Experimental Psychology

3rd edition
Experimental Psychology
a methodological approach

F. J. McGUIGAN
University of Louisville

PRENTICE-HALL, INC., Englewood Cliffs, New Jersey 07632

Library of Congress Cataloging in Publication Data

McGUIGAN, FRANK J.
 Experimental psychology.

 Bibliography: p.
 Includes index.
  1. Psychology, Experimental. I. Title. [DNLM:
 1. Psychology, Experimental. BF181 M148e]
 BF181.M24 1978   150'.7'24   77-25206
 ISBN 0-13-295162-2

Printed in the United States of America

10   9   8   7   6   5   4   3   2   1

Prentice-Hall International, Inc., London
Prentice-Hall of Australia Pty. Limited, Sydney
Prentice-Hall of Canada, Ltd., Toronto
Prentice-Hall of India Private Limited, New Delhi
Prentice-Hall of Japan, Inc., Tokyo
Prentice-Hall of Southeast Asia Pte. Ltd., Singapore
Whitehall Books Limited, Wellington, New Zealand

To two charming ladies

Constance and Joan

Contents

8
Experimental Design: The Case of Two Matched Groups   *197*

9
Experimental Design:
The Case of More Than Two Randomized Groups   *230*

10
Experimental Design: The Factorial Design   *280*

11
Experimental Design: Within-Groups Designs   *326*

12

Quasi-Experimental Design: Seeking Solutions to Society's Problems

13

The Logical Bases of Experimental Inferences

14

The Inductive Schema: An Overview of Some Characteristics of Science

15

Generalization, Explanation, and Prediction in Experimentation

16

Miscellany

Preface

Preface to First Edition

Experimental psychology was born with the study of sensory processes; it grew as additional topics, such as perception, reaction time, attention, emotion, learning, and thinking, were added. Accordingly, the traditional course in experimental psychology was a course the content of which was accidentally defined by those lines of investigation followed by early experimenters in those fields. But times change, and so does experimental psychology. The present trend is to define experimental psychology not in terms of specific content areas, but rather as a study of scientific methodology generally, and of the methods of experimentation in particular. There is considerable evidence that this trend is gaining ground rapidly.

This book has been written to meet this trend. His methods no longer confined to but a few areas, the experimental psychologist conducts research in almost the whole of psychology—clinical, industrial, social, military, and so on. To emphasize this point, we have throughout the book used examples of experiments from many fields, illustrative of many methodological points.

In short, then, the point of departure for this book is the relatively new conception of experimental psychology in terms of methodology, a conception which represents the bringing together of three somewhat distinct aspects of science: experimental methodology, statistics, and philosophy of science. We have attempted to perform a job analysis of experimental psychology, presenting the important techniques that the experimental psychologist uses in his everyday work. Experimental methods are the basis of experimental psychology, of course; the omnipresence of statistical presentations in journals attests the importance of this aspect of experimentation. An understanding of the philosophy of science is important to an understanding of what science is, how the scientific method is used, and particularly of where experimentation fits into the more general framework of scientific methodology. With an understanding of the goals and functions of scientific methodology, the experimental psychologist is prepared to function efficiently, avoiding scientifically unsound procedures and fruitless problems.

Designed as it is to be practical in the sense of presenting information on those techniques actually used by the working experimental psychologist, it is hoped for this book that it will help maximize transference of performance from a course in experimental psychology to the type of behavior manifested by the professional experimental psychologist.

Preface to Third Edition

As I reflect back over the years since I first started working on the first edition of this book (that was about 1955) I think the real beginning must have been my first course in (traditional) experimental psychology. As a young student my aspirations were to become an experimental psychologist, but my experimental psychology course and I did not prosper all that much together. The class was normal for its time, and I recall many hours of laboratory work in which we plotted the blind spot, measured the knee jerk, were ushered into rooms with sundry components where we were instructed to "put it together," etc. One assemblage of apparatus, I remember, turned out to be the Dunlap Rotating Chair developed in the early years of this century by Professor Dunlap for (among other things) inducing nystagmus. In no sense am I disparaging these important topics, for these efforts were critical to the evolution of experimental psychology, but, in Leon Festinger's terms, I must have been experiencing cognitive dissonance: either *I* had to change my goal of becoming an experimental psychologist, or experimental psychology should be defined otherwise. Some twenty years later I attempted to change that definition from one of traditional content to one of methodology.

My initial efforts were not met with great enthusiasm, but perseverence is the hallmark of our trade, and with the encouragement of Ed Thomas (then Prentice-Hall psychology editor) and Kenneth Spence (then psychology advisory editor) the 1960 edition was completed. When it was through Spence accepted it (somewhat reluctantly I thought at the time, but apparently not, because he later "adopted" me as an honorary Iowa student). In any event, a methodological definition which was born with that first maverick edition now seems to flourish. Numerous methodologically oriented experimental psychology books now abound—the strategy of asking questions and seeking sound answers regardless of the area of behavior now replaces books based on content definitions. It is indeed heart-warming to observe scientists who refer to themselves as "experimental psychologists" working in such disparate areas as industrial psychology, clinical psychology, educational psychology, and social psychology. Although the traditional content of psychology is obviously of primary importance, it should be assigned to such specific courses as "learning," or "sensation and perception."

One still debated question concerns when and how statistics should be introduced to psychology majors. A strong independent course in statistics continues to be a necessity, but I am also more convinced than ever that it is highly desirable to integrate statistics with experimental methods. Whether the statistics course should come before, during or after the experimental psychology course is a difficult question—through my work with students over the years I have concluded that there is probably about as much course transfer from experimental psychology to statistics as vice versa, so I haven't been rigid about recommending either as a requirement for the other. I do know that "statistics" doesn't make too much sense to many of our students until they actually have to use statistical methods in answering questions they have posed in their own experiments. I thought that Levy (1973) used this formula rather nicely in his "Psychological Statistics: A Teaching Paradigm": "In conclusion, we should avoid giving our students the impression that data analysis is concerned with mere technique which is distinct from, or auxiliary to, the psychologist's own thinking as an experimenter" (p. 12).

It is also gratifying to note the variety of uses to which *Experimental Psychology* has been put, as in educational research courses. Apparently the book can be adapted to either a one-semester or a one-year course, as Edward Simmel suggested: "Through the arrangement of the design chapters in order from simpler to more complex, the instructor can stop at any point which suits his needs without sacrificing continuity . . . it is now clearly usable for a year course in experimental psychology, but would also be useful for the first-year graduate students, either as a review or for those who have had only a single-semester course as undergraduates." I thoroughly agree with Dr. Simmel, and in fact over the years I have built a first-year graduate research practicum largely around the philosophy of the book. After falsely assuming for some years that graduate students had acquired and *retained* basic principles of scientific research, I reluctantly concluded that they at least needed a refresher course.

I have also watched developments in the philosophy of science (and of scientific psychology) with special interest over the decades; with others I am somewhat alarmed that the strong dose of philosophy of science that our students used to seek has been diminished or eliminated. The basic principles of logical positivism remain the foundation of our science, though they have been modified and extended in a number of ways. Perhaps now more than ever, an important function of the book is to introduce our students to philosophy of science, especially in chapters 2 and 3 (on "testability") and in later chapters concerning generalization, explanation, and prediction from empirical findings.

A number of changes were compelled since the 1968 edition. Among these was the thorough rewriting of the "how to" section on preparing an article for publication; there have been numerous changes on this matter, and special thanks go to the American Psychological Association for reading and making some valuable suggestions about Chapter 4. I have also incorporated several new topics, such as "research ethics," replication, a guide to use of statistical analyses, review items for each chapter, and a glossary. But most importantly, a new chapter was added that extends the presentation of quasi-experimental designs to the solution of society's problems. This effort reflects to a large extent the interest and work of B. F. Skinner and Donald Campbell, as acknowledged herein.

Finally, over the years I have very much enjoyed hearing from students and teachers on positive and negative aspects of the book, and it would be my pleasure to hear from others in the years that follow.

F. J. M.
Louisville, Kentucky

1

An Overview of Experimentation

The Nature of Science

The questions that concern psychologists are singularly challenging and complex. For this reason it is necessary for us to use the most effective research methods that society can make available to study these problems. Our experience of many centuries has clearly indicated that scientific methods have yielded the soundest knowledge.

Definitions of "science" vary widely, but they can generally be categorized in two (overlapping) classes: content definitions and process definitions. A typical content definition would be that "science is an accumulation of integrated knowledge," while a process definition would state that "science is that activity of discovering important variables in nature, of relating those variables, and of explaining those relationships (laws)." A definition that incorporates content *and* processes is: "Science is an interconnected series of concepts and conceptual schemes that have developed as a result of experimentation and observations" (Conant, 1951, p. 25). A similar definition would be that science is "a systematically organized body of knowledge about the universe obtained by the scientific method."[1]

It is likely that there is no completely adequate definition of "science." To arrive at a definition that will help us to achieve an understanding of some of the basic characteristics of science, and that will facilitate systematic presentation of those characteristics, we might consider the various sciences as a group; we can then abstract the salient characteristics that distinguish sciences from other disciplines. Figure 1-1 is a schematic representation of the disciplines we study, rather crudely categorized into three groups (excluding the formal disciplines, mathematics, and logic). Within the inner circle we have represented what are commonly called the sciences. The next circle embraces various disciplines that are not usually thought of as sciences, such as the arts and some of the humanities. Outside that circle are yet other disciplines which, for lack of a better term, are designated as metaphysical disciplines.

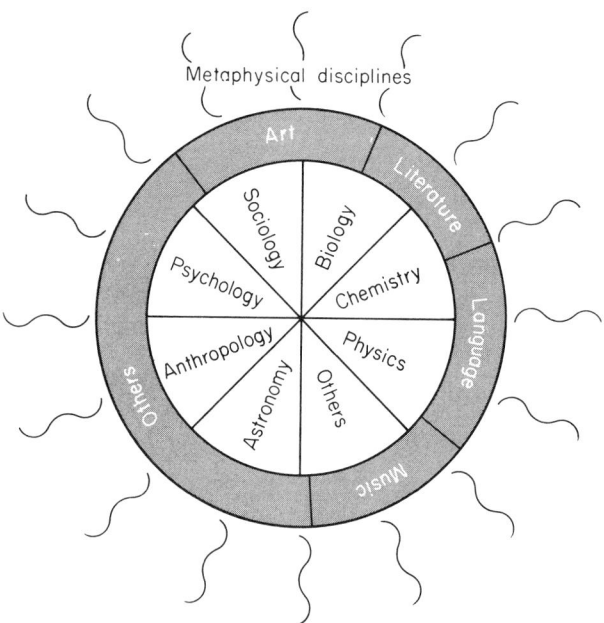

Figure 1-1. Three groups of disciplines which we study. Within the inner circle are the sciences. The second circle contains the arts and some of the humanities; the metaphysical disciplines fall outside the circles.

---

[1] Thanks to Eleanor Simon for this precise and general statement, and also for her good suggestions on a number of other matters throughout the book.

The sciences in the inner circle certainly differ among themselves in a number of ways. But in what important ways are they similar to each other? Likewise, what are the similarities among the disciplines in the outer circle? What do the metaphysical disciplines outside the circle have in common? Furthermore, in what important ways do each of these three groups differ from each other? Answers to these questions should enable us to arrive at an approximation to a general definition of "science."

One common characteristic of the sciences is that they all use the same general approach in solving problems—"the scientific method." The "scientific method" is a serial process by which all the sciences obtain answers to their questions. Neither of the other two groups explicitly uses this method.

The disciplines within the two circles differ from the metaphysical disciplines with regard to the type of problem studied. Individuals who study the subject matter areas within the two circles attempt to consider only problems that can be solved; those whose work falls outside of the circle generally study unsolvable problems. Briefly, a *solvable* problem is one that poses a question that can be answered with the use of our normal capacities. An *unsolvable* problem raises a question that is essentially unanswerable. Unsolvable problems usually concern supernatural phenomena or questions about ultimate causes. For example, the problem of what caused the universe is unsolvable and is typical of studies in religion and classical philosophy.[2] Ascertaining what is and what is not a solvable problem is an extremely important topic and will be taken up in greater detail in chapter 2.

It is important to emphasize that "solvable" and "unsolvable" are technical terms and certain vernacular meanings should not be read into them. It is not meant, for instance, to establish a hierarchy of values among the various disciplines by classifying them according to the type of problem studied. We are not necessarily saying, for example, that the problems of science are "better" or more important than the problems of religion. The distinction is that solvable problems may be attacked by studying observable events in the world around us—they are susceptible to empirical solution, but this is not the case with unsolvable problems. Individuals whose work falls within the two circles (particularly within the inner one) simply believe they must limit their

---

[2] Crude categorizations are dangerous. We merely want to point out *general* differences among the three classes of disciplines. A number of theological problems, for example, *are* solvable (e.g., "does praying beneficially affect our future behavior?"). Hence, while it *is* possible to develop at least a limited science of religion, most theologians are not interested in answering their questions in an empirical manner.

study to problems that they are capable of solving. Of course, some scientists also devote part of their lives to the consideration of supernatural phenomena. But it is important to realize that when they do, they have "left the circle" and are, for that time, no longer behaving as scientists.

In summary, first, the sciences use the scientific method, and they study solvable problems. Second, the disciplines in the outer circle do not use the scientific method, but their problems are typically solvable. And third, the disciplines outside the circles neither use the scientific method nor do they pose solvable problems. These considerations lead to the following definition: *"Science" is the application of the scientific method to solvable problems.* It can be seen that this definition incorporates both the process (method) and content definitions, in that the study of solvable problems results in systematic knowledge. Generally, neither of the other two groups of disciplines have *both* these features in common.

The consequences of this very general definiton are enormous and lead us to specify several of the more important concepts of science. The classical behaviorists, led by John B. Watson in the early part of the century, were important in the development of psychology as a science. Watson's program for a transition from a nonscience to a science was as follows: "If psychology is ever to become a science, it must follow the example of the physical sciences; it must become materialistic, mechanistic, deterministic, objective" (Heidbreder, 1933, p. 235). Watson's demand that we be materialistic states what is now obvious, namely that we must study only physical events[3] like observable responses, rather than ghostly "ideas" or a "consciousness" of a nonmaterial mind. Materialism is interrelated with objectivity, for it is impossible to be objective when seeking to study unobservable phenomena. We are *objective* as a result of our application in science of a principle of intersubjective reliability. That is, we all have "subjective" experiences when we observe an event. "Intersubjective" means that two or more people may share the same experience. When they verbally report the same subjective experience, we conclude that the event really (reliably) occurred (was not an hallucination). In short, the data of science are public in that they are gathered objectively — scientifically observed events are reliably reported through the subjective perceptions of a number of observers, not just one. Watson's request that we be *deterministic* was not new in psychology, but is

---

[3] Our everyday language sometimes leads us to unfortunate habits, such as this redundancy — "physical events" implies that there may be nonphysical events, which is precisely what Watson and his colleagues tried to eliminate from early psychology.

critical for us. Determinism is the assumption that there is lawfulness in nature. If there is lawfulness, we are able to ascertain causes for the events that we seek to study. To the extent to which nature is nondeterministic, it is chaotic with events occurring spontaneously (without causes). We have, incidentally, no assurance that all events are determined. However, we must assume that those that we study *are* lawful if we ever hope to discover laws for them.[4]

With these considerations and our general definition of science in hand let us consider the scientific method, primarily as it is applied in psychology. The more abstruse and enigmatic a subject is, the more rigidly we must adhere to the scientific method and the more diligently we must control our variables. Chemists work with a relatively limited set of variables, while psychologists must study considerably more complex phenomena. We cannot afford to be sloppy in our research. And since the most powerful application of the scientific method is experimentation, we shall focus principally on how experiments are conducted. The following brief discussion will serve as an overview of the rest of the book. It will include a general picture of how the experimental psychologist proceeds as an orientation to experimentation. Because this overview is so brief, however, complex matters will necessarily be oversimplified. Possible distortions resulting from this oversimplification will be corrected in later chapters.

Psychological Experimentation: An Application of the Scientific Method[5]

A psychological experiment starts with the formulation of a problem, which is usually best stated in the form of a question. The only requirement that the problem must meet is that it be solvable— the question that it raises must be answerable with the tools that are available to the psychologist. Beyond this, the problem may be concerned with any aspect of behavior, whether it is judged to be important or trivial. One lesson of history is that we must not be hasty in judging the importance of the problem on which a scientist works, for

[4] Watson's "mechanism" refers to the assumption that we behave in accordance with "mechanical" principles (those of physics and chemistry). But since mechanism (as opposed to vitalism) is now of historical interest only, we shall not dwell on the issue as it was formulated in years past, principally in biology.

[5] There are those who hold that psychologists do not formally go through the following steps of the scientific method in conducting their research. We would agree with this statement for many researchers. However, a close analysis of the actual work of such people would suggest that they at least informally approximate the following pattern, regardless of how they verbalize it.

many times what was momentarily discarded as being of little importance contributed sizably to later scientific advances.

The experimenter generally expresses a tentative solution to the problem. This tentative solution is called a hypothesis; it may be a reasoned potential solution or only a vague guess (it is an empirical hypothesis, not a null hypothesis, which will be discussed in chapter 5). Following the statement of the hypothesis, the experimenter seeks to determine whether the hypothesis is (probably) true or (probably) false, i.e., does it solve the problem the psychologist has formulated? To answer this question we must collect data, for a set of data is our only criterion. Various techniques are available for data collection, but, as we said, experimentation is the most powerful.

One of the first steps that the experimenter will take in actually collecting data is to select a group of participants with which to work. The type of participant studied will be determined in part by the nature of the problem. If the concern is with psychotherapy, one may select a group of mentally disturbed patients. A problem concerned with the function of parts of the brain would entail the use of animals (for few humans volunteer to serve as participants for brain operations). Learning problems may be investigated with the use of college sophomores, chimpanzees, rats, etc. But whatever the type of participant, the experimenter typically assigns them to groups. We shall consider here the basic type of experiment, namely, one that involves only two groups. Incidentally, you will note that we refer to people who collaborate in an experiment for the purpose of allowing their behavior to be studied as *participants,* rather than by the traditional term *subjects.* As Gillis (1976) pointed out, "participants" is socially more desirable because "subjects" suggest that the participants are "being used," or that there is a status difference between the experimenter and the subject (as a king and his subjects). It is important that individuals who participate in an experiment be well respected, as suggested by the use of the word "participants" in the American Psychological Association's (1973) *Ethical Principles in the Conduct of Research with Human Participants* (see chapter 16). Experimental participants should have a prestigious status, for they are critical in the advancement of our science. Other terms ("children," "students," "animals") are useful alternatives.

The assignment of participants to groups must be made in such a way that the groups will be approximately equivalent at the start of the experiment; this is accomplished through *randomization,* a term to be discussed. The experimenter next typically administers an experimental treatment to one of the groups. The experimental treatment is what one wishes to evaluate, and it is administered to the *experimental group.* The other group, called the *control group,* usually receives a normal or standard treatment. It is important, here, to understand clearly just what the terms "experimental" and "normal" or "standard treatment" mean.

In the study of behavior, the psychologist generally seeks to establish empirical relationships between aspects of the environment, broadly conceived, and aspects of behavior. These relationships are known by a variety of names, such as hypotheses, theories, or laws. Such relationships in psychology essentially state that if a certain environmental characteristic is changed, behavior of a certain type also changes.[6]

The aspect of the environment that is experimentally studied is called the *independent variable;* the resulting change in behavior is called the *dependent variable.* Roughly, a variable is anything that can change in value. It is a quality that can exhibit differences in value, usually in magnitude or strength. Thus it may be said that a variable generally is anything that may assume different numerical values. Anything that exists is a variable, according to E. L. Thorndike, for this prominent psychologist asserted that anything that exists, exists in some quantity. Let us briefly elaborate on the concept of a "variable," after which we shall distinguish between independent and dependent variables.

Psychological variables change in value from time to time for any given organism, between organisms, and according to various environmental conditions. Some examples of variables are the height of women, the weight of men, the speed with which a rat runs a maze, the number of trials required to learn a poem, the brightness of a light, the number of words a patient says in a psychotherapeutic interview, and the amount of pay a worker receives for performing a given task.

Figure 1-2 schematically represents one of these examples, the speed with which a rat runs a maze. It can be seen that this variable can take on any of a large number of magnitudes, or, more specifically, it can exhibit any of a large number of time values. In fact, it

---

[6] By saying that the psychologist seeks to establish relationships between environmental characteristics and aspects of behavior, we are being unduly narrow. Actually we are also concerned with processes that are not directly observed (variously called logical constructs, intervening variables, hypothetical constructs, etc.). Since, however, it is unlikely that work of the young experimentalist will involve hypotheses of such an abstract nature, they will not be emphasized here. The highly arbitrary character of defining and differentiating among the various kinds of relationships should be emphasized — frequently the grossly empirical kind of relationship that we are considering under the label "hypothesis," once it is confirmed, is referred to as an empirical or observational law; or before it is tested, merely as a "hunch" or "guess."

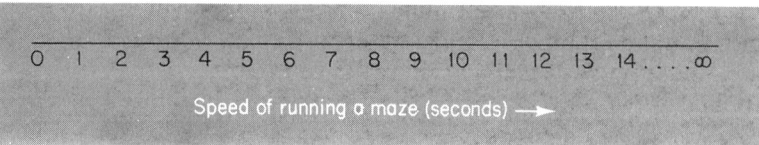

Figure 1-2. Diagrammatic representation of a continuous variable.

may "theoretically" assume any of an infinite number of such values, the least being zero seconds, and the greatest being an infinitely large amount of time. In actual situations, however, we would expect it to exhibit a value of a number of seconds, or at the most, several minutes. But the point is that there is no limit to the specific time value that it may assume, for this variable may be expressed in terms of any number of seconds, minutes, hours, including any fraction of these units.

For example, we may find that a rat ran a maze in 24 seconds, in 12.5 seconds, or in 2 minutes and 19.3 seconds. Since this variable may assume any fraction of a value (it may be represented by any point along the line in Figure 1-2), it is called a *continuous* variable. A continuous variable is one that is capable of changing by any amount, even an infinitesimally small one. A variable that is not continuous is called a *discontinuous* or *discrete* variable. A discrete variable can assume only numerical values that differ by clearly defined steps with no intermittent values possible. For example, the number of people in a theater would be a discrete variable, for, barring an unusually messy affair, one would not expect to find a part of a person in such surroundings. Thus, one might find 1, 15, 299, or 302 people in a theater, but not 1.6 or 14.8 people. Similarly, gender (male or female) and eye color (brown, blue) are frequently cited as examples of discrete variables.[7]

We have said that the psychologist seeks to find relationships between independent and dependent variables. There are an infinite (or at least indefinitely large) number of independent variables available in nature for the psychologist to examine. But we are interested in discovering those *relatively* few that affect a given kind of behavior. In short, we may say that an independent variable is any variable that is investigated for the purpose of determining whether it influences behavior. Some independent variables that have been scientifically inves-

[7]We may note that some scientists question whether there are actually any discrete variables in nature. They suggest that we simply "force" nature into "artificial" catagories. Color, for example, may more properly be conceived of as a continuous variable—there are many gradations of brown, blue, etc. Nevertheless, scientists find it useful to categorize variables into classes as discrete variables, and to view such categorization as an approximation.

tigated are water temperature, age, hereditary factors, endocrine secretions, brain lesions, drugs, loudness of sounds, and home environments.

Now, with the understanding that an experimenter seeks to determine whether an independent variable affects a dependent variable (either of which may be continuous or discrete), let us return to our consideration of experimental and control groups. To determine whether a given independent variable affects behavior the experimenter administers one value of it to the experimental group and a second value of it to the control group. The value administered to the experimental group is, as we have said, the "experimental treatment," while the control group is usually given a "normal treatment." Thus, the essential difference between the "experimental" and "normal" treatment is the specific value of the independent variable that is assigned to each group. For example, the independent variable may be the intensity of a shock (a continuous variable). The experimenter may subject the experimental group to a high intensity and the control group to a lower intensity or zero intensity.

To elaborate on the nature of an independent variable, let us consider another example of how one might be used in an experiment. Visualize a continuum similar to Figure 1-2, composed of an infinite number of possible values that the independent variable may take. If, for example, we are interested in determining how well a task is retained as a result of the number of times it is practiced, our continuum would start with zero trials and continue with one, two, three, etc., trials (this would be a discrete variable).

Let us suppose that in a certain industry workers are trained by performing an assembly line task 10 times before being put to work. After a while, however, it is found that the workers are not assembling their product adequately, and it is judged that they have not learned their task sufficiently well. Some corrective action is indicated, and the foreman suggests that the workers would learn the task better if they were able to practice it 15 times instead of 10. Here we have the makings of an experiment of the simplest sort.

We may think of our independent variable as the "number of times that the task is performed in training," and will assign it two of the possibly infinite number of values that it may assume—10 trials and 15 trials (see Figure 1-3). Of course, we could have selected any number of other values, one trial, five trials, or 5,000 trials, but because of the nature of the problem with which we are concerned, 10 and 15 seem like reasonable values to study. We will have the experimental group practice the task 15 times, the control group 10 times. Thus, the control group receives the normal treatment (10 trials), and

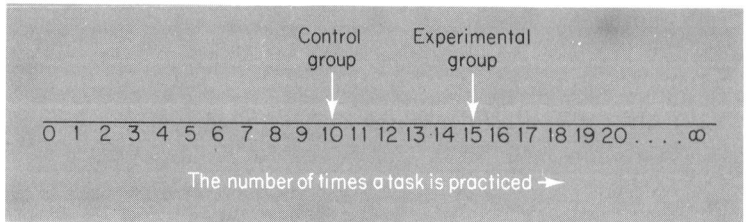

Figure 1-3. Diagrammatic representation of an independent variable. The value of the independent variable assigned to the control group is 10 trials; that assigned to the experimental group, 15 trials.

the experimental group is assigned the experimental or new treatment (15 trials). In many cases, of course, it is arbitrary which group is labeled the control group and which is called the experimental group. Sometimes both treatments are novel ones, in which case it is impossible to label the groups in this manner—they might simply be called "Group 1," and "Group 2." In another instance, if one group is administered a "zero" value of the independent variable, and a second group is given some positive amount of that variable, then the zero treatment group would be called the "control group" while the other would be the "experimental group."

The *dependent* variable is usually some well-defined aspect of behavior (a response) which the experimenter measures in the experiment. It may be the number of times the participant says a certain word, the rapidity with which a participant learns a given task, the number of items a worker on a production line can turn out in an hour, and so on. The value obtained for the dependent variable is the criterion of whether or not the independent variable is effective. It is in this sense that it is called a *dependent* variable—the value that it assumes is expected to be dependent on the value assigned to the independent variable.[8] Thus, an experimenter will vary the independent variable and note whether the dependent variable changes. If the dependent variable changes in value as the independent variable is manipulated, then it may be asserted that there is a relationship between the two. (The psychologist has discovered an empirical law.) If the dependent variable does not change, however, it may be asserted that there is a lack of relationship between them. For example, let us assume that a light of high intensity is flashed into the eyes of each participant of the experimental group, while those of the control group are subjected to a low intensity light. The dependent variable might be the amount of contraction of the iris diaphragm (the pupil) of the

---

[8] It may be added that the dependent variable is also dependent on some of the *extraneous* variables, discussed shortly, that are always present in an experiment.

eye which, it may be noted, is an aspect of behavior—a response. If we find that the average contraction of the pupil of the experimental participants exceeds that of those in the control group, we may conclude that intensity of light is an effective independent variable. We can tentatively assert the following relation: The greater the intensity of a light that is flashed into a participant's eyes, the greater the contraction of the pupil. If, on the other hand, we find no difference in the amount of contraction of the pupil between the two groups, we would assert that there is a lack of relationship between these two variables.

Perhaps the most important principle of experimentation, stated in an ideal form, is that the experimenter must hold constant all of the variables that may affect the dependent variable, except the independent variable(s) whose effect is being evaluated.[9] Obviously, there are a number of variables that may affect the dependent variable, but the experimenter is not immediately interested in these. For the moment, the interest is in only one thing—the relationship, or lack of it, between the independent and the dependent variables. If the experimenter allows a number of other variables to operate freely in the experimental situation (call them *extraneous* variables) the experiment is going to be contaminated. For this reason one must *control* the extraneous variables in an experiment.

A simple illustration of how an extraneous variable might contaminate an experiment, and thus make the findings unacceptable, might be made using the last example. Suppose that, unknown to the experimenter, all the participants of the experimental group had that morning received a routine vaccination. But the serum contained a substance that affected the contraction of the pupil of the eye. In this event, the dependent variable measures that the experimenter collected would have, to be generous, little value. For example, if the effect of the serum were such as to cause the pupil not to contract, the participants of the experimental group would show about the same lack of contraction as those in the control group. It would thus be concluded that the independent variable did not affect the response being studied. The findings would assert that these two variables of light intensity and pupillary contraction are not related, when in fact they are. It was that the dependent variable was affected by an extraneous variable (the serum); and the effects of this extraneous variable obscured the influence of the independent variable. This topic of controlling extraneous variables that might invalidate an experiment is of sufficiently great importance that an entire chapter will be devoted to

[9] Chapter 6 will show why this is an ideal statement and the ways in which it needs to be modified for any given experiment.

it. In chapter 6 we shall study various techniques for handling unwanted variables in an experiment.

With this discussion of independent and dependent variables behind us, let us return to our general discussion of the scientific method as applied to experimentation. We have said that a scientist starts an investigation with the statement of a problem, after which a hypothesis is advanced as a tentative solution to that problem. One may then conduct an experiment to collect data—data which should indicate the probability that the hypothesis is true or false. The scientist may find it advantageous to use certain types of apparatus and equipment in the experiment. The particular type of apparatus used will naturally depend on the nature of the problem. In general, apparatus is used in an experiment for two main reasons: First, to administer the experimental treatment; and second, to allow, or to facilitate, the collection of data.

The hypothesis that is being tested will predict the way in which the data should point. It may be that the hypothesis will predict that the experimental group will perform better than the control group. By confronting the hypothesis with the dependent variable values of the two groups the experimenter can determine if the hypothesis accurately predicted the results. But it is difficult to tell whether the (dependent variable) values for one group are higher or lower than the values for the second group simply by looking at a number of unorganized data for the two groups of participants. Therefore, the experimenter must reduce the data to numbers that can be reasonably handled, numbers that will provide an answer—for this reason, we must resort to statistics.

For example, we may compute an average (mean) score for both the experimental group and the control group. We might find that the experimental group has a higher mean score (say, 100) than the control group (say, 99). While we note that the experimental group has a higher mean score, we also note that the difference between the two groups is very small. Is this difference, then, a "real" difference, or is it only a chance difference? What are the odds that if we conduct the experiment again, we would obtain the same results? If the difference is a "real," reliable difference, then the experimental group should obtain a higher mean score than the control group almost every time the experiment is repeated. But if there is no "real" difference between the two groups, we would expect to find the experimental group receiving the higher score half of the time and the control group being superior the other half of the time. To tell whether the difference between the two groups in one experiment is "real," rather than simply due to random fluctuations (chance), the experimenter resorts to any of a variety of statistical tests. The particular statistical test(s) used will be determined by the type of data obtained and the general design of the experiment. But the point is that, on the basis of such tests, it can be determined whether the difference between the two groups is likely to be "real" and reliable or merely "accidental." The tests indicate whether or not the difference is statistically reliable. If the difference between the dependent variable scores of the groups is reliable, the difference is very probably not due to random fluctuations; and it is concluded that the independent variable is effective (providing that the extraneous variables have been properly controlled).

When you read psychological journals you will often note that "significant" is used to mean "reliable." However, to say that you have a significant difference is sometimes unfortunate for it may suggest that your reliable difference is an important one, which of course it might not be at all. It is indeed confusing when psychologists try to communicate to a newspaper reporter, for instance, that a significant statistical test that they conducted was not an important finding. As Porter (1973) pointed out

> . . . the technical jargon of statistics itself has a word and concept that fits the situation: *reliable.* A reliable outcome is one that can be expected to reappear on reexamination. A reliable difference will be found again if the experiment is repeated. An F, a z, or whatever is significant in that it signifies the reliability of whatever observation is under test. An extremely reliable difference can be every bit as trivial as its most untrustworthy counterpart; there is no need to mislead one's audience nor to delude oneself with *highly significant* (pp. 188–89).

Thus, just as we will often continue to use "subjects" for "participants" in psychological writings, "significant" will continue to be used, though "reliable" is preferable.

By starting with two equivalent groups, administering the experimental treatment to one but not to the other, and collecting and analyzing the (dependent variable) data thus obtained, suppose we find a reliable difference between the two groups. We may legitimately assume that the two groups eventually differed because of the experimental treatment. Since this is the result that was predicted by our hypothesis, the hypothesis is supported, or confirmed. In other words, when a hypothesis is supported by experimental data, the probability that the hypothesis is true is increased. On the other hand, if, in the above example, the control group is found to be equal or superior to the experimental group, the hypothesis is not supported by the data and we may conclude that it is probably false. Naturally, this step of the scientific method in which the hypothesis is tested is considerably oversimplified in our brief presentation. It will be necessary to consider the matter more thoroughly later (see chapter 13).

Closely allied with testing of the hypothesis is an additional step of the scientific method, "generalization." After completing the phases outlined above, the experimenter may feel quite confident that the hypothesis is true for the specific conditions under which it was tested. We must underline *specific* conditions, however, and not lose sight of how specific they are in any given experiment. But the scientist *qua* scientist is not concerned with truth under a highly restricted set of conditions. Rather, we usually want to make as *general* a statement as we possibly can about nature. And herein lies much of our joy and grief, for the more we generalize our findings, the greater are the chances for error. Suppose that one has used college students as the participants of an experiment. This selection does not mean that the researcher is interested *only* in the behavior of college students. Rather, the interest is probably in the behavior ot *ll* human beings and perhaps even of *all* organisms. Because it has been found that the hypothesis is probably true for a particular group of participants, can we now say that it is probably true for all humans? Or must we simply restrict the results to college students? Or must the focus be narrowed even further, limiting it to those attending the college at which the experiment was conducted? This, essentially, is the question of generalization — how widely can the experimenter generalize the results obtained? We want to generalize as widely as possible, yet not so widely that the hypothesis "breaks down." The question of how widely we may safely generalize a hypothesis will be discussed in chapter 15. The broad principle to remember is that we should state that a hypothesis is applicable to as wide a set of conditions (e.g., to as many classes of participants) as the nature of the experiment warrants.

The next step in the scientific method, closely related to the preceding steps, concerns making predictions on the basis of the hypothesis. By this we mean that a hypothesis may be used to predict certain events in new situations — to predict, for example, that a different group of participants will act in the same way as a group studied in an earlier experiment.[10] The step in the scientific method of making a prediction is closely connected with another one, that of *replication*. By *replication* we mean that an additional experiment is conducted in which the *method* of the first experiment is precisely repeated. The confirmed hypothesis may thus be used as the basis for predicting that a new sample of participants will behave as did the original sample. If

[10] A case can be made for not including prediction as a part of the scientific method, at least in some sciences (cf. Scriven, 1959).

the prediction made by the use of the previously confirmed hypothesis is found to hold in the new situation, the probability that the hypothesis is true is tremendously increased. The difference between replicating a previous experiment and supporting the results or conclusions of that experiment should be emphasized. In a replication, the methods of an experiment have been repeated, but the results may confirm or disconfirm the previous experiment. Sometimes researchers erroneously state that they have "replicated an experiment" when they mean to say that they have "confirmed the findings of that experiment."

The relationship between the independent and dependent variables may be formulated as an empirical law, particularly if the relationship has been confirmed in a replication of the experiment (in accordance with the experimenter's prediction). The final step in the scientific method (by our formulation) is that of explanation, which involves the use of theory. We seek to explain our results by means of some appropriate theory. For instance, Galileo's experiments on falling bodies resulted in his familiar law of $S = 1/2\ gt^2$, which was later explained by the theories of Newton (see chapter 14).

In summary let us set down the various steps in the scientific method. (Be advised, however, that there are no rigid rules to follow in doing this. In any process that one seeks to classify into a number of arbitrary categories, some distortion is inevitable. Another source might offer a different classification, while still another one might refuse, quite legitimately, to attempt such an endeavor.)

First, the scientist selects an area of research and states a problem for study. Next, a hypothesis is formulated as a tentative solution to the problem. Third, one collects data relevant to the hypothesis. Following this, a test is made of the hypothesis by confronting it with the data—we organize the data through statistical means and make appropriate inferences to determine whether the data support or refute the hypothesis. Fifth, assuming that the hypothesis is supported, we may wish to generalize to all things with which the hypothesis is legitimately concerned, in which case we should explicitly state the generality with which we wish to advance the hypothesis. Sixth, we may wish to make a prediction to new situations, to events not studied in the original experiment. In making a prediction we may test the hypothesis anew in the novel situation; that is, we might replicate (conduct the experiment with a new sample of participants to determine whether the estimate of the probability of the hypothesis can legitimately be increased). Finally, we should attempt to explain our findings by means of a more general theory.

An Example
of a Psychological Experiment

To make the preceding discussion more concrete, consider an example of how an experiment might be conducted from its inception to its conclusion. This example is taken from the area of clinical psychology which, like any applied area, is a methodologically difficult one in which to conduct sound research. Our purpose, however, is primarily to illustrate the application of the preceding principles. Let us assume that a clinician has some serious questions about the effect of traditional psychotherapy as a "cure" for clients. Traditional psychotherapy has been conducted primarily at the verbal level in which the client (or patient) and therapist discuss in some manner the client's problems. Psychoanalysis emphasized the value of "verbal outpouring" from the patient for the purpose of catharsis, originally referred to by Freud and Brever as "chimney sweeping." In our example the therapist is not sure whether interchange strictly at a verbal level is effective. Rather, would directly dealing with the client's behavior (as in clinical progressive relaxation or behavior modification) be the more effective approach? The problem may be stated as follows: Should a clinical psychologist engage in verbal psychotherapy and talk with clients about their problems, or should the psychologist attempt to modify behavior, ignoring interaction at a strictly verbal level? Assume that the therapist believes the latter to be the better procedure. The reasons for or against this opinion need not detain us here. We simply note the hypothesis: If selected responses of a client undergoing therapy are systematically manipulated in accordance with principles of behavior theory, then recovery will be more efficient than if the therapist engages in strictly verbal discourse about the client's problems. We might identify the independent variable as "the amount of systematic manipulation of behavior" and assign two values to it: first, a maximal amount of systematic manipulation, and second, a zero amount of systematic manipulation of behavior in which clients are left free to do whatever they wish. In this zero amount of the experimental treatment, presumably clients will wish to talk about their problems, in which case the therapist would merely serve as a "sounding board" as in Carl Rogers' nondirective counseling procedures.

Suppose that the clinical psychologist has ten clients, and that they are randomly assigned to two groups of five each. A large amount of systematic manipulation of behavior will then be given to one of the groups, and a zero (or minimum) amount will be administered to the second group. The group that receives a zero or minimum amount of systematic manipulation will be called the control group; the group that receives the maximum amount will be called the experimental group.[11]

Throughout the course of therapy, then, the therapist administers the two different treatments to the experimental and control groups. During this time it is important to prevent extraneous variables from acting differently on the two groups. For example, we would want the clients from both groups to undergo therapy in the same place (the office, for instance) to eliminate the possibility that the progress of one group might differ from that of the other group because of the immediate surroundings in which the therapy takes place.

The dependent variable here may be specified as the progress toward recovery. Such a variable is obviously rather difficult to measure, but for illustrative purposes we might use a time measure. Thus, we might assume that the earlier the client is discharged by the therapist, the greater is the progress toward recovery. The time of discharge might be determined when the client has no further complaints of difficulties. Assuming that the extraneous variables have been adequately controlled, the progress toward recovery (the dependent variable) depends on the particular values of the independent variable used, and on nothing else.

As therapy progresses the psychologist collects the data. Specifically, the amount of time each client spends in therapy before discharge is determined. After all the clients are discharged, the therapist compares the times for the experimental group against those for the control group. Let us assume that the mean amount of time in therapy of the experimental group is lower than that of the control group, and, further, that a statistical test indicates that the difference is significant. That is, the group that received a minimum amount of systematic behavioral manipulation had a significantly longer time-in-therapy (the dependent variable) than did the group that received a large amount. This is precisely what the therapist's hypothesis predicted. Since the results of the experiment are in accord with the hypothesis, we may conclude that the hypothesis is confirmed.

Now the psychotherapist is happy, since the problem has been solved and the better method of psychotherapy has been determined. But has "truth" been found only for the psychologist, or are the results applicable to other situations—can other therapists also benefit by these results? Can the findings be extended, or generalized, to all

---

[11] Since it may not be possible to completely avoid guiding or influencing the behavior of the clients, this example well illustrates that *frequently* it is not appropriate to say that a zero amount of the independent variable can be administered to a control group.

therapeutic situations of the nature that were studied? How can the findings be explained in terms of a broader principle (a more general theory)? After serious consideration of these matters, the psychologist formulates an answer and publishes the findings in a psychological journal.

Inherent in the process of generalization is that of prediction (although there can be generalizations that are not used to make predictions). Here, in effect, what the therapist does by generalizing is to predict that the same results would be obtained if the experiment were repeated in a new situation. In this simple case the therapist would essentially say that for other subjects, systematic manipulation of behavior will result in more rapid recovery than mere verbal psychotherapy. To test this prediction, another therapist might conduct an experiment as outlined above (the experiment is replicated). If the new findings are the same, the hypothesis is again supported by the data. With this independent confirmation of the hypothesis as an added factor, it may be concluded that the probability of the hypothesis is increased. That is, our confidence that the hypothesis is true is considerably greater than before.[12]

With this overview before us, let us now turn to a detailed consideration of the phases of the scientific method as it applies to psychology. The first matter on which we should enlarge is "the problem."

Review of Chapter 1

In studying this book, as with most of your studies, it is recommended that you employ the whole (rather than the part) method of learning. To apply the whole method here you would first read through the table of contents and "thumb" through the entire book attempting to get a general picture of the task at hand. Then employ the naturally developed units of learning presented in the form of chapters. You would thus regard chapter 1 as a whole unit in which you would engage in several trials — you would first quickly breeze through the chapter noting the important topics. Then you might read through the chapter quite hastily, adding somewhat more to

---

[12] The oversimplification of several topics in this chapter is especially apparent in this fictitious experiment. For one, adequate control would have to be exercised over the important extraneous variable of the therapist's own confidence in, and preference for, one method of therapy. For another, it would have to be demonstrated that the participants used in this study are typical of those elsewhere before a legitimate generalization of the findings could be asserted. But these problems will be handled in due time.

your understanding of each topic. Then, at some later time when you *really* get down to business, you should read for the next "trial" for great detail, perhaps even outlining or writing down critical concepts and principles. Finally, perhaps when an examination is imminent, you would want to review your outline or notes of this chapter along with those from other chapters of the book. To help you start off, note that we have a glossary at the end of the book. You might also ask yourself questions such as the following.

1. What is the major difference between scientific and metaphysical endeavors?
2. Do you understand what Watson meant by "materialism," "determinism," and "objectivity?"
3. What is meant by "empiricism" and empirical laws?
4. Are there firm, well-established steps in the scientific method that are accepted by all scientists? Why or why not?
5. List the steps of the scientific method as presented in this book.
6. Why do you think that the problems of psychology are held to be the most challenging and complex that we face? Maybe you disagree with this.
7. Can you define the following? These are terms that you should emphasize as you proceed on through the remaining chapters of the book:
   randomization
   null hypothesis
   experimental group
   control group
   independent variable
   dependent variable
   continuous versus discrete variables
   extraneous variables
   control of extraneous variables
   statistical significance and statistical reliability
8. Edward L. Thorndike's complete statement, referred to on page 7, was that "If a thing exists, it exists in some amount. If it exists in some amount, it can be measured." Do you accept this?
9. You might wish to look over the oversimplified example of an experiment given about clinical psychology, and think of a psychological problem that especially interests you. For instance, you might be concerned about developing a more effective penal system, of controlling drug abuse, or of ascertaining the systematic effect of amount of reinforcement on a pigeon's behavior. How would you design an experiment to solve your problem?

   Finally, don't forget that your library is full of sources that you can use to elaborate on items covered in each chapter. For instance, you might wish to read further on the exciting history of the development of the concept of "materialism". By getting a good start on the study of this first chapter, your learning of the rest of the book should be materially enhanced.

2
The Problem

What Is a Problem?

A scientific inquiry starts when we have already collected a certain amount of knowledge and that knowledge indicates that there is something we don't know. It may be that we simply do not have enough information to answer a question, or it may be that the knowledge that we have is in such a state of disorder that it cannot be adequately related to the question. In either case a problem exists. The formulation of a problem is especially important for it guides us in the remainder of our inquiry. Great creativity is required here if our research is to be valuable for society. Some people see only trivial problems to research. A certain amount of genius is required to formulate an important problem with far-reaching consequences. The story is told of Isaac Newton's request for research support from the king, phrased for illustrative purposes in terms of gravitational pull on apples to the earth. The king's research grant committee rejected Newton's proposed research on gravitational theory, but asked if he would revise his grant proposal so as to solve the problem of preventing apples from bruising when they fall and hit the ground. Let us now see, in a more specific way, how we become aware of a problem.

Ways in Which a Problem Is Manifested

First, it should be obvious that studying past research helps you to become aware of problems, from among which you can select and formulate those of special interest to you. To study past research, we are fortunate to have a number of important psychological journals available in our libraries (or professors' offices for reliable borrowers). These journals cover a wide variety of researchable topics so that you can select those concerned with problems of social psychology, clinical psychology, learning, or whatever interests you. To get an overall view of the entire field of psychology, and even of research in related fields, you might wish to survey the numerous condensations that periodically appear in the journal entitled *Psychological Abstracts.* By studying our journals, we can note that the lack of sufficient knowledge that bears on a problem is manifested in at least three, to some extent overlapping, ways: first, when there is a noticeable gap in the results of investigations; second, when the results of several inquiries disagree; and third, when a "fact" exists in the form of a bit of unexplained information. As you think through these ways in which we become aware of a problem, you might start to plan the introduction of your first experimental write-up. It is in the introduction that you introduce your reader to the problem that you are seeking to solve, and it is there that you help the reader to understand why you think the problem an important one. Let us now consider in greater detail each of the three ways in which we become aware of a problem.

A Gap in Our Knowledge

Probably the most apparent way in which a problem is manifested is when there is a straightforward absence of information; we are aware of what we know and there is simply something that we do not know. If a community group plans to establish a clinic to provide psychotherapeutic services, two natural questions for them to ask are, "What kind of psychotherapy should we offer?" and "Of the different systems of psychotherapy, which is the most effective?" Now these questions are extremely important, but there are few scientifically acceptable studies that provide answers. Here is an apparent gap in our knowledge. Collection of data with a view toward filling this gap is thus indicated.

Students most often conduct experiments in their classes to solve problems of this type. They become curious about why a given kind of behavior occurs, about whether a certain kind of behavior can be produced by a given stimulus, about whether one kind of behavior is related to another kind of behavior, and so forth. Frequently, some casual observation serves as the basis for their curiosity and leads to the formulation of this kind of problem. For example, one student had developed the habit of lowering her head below her knees when she came to a taxing question on an examination. She thought that this kind of behavior facilitated her problem solving ability, and her reasoning was that she thereby "got more blood into her brain." Queer as such behavior might strike you, or queer as it struck her professors (who developed their own problem of trying to find where she hid the crib notes that she was so obviously studying), such a phenomenon is possible. And there were apparently no relevant data available. Consequently, the students in the class conducted a rather straightforward, if somewhat unusual, experiment: They asked participants to solve auditorily presented problems as their bodily positions were systematically maneuvered through space.

Similar problems that have been developed by students are: What is the effect of consuming a slight amount of alcohol on motor performance and on problem solving ability? Can the color of the clothes worn by a roommate be controlled through the subtle administration of verbal reinforcements? Do students who major in psychology evidence a higher amount of situational anxiety than those whose major is a "less dynamic" subject? Such problems as these are rather typically selected for the early experiments in a course in experimental psychology, and they are quite valuable, at least in helping the student to learn appropriate methodology. As students read about previous experiments that have been conducted in areas related to the problem that they have chosen, their storehouse of scientific knowledge grows, and their problems become more sophisticated. One cannot help being impressed by the high quality of research conducted by undergraduate students toward the completion of their course in experimental methodology. Fired by their enthusiasm for conducting their own original research, it is not infrequent for them to attempt to solve problems made manifest by contradictory results or by the existence of phenomena for which there is no satisfactory explanation.

Contradictory Results

To understand how the results of different attempts to solve the same problem may differ, consider three separate experiments that have been reported in psychological journals. All three experiments were very similar, and they all addressed themselves to the following question: "When a person is learning a task, are rest pauses more beneficial if concentrated during the first part of the total practice session or if concentrated during the last part?" For instance, if a person is to spend ten trials in practicing a given task, would learning be more efficient if rest pauses were concentrated between the first five trials (early in learning) or between the last five trials (late in learning)? The general design of all three experiments was as follows: One group of participants practiced a task with rest pauses concentrated during the early part of the practice session. As these individuals continued to practice the task on additional trials, the length of the rest pauses between trials progressively decreased. A second group of participants practiced the task with progressively increasing rest pauses between trials; as the number of trials on which they practiced the task increased, the amount of rest between trials became larger.

The first experiment indicated that progressively increasing rest periods are superior; the second experiment showed that progressively decreasing rest periods led to superior learning; while the third experiment indicated that the effects of progressively increasing and progressively decreasing rest periods are about the same. Why do these three studies provide us with conflicting results?

One possible reason for conflicting results is that one or more of the experiments was poorly conducted — certain principles of sound experimentation may have been violated. Perhaps the most common error in experimentation is the failure to control important extraneous variables. To demonstrate briefly how such a failure may produce conflicting results, let us assume that one important extraneous variable is not considered by the experimenter. Unknown to the experimenter, this variable is actually influencing the dependent variable. In one experiment it happens to assume one value, while in a second experiment on the same problem, it happens to assume a different value. Thus, it leads to different values for the dependent variable in the two experiments. The publication of two independent experiments with conflicting conclusions then presents the psychological world with a problem. The solution to this particular problem can be achieved by systematically varying the extraneous variable in a repetition of the two experiments. Let us illustrate by considering a set of experiments having to do with language suppression. In one experiment (Webster and Weingold, 1965) two pronouns were selected and repeatedly exposed in a variety of sentences to the participants of an experimental group. Control participants were exposed to the same sentences except that other pronouns were substituted for the special two. From a larger list of pronouns (that contained those two of special interest) both groups of participants selected a pronoun to use in a sentence. More specifically, the participants were told to compose sentences using any of the pronouns from the list. It was found that the experimental group tended to avoid one of those pronouns to which they had previously been exposed, relative to the frequency of their selection by the control group. It was concluded that prior verbal stimulation produces a satiation effect so that there is a suppression of pronoun choice. Regardless, however, of the theoretical issues involved, our purpose is satisfied by focusing on one of the extraneous variables in the experiment, viz., the location of the experimenter who presented the verbal materials. In this experiment the experimenter sat outside the view of the participants. Albrecht (1965) undertook to repeat the experiment, though she sat in a position that allowed the participants to see her; quite possibly the participants could thus receive additional cues, such as those that occurred when the experimenter recorded response information. The results of this repetition, needless to say, did not show a suppression effect of the two pronouns by the experimental group. Not to be discouraged, however, Albrecht and Webster (1966) again repeated the original experiment except that this time it was made certain that the experimenter was hidden from the participant's view. And this time the results confirmed Webster and Weingold's original findings.

It would thus appear that the extraneous variable of experimenter location was sufficiently powerful to influence the dependent variable values. The fact that it was different in the second experiment led to results that conflicted with those of the first experiment, thus creating a problem. The problem was apparently solved, however, by controlling this variable, and thus the reason for the conflicting results was established. We may only add that it would have been preferable to have repeated the first two experiments simultaneously, in place of the third, systematically varying experimenter location by means of a factorial design. But this point will be delayed until we take up the factorial design in chapter 10.

Explaining a Fact

The third way in which we become aware of a problem is when we are in possession of a "fact," and we ask ourselves "Why is this so?" A fact, existing in isolation from the rest of our knowledge, demands *explanation*. A science consists not only of knowledge, but of *systematized* knowledge. The greater the systematization, the greater is the scientist's understanding of nature. Thus, when a new fact is acquired, the scientist seeks to relate it to the already existing body of knowledge. But one does not know exactly where in the framework of knowledge the new fact fits, or even that it will fit. If, after sufficient reflection, we are able to appropriately relate the new fact to existing knowledge, it may be said that we have *explained* it. That fact presents no further problem. On the other hand, if the fact does not fit in with existing knowledge, a problem is made apparent. The collection of additional information is necessary so that eventually, the scientist hopes, the new fact will be related to additional information in such a manner that it will be "explained." By this gradual process, the scientist's understanding and control of nature is extended. Some problems of this kind will lead to little that is of significance for science, while others may result in major discoveries. Examples of new portions of knowledge that have had revolutionary significance are rare in pscyhology since it is such a new science, but they are relatively frequent in other sciences.[1] To illustrate how the discovery of a new fact created a problem, the solution of which has had important consequences, consider the following example.

One day the Frenchman, Henri Becquerel, found that a piece of photographic film had been fogged. He could not immediately explain this, but in thinking about it he noticed that a piece of uranium had been placed near the film before the fogging. Existing theory did not suggest that there was any connection between the uranium and the fogged film. But Becquerel suggested that the two events were connected to each other. To relate the events more specifically, he had to postulate that the uranium gave off some unique kind of energy. Working along these lines, he eventually determined that the metal gave off radioactive energy which caused the fogging, for which finding he received the Nobel Prize. This discovery led to a whole series of developments that have resulted in present-day theories of radioactivity.

The explanation of a fact constitutes a hypothesis (or theory), and it is characteristic of hypotheses that they also apply to other phenomena. That is, most hypotheses are sufficiently general that they are possible explanations of several facts. Hence, the development of a hypothesis that accounts for one fact may be a fertile source of additional problems in the sense that one may ask: "What other phenomena can it explain?" One of the most engaging aspects of the scientific enterprise is to tease out the implications of a general hypothesis and to subject those implications to additional classical empirical tests. An illustration is Hull's (1943) principles of inhibition. To oversimplify

---

[1] Wertheimer's attempts to explain the phi phenomenon may be one such case in psychology.

the matter, Hull was presented with the fact in Pavlovian conditioning of spontaneous recovery—that with the passage of time a response that had been extinguished will recover some of its strength and will again be evoked by a conditional stimulus. To explain this fact Hull postulated that there is a temporary inhibition factor that is built up each time an organism makes a response. He called this factor "reactive inhibition" and held that it is a tendency to *not* make a response, quite analogous to fatigue. When the amount of inhibition is sufficient in quantity, the tendency *not* to respond is sufficiently great that the response is extinguished. But with the passage of time, reactive inhibition (being temporary, like fatigue) dissipates, and the tendency to not respond is reduced. Hence, the strength of the response increases and it thus can reoccur—the response "spontaneously recovers."

Our point is not, of course, to argue the truth or falsity of Hull's inhibitory principles, but merely to show that a hypothesis that can explain one behavioral phenomenon can be tentatively advanced as an explanation of other phenomena. For example, the principle of reactive inhibition has also been extended to explain why distributed practice is superior to massed practice, to account for an observed superiority of a whole method over a part method of learning, and so forth (e.g., Calvin et al., 1961, Ch. 26). Each such attempt to apply this principle to other phenomena has constituted an additional problem. This, as well as others of Hull's principles, were extremely fruitful in generating new problems in our recent history that were susceptible to experimental attack.

We can thus see that the growth of our knowledge progresses as we acquire a bit of information, as we advance tentative explanations of that information, and as we explore the consequences of those explanations. In terms of problems, science is a mushrooming affair. As Dubs (1930) correctly noted, every increase in our knowledge results in a greater increase in the number of our problems. We can, therefore, judge a science's maturity by the number of problems that it has; the more problems that a given science faces, the more advanced it is.

We will conclude this section with a special thought for the undergraduate student who might be worrying about how to find a problem on which to experiment. This difficulty is not unique for the undergraduate, for we often see Ph.D. students in a panic to select a problem, fearing that they will choose a topic inappropriate for the Nobel Prize. We would like to help undergraduate *and* graduate students relax on this point—do the best that you can in selecting a problem, and don't worry excessively about its ultimate importance. However, once you have selected your problem, make as sure as you can that you study that problem with sound research methodology. You should not expect more than this from yourself.

A Problem Must Be Solvable

Not all questions that people are capable of asking can be answered by science.[2] As noted in chapter 1, a problem can qualify as the object of scientific inquiry only if it is solvable, as distinguished from an unsolvable problem. And since a hypothesis is a tentative solution of a problem, science deals only with hypotheses that are *testable*. One of the most important activities related to science, and also one of the most complex, is the determination of a criterion of testability. Such a criterion should enable us to determine whether or not a problem is solvable, i.e., a problem is solvable if, and only if, it is possible to advance a testable hypothesis as a tentative solution of it. Because of the numerous discordant views on testability, it would be impossible to offer a presentation that would satisfy all. As a result we shall get into this matter only insofar as it affects the everyday life of the experimental psychologist.[3]

The Truth Theory of Testability

Briefly, we shall say that a problem is solvable if there can be empirical reasons for answering it in a "yes" or "no" fashion. There are two important stages in the development of the theory of testability that we are following here. The first stage is the statement of truth (or verifiability) theory of testability. The second stage, an improvement on the truth theory, is the statement of a probability theory of testability.

The main principle of the truth theory with which we shall be concerned may be expressed as follows: *A proposition is testable if, and only if, it is possible to determine that the proposition is either true or false.* It follows that only a proposition (a statement or sentence) can be testable. Hypotheses are propositions, so we can use the truth theory to determine whether or not hypotheses are testable. For if it is possible to determine that a hypothesis is true or false, then the hypothesis is

---

[2] To be more precise, if a "question" can't be answered, it's really not a question at all. Hence, what might appear to be questions or problems merely because of their grammatical construction are more appropriately called "pseudoquestions" or "pseudoproblems." An example that nobody would waste time on is, "Is he either and/or I during somebody?" but much effort has been expended on pseudoquestions of the sort, "Does Ra (the Egyptian Sun God) have a green beard?"

[3] Historically, the problem has often been approached by the use of the terms *meaning* and *meaningfulness*. We shall here prefer the more neutral words *solvable* when referring to a problem and *testable* when referring to a hypothesis. For more advanced treatments of the topic refer to Reichenbach (1938); Frank (1956); Feigl and Scriven (1956), especially the chapter in the latter by Rudolph Carnap entitled "The Methodological Character of Theoretical Concepts;" and Hempel (1965). Here we shall largely follow Reichenbach's development.

testable. But if it is not possible to determine that the proposition is either true or false, then the hypothesis is not testable and should be discarded as being worthless to science.

Second, it follows that knowledge can only be expressed in the form of propositions. The following statements are examples of what we call knowledge: "That table is brown." "Intermittent reinforcement schedules during acquisition result in increased resistance to extinction." "$E = MC^2$." It is important to note, then, that events, observations, objects, or phenomena per se are not knowledge and it is irrelevant here whether events are private or external to a person. For example, external phenomena such as the relative location of certain stars, a bird soaring through the air, or a painting are not knowledge; such things are neither true nor false, nor are our perceptions of them true or false. Similarly, a feeling of pain in your stomach or your aesthetic experience when looking at a painting are not in themselves instances of knowledge. *Statements* about events and objects, however, are candidates for knowledge. For example, the statements, "He has a stomach pain," and "I have a stomach pain," may be statements of knowledge, depending on whether or not they are true. In short the requirement that knowledge can occur only in the form of a statement is critical for the process of testability. Problems are best stated in the form of questions, and such questions must be answerable if they are to be subjected to scientific inquiry. When we say that a problem (stated as a question) is solvable, it must be possible to state a hypothesis as a potential answer to the problem, and it must be possible to determine that the hypothesis is either true or false. In short, a solvable problem is one for which a hypothesis that is testable by the truth criterion can be stated.[4] If we determine that the statement of the hypothesis is true, then that statement is an instance of what we define as knowledge.

The Probability Theory of Testability

The words "true" and "false" have been used frequently in the above discussion. Strictly speaking, these words have been used only as approximations, for it is impossible to determine beyond all doubt that any given empirical proposition is true or false. The kind of world

---

[4] Of course the hypothesis must be *relevant* to the problem. For instance, if our problem is "What is the average height of Pygmies?" an irrelevant (but probably true) hypothesis would be "If a person smokes opium, then he will develop hallucinations." By relevance we shall mean that an inference can be made from the hypothesis to the problem, and the results of that inference constitute a solution to the problem.

that we have been given for study is simply not of this nature. The best that we can do is to say that a certain proposition has a determinable degree of probability.[5] Thus, we cannot say in a strict sense that a certain proposition is true—but the best that we can say is that it is *probably* true. Similarly, we cannot say that another proposition is false; rather, we must say that it is highly improbable. Strictly speaking, then, the truth theory is inadequate for our purposes, for according to it, no empirical proposition would ever be known to be testable since no empirical proposition can ever be (absolutely) true or false. Hence we shall substitute the probability theory for the truth theory, the essential difference between the two being that the words "a degree of probability" are substituted for "true" and "false." The main principle of the probability theory with which we shall be concerned is: *A proposition is testable if, and only if, it is possible to determine a degree of probability for it.*

When we say that a proposition is testable, then, we understand that it is testable as defined by the probability theory of testability. In short a problem is solvable if: (1) a relevant hypothesis can be advanced as a tentative solution for it, and (2) it is possible to test that hypothesis by determining a degree of probability for it.

Kinds of Possibilities

Let us now enlarge our understanding of the probability theory of testability. In particular, let us focus on the word "possible" contained in our statement of it. What does "possible" refer to? Does it mean that we can test the hypothesis *now*, or at some time in the future? Consider the question, "Is it possible for us to fly to Uranus?" If by "possible" here we mean that one can step into a rocket ship today and set out on a successful journey, then clearly such a venture is not possible. But if we mean that such a trip is likely to be possible sometime in the future, then the answer is in the affirmative. In the following simplified discussion we shall consider two interpretations of "possible." The first interpretation we shall call *presently attainable,* and the second *potentially attainable.*

Presently Attainable. This interpretation of "possible" concerns those possibilities that lie within the powers of people at the present time. If a certain task can be accomplished with the equipment that is immediately available, we would say that the solution to the task is presently attainable. But if the task cannot be accomplished with tools

---

[5] By "degree of probability" we mean that the proposition is true with a probability somewhere between 0 (absolutely false) and 1 (absolutely true). For instance, if a proposition has a probability of .5, then it is just as likely to be true as false.

that are presently available, the solution is not presently attainable. For example, building a bridge over the Suwannee River is presently attainable, but living successfully on Venus is not presently attainable.

Potentially Attainable. This interpretation concerns those possibilities that *may* come within the powers of people at some future time, but which are not possessed at the present. Whether or not they actually will be possessed in the future is a difficult matter to decide now. In that technological advances are sufficiently successful that we actually come to possess the powers, then the potentially attainable becomes presently attainable. For example, a trip to Uranus is not presently attainable, but we fully expect such a trip to be technologically feasible in the future. Successful accomplishment of such a venture is "proof" that the task should be shifted into the presently attainable category. Less stringently, when we can specify the procedures for solving a problem, and when it has been demonstrated that those procedures can actually be used, then we may shift the problem from the potentially to the presently attainable category.

Classes of Testability

With these two interpretations of the word "possible" in hand we may now consider two classes of testability, each based on our two interpretations.

Presently Testable. If the determination of a degree of probability for a proposition is presently attainable, then the proposition is presently testable. This statement allows considerable latitude, which we must have in order to justify work on problems which have a low probability of being satisfactorily solved as well as on straightforward, cut-and-dried problems. If one can conduct an experiment in which the probability of a hypothesis can be ascertained with the tools that are presently at hand, then clearly the hypothesis is presently testable. If we cannot now conduct such an experiment, the hypothesis is not presently testable.

Potentially Testable. A proposition is potentially testable if it may be possible to determine a degree of probability for it at some time in the future, if the degree of probability is potentially attainable. Although such a proposition is not presently testable, improvement in our techniques and the invention of new ones may make it possible to test it later. Within this category we also want to allow wide latitude. There may be statements for which we know with a high degree of certainty how we will eventually test them, though we simply cannot do it now. At the other extreme are statements for which we have a good deal of trouble imagining the procedures by which they will eventually be tested, but we are not ready to say that someone will not some day design the appropriate tools.

A Working Principle for the Experimenter

On the basis of the above considerations, we may now formulate our principles of action with regard to hypotheses. First, since psychologists conducting experiments must work only on problems that have a possibility of being solved with the tools that are immediately available, we must apply the criterion of present testability in our everyday work. Therefore, only if it is clear that a hypothesis is presently testable should it be considered for experimentation. The psychologist's problems, which are not presently but are potentially testable, should be set aside in a "wait and see" category. When sufficient advances have been made so that the problem can be investigated with the tools of science, it becomes presently testable and can be solved. If sufficient technological advances are not made, then the problem is maintained in the category of potential testability. On the other hand, if advances show that the problem that is set aside proves not to be potentially testable, it should be discarded as soon as this becomes evident, for no matter how much science advances, no solution will be forthcoming.

Applying the Criterion of Testability

In our everyday research we apply the preceding principles essentially as follows. First, we formulate a problem that we seek to solve, and then a hypothesis that is a potential solution to the problem. As we will note in the next chapter, the hypothesis is typically a statement that is general in scope in that it refers to a wide variety of events with which the problem is concerned. We then observe a sample of those events in our effort to collect data and confront the hypothesis with those observations. Next, we test the hypothesis, a process by which we conclude that the hypothesis is confirmed (supported) by the data or disconfirmed (not supported). More particularly, if our summary statements of the observations are in accord with our hypothesis, we then say that the hypothesis is confirmed — otherwise it is disconfirmed. This extremely complex process of testing hypotheses will be elaborated on throughout the book, but for now it is important to note that there are two specific criteria in order for a hypothesis to be tested (and thus to be confirmed or disconfirmed):

1. Do all of the variables contained in the hypothesis actually refer to empirically observable events?
2. Is the hypothesis *formulated* in such a way that it is possible to relate it to empirically observable events?

If all of the events referred to in the hypothesis are publicly observable, then the first criterion is satisfied. Ghosts, for instance, are not typically considered to be reliably observable by people in general, so that problems formulated about ghosts are unsolvable and corresponding hypotheses about them are untestable. Then, if a hypothesis is well formed in accordance with our rules of language, and if we can unambiguously specify how we might relate the hypothesis to empirically observable events in order to render a confirmed-disconfirmed decision, then our second criterion is satisfied. The effort to formulate a hypothesis might make reference to events and objects that are readily observable, such as "dogs smell many things," but the words might not be put together in a sensible fashion ("smell do dogs," "dogs smell do many things," etc.). Stated in such extreme forms, you might think that reasonable scientists would never formulate unsolvable problems or corresponding untestable hypotheses. Unfortunately, however, we *are* frequently victimized by precisely these errors, though in more subtle form. It is, in fact, often difficult for us to sift out statements that are testable from those that are untestable by applying the above criteria of testability. Those statements that merely pretend to be hypotheses are called "pseudostatements" or "pseudohypotheses." Pseudostatements (like "ghosts can solve problems") are meaningless because it is not possible to determine a degree of probability for them. The task of identifying some pseudostatements in our science is quite easy while others are often difficult and exacting. Since the proper formulation of, and solution to, a problem is basic to the conduct of an experiment, it is essential that the experimenter be agile in formulating solvable problems and testable hypotheses.

The Unstructured Problem

First, let us consider some problems that are obviously unsolvable as they are formulated. The student just learning how to develop, design, and conduct experimental studies usually has difficulty in isolating pseudoproblems from solvable problems. This discussion about unsolvable problems, therefore, is to give the student some perspective, so that he or she can become more proficient at recognizing and stating solvable problems. Your psychology instructor with years of experience, however, must face the fact that the vague, inadequately formulated problem will be asked by introductory students for many generations to come. How, for instance, can one answer such questions as: "What's the matter with his (her, my, your) mind?" "How does the mind work?" "Is it possible to change human nature?" and so forth. These problems are unsolvable because the intent is unclear and the domain to which they refer is so amorphous that it is impossible to specify what the relevant observations would be, much less to relate observations to such vague formulations. After lengthy discussion with the asker, however, it *might* be possible to determine what the person is trying to ask and to thereby reformulate the question so that it does become answerable. Perhaps, for example, suitable dissection of the question "What's the matter with my mind?" might lead to a reformulation such as "Why am I compelled to count the number of door knobs in every room that I enter?" Such a question is still difficult to answer, but at least the chances of success are increased because the question is now more precisely stated and refers to events that are more readily observable. Whether the game is worth the candle is another matter. For the personal education of the student, it probably is. Reformulations of this type of question, however, are not very likely to advance science.

Inadequately Defined Terms and the Operational Definition

Vaguely stated problems like the above typically contain terms that are inadequately defined, which contributes to their vagueness. However there may be problems that are solvable if we but knew what was meant by one of the terms contained in their statement. Consider, for example, the topical question, "Can machines think?" This is a contemporary analogue of the question that Thorndike took up in great detail early in the century: "Do lower animals reason?" Whether or not these problems are solvable depends on how "think" and "reason" are defined. Unfortunately, much energy has been expended in arguing such questions in the absence of clear specifications of what is meant by the crucial terms. Historically, the disagreements between the disciples, Jung and Adler, and the teacher, Freud, are a prime example. Just what *is* the basic driving force for humans? Is it the libido, with a primary emphasis on sexual needs? Is it Jung's more generalized concept of the libido as "any psychic energy?" Or is it, as Adler held, a compensatory energy, a "will to power?" This problem, it is safe to say, will continue to go unsolved until these hypothesized concepts are adequately defined, if in fact they ever are.

A question that is receiving an increasing amount of attention from many points of view is "How do children learn language?" In their step by step accounts of the process, linguists and psychologists frequently include a phase in language development which may be summarized as "Children then learn to imitate the language production of adults around them." The matter is usually left there with the feeling that this highly complex process is well accounted for. A closer analysis of "Do children learn language by imitation?" however, leads us to be not so hasty. Because we don't know what the theorist means by "imitation"—its sense may vary from a highly mystical interpretation to a concrete, objectively observable behavioral process—the question is unsolvable at this stage of its formulation.

One of the main reasons that such problems as those considered above are unsolvable as they have been stated is that many of the terms have been imported from everyday language. Our common language is replete with ambiguities, as well as with multiple definitions for any given word. If we do not give cognizance to this point, we can expend our argumentative (and research) energies in vain. Everyone can recall, no doubt, at least several lengthy and perhaps heated arguments which, on more sober reflection, were found to have resulted from a lack of agreement on the definition of certain terms that were basic to the discussion. To illustrate, suppose a group of people carried on a discussion about happiness. The discussion would no doubt take many turns, produce many disagreements, and probably result in considerable unhappiness on the part of the disputants. It would probably accomplish little, unless at some early stage in the discussion the people involved were able to agree on an unambiguous definition of "happiness." Although it is impossible to guarantee the success of a discussion in which the terms are adequately defined, without such an agreement there would be no chance of success whatsoever.

The importance of adequate definitions in science cannot be too strongly emphasized. The main functions of good definitions are: (1) to clarify the phenomenon under investigation; and (2) to allow us to communicate with each other in an unambiguous manner. These functions are accomplished by *operationally defining* the empirical terms with which the scientist deals.

When we face the problem of how to define a term operationally we, in large part, address ourselves to the question of whether or not our problem is solvable. That is, with reference to the two criteria presented above for ascertaining whether or not a problem is solvable, we made the point that the events referred to in the statement of the problem should all be publicly observable. If the terms contained in the statement of the problem can be operationally defined, then it is clear that they are empirically observable by a number of people, and the scientist has moved a long way toward rendering the problem solvable.

Essentially, an operational definition is one that indicates that a certain phenomenon exists and does so by specifying precisely how (and preferably in what units) the phenomenon is measured. That is, an operational definition of a concept consists of a statement of the operations necessary to produce the phenomenon. Once the method of recording and measuring a phenomenon is specified, that phenomenon is said to be operationally defined. The precise specification of the defining operations obviously accomplishes the intent of the scientist — by performing those operations, a phenomenon is produced and a number of observers can agree on the existence and characteristics of the phenomenon. Hence, a phenomenon that is operationally defined is reproducible by other people. Because we operationally define a concept, the definition of the concept consists of the objectively stated operations performed in producing it. Others can then reproduce the phenomenon by repeating these operations. For example, when we define air temperature, we mean that the column of mercury in a thermometer rests at a certain point on the scale of degrees. Consider the psychological concept of hunger drive. One way of operationally defining this concept is in terms of the amount of time that an organism is deprived of food. Thus, one operational definition of *hunger drive* would be a statement about the number of hours of food deprivation. Accordingly, we might say that an organism that has not eaten for 12 hours is more hungry than one that has not eaten for 2 hours.

A considerable amount of work has been done in psychology on steadiness. There are a number of different ways of measuring steadiness, and accordingly there would be a number of different operational definitions of the concept. For example, let us consider the Whipple Steadiness Test. The apparatus in this test consists of a series of holes, varying from large to small in size, and a stylus. The participant attempts to hold the stylus as steadily as possible in each hole, one at a time, without touching the sides. The number of contacts made is automatically recorded. Presumably, the steadier participant will make fewer contacts than the unsteady participant. Therefore, we would operationally define steadiness as the number of contacts made by an individual when taking the Whipple Steadiness Test. Let us add that if we measured steadiness by using other types of apparatus, then we would have additional operational definitions of steadiness.

We can now see that the first step in solving a problem is to ask whether or not the critical empirical terms can be operationally defined. What we are basically requiring is a specification of the laboratory methods and techniques for producing stimulus events and for recording and measuring response phenomena. We must be able to refer to ("point" to) some event in the environment that corresponds to each empirical term in the statement of problems and hypotheses. If no such operation is possible for all these terms, we must conclude that the problem is unsolvable and that the hypothesis is untestable. In short, by subjecting the problem to the criterion of operational definition of its terms, we render a solvable-unsolvable decision, on the basis of which we either continue or abandon our research on that question.

The movement known as operationism was initiated in 1927 by P. W. Bridgeman. The prime assumption of operationism is that the adequate definition of the variables with which a science deals is a prerequisite to advancement. In the years that have followed much has been written concerning operationism, writings that have led to many arguments. An advanced discussion of operational definitions can thus lead into matters far beyond what we need to take up here. For instance, operationism has been criticized because the operational definitions are often specific to the particular empirical investigation in which they are used. Variables specified in the statement of problems may be operationally defined in different ways by different experimenters, even though they are identified by the same word — the numerous different definitions of "anxiety" being a case in point. "Anxiety" may be operationally defined by one experimenter through the use of the Taylor Scale of Manifest Anxiety, while a different researcher may define it in terms of the operations of the Palmar Perspiration Index. Unfortunately, these two different measures of anxiety correlate approximately zero with each other (Calvin et al., 1956a). While the problem of different operational definitions of the same term is irritating, it is not at all insurmcuntable. We simply have a number of different definitions of anxiety which we might label, $Anxiety_1$, $Anxiety_2$ ... $Anxiety_N$. As we advance in our studies, we might arrive at a fundamental definition of anxiety that would encompass all the specific definitions, so that there would be one general definition that would fit all experimental usages. In the meantime, however, it is critical that we continue to use operational definitions in experimentation for at least they communicate clearly just what the researcher did in measuring and recording the events studied in the research being reported.

Operationism has also been criticized because it demands that all the phenomena with which we deal must be strictly observable, operationally definable. This requirement, when rigidly adhered to, would lead us to prematurely exclude certain phenomena from scientific investigation. "Images," for instance, were forbidden in the vocabulary of many psychologists some years ago on the basis that it was not possible to operationally define them. Still it is important from a broad perspective that we maintain some concepts, like we did "images," even though we are not at the moment able to specify how we would operationally define them. Eventually such phenomena might be subjected to fruitful scientific study, once advances in techniques for measuring them are made. Such concepts can thus be maintained in our "potentially solvable" category of problems. Within recent years the topic of imagery and images has reentered psychology in a most impressive manner so that we are now vigorously studying images in a number of different ways. Meehl (1973) pointed out a similar example in physics: In 1931 Pauli developed the notion of a neutrino, solely to preserve the laws of conservation even though the neutrino hypothesis was probably not testable—the imagined new particle was presumably not real and therefore could not be observed because it had zero charge and zero rest mass. It was not until 45 years later that experimenters were successful in detecting the neutrino more directly. It can thus be seen that we can maintain an operational approach and still keep some concepts in our science that are not immediately susceptible to operational definition, for eventually those concepts may turn out to be of considerable importance. (A major international issue has more recently concerned whether or not the U.S. should deploy a neutrino bomb.)

Impossibility of Collecting Relevant Data

Sometimes we have a problem that is sufficiently precise and whose terms are operationally definable, but we are at a loss to specify how we would collect the necessary data. As an illustration, consider the question of the effect of psychotherapy on the intelligence of a clinical patient who cannot talk. Note that we can adequately define the crucial terms such as "intelligence" and "therapy." The patient, we observe, scores low on an intelligence test. After considerable clinical work the patient's speech is improved; on a later intelligence test the patient registers a significantly higher score. Did the intelligence of the patient actually increase as a result of the clinical work? Alternatives are possible: Was the first intelligence score invalid because of the difficulties of administering the test to the nonverbal patient? Did the higher score result from merely "paying attention" to the patient? Was the patient going through some sort of transition period such that merely the passage of time (with various experiences) provided the opportunity for the increased score? Clearly it is impossible to decide among these possibilities and the problem is unsolvable as stated.

"If you attach the optic nerve to the auditory areas of the brain, will you sense visions auditorily?" Students will probably continue arguing this question until neurophysiological technology progresses to the point that we can change this potentially solvable problem to the presently solvable category. A similar candidate for dismissal from the presently solvable category is a particular attempt to explain reminiscence, a phenomenon that may appear under certain very specific conditions. To illustrate reminiscence briefly, let us say that a person practices a task such as memorizing a list of words, though the learning is not perfect. The person is tested immediately after a certain number of practice trials. Then, after a period of time during which there is no further practice, there is a test on the list of words again. On this second test suppose that it is found that the participant recalls more of the words than on the first test. This is reminiscence. We may say that reminiscence occurs when the recall of an incompletely learned task is more complete after a period of time has elapsed following the learning period than it is immediately after the learning period. The problem is how to explain this phenomenon.

One possible explanation of reminiscence is that although there are no formal practice trials following the initial learning period, the participant continues covertly to practice the task. That is, the individual "rehearses" the task "to himself or herself" following the initial practice period, and before the second test. This informal rehearsal could well lead to a higher score on the second test. Our purpose is not to take issue with this attempt to explain reminiscence. Rather, we wish to examine a line of reasoning that led one psychologist to reject "rehearsal" as an explanation of the phenomenon.

The psychologist listed several types of evidence which suggest that rehearsal cannot account for reminiscence. Among this evidence is the following statement: "Rats show reminiscence in maze learning (Burch & Magsdick, 1933), and it is not easy to imagine rats rehearsing their paths through a maze between trials" (Deese, 1952, p. 175). Such a statement cannot seriously be considered as bearing on the problem of reminiscence—there is simply no way at present to determine whether rats do or do not rehearse, assuming the common definition of "rehearse." Hence, the hypothesis that rats show reminiscence but do not rehearse is not presently testable. If we are successful in developing an effective "thought reading machine" (cf. McGuigan, in press) then we might be able to apply it to the subhuman level too. (This does not mean, of course, that the rehearsal hypothesis or other explanations of reminiscence are untestable).

To take another example of an unsolvable problem, let us consider testing two theories of forgetting: the disuse theory, which says that forgetting occurs strictly because of the passage of time; and the interference theory, which says that forgetting is the result of competition from other learned material. Which theory is more probably true? A classic experiment by Jenkins and Dallenbach (1924) is frequently cited as evidence in favor of the interference theory, and this is scientifically acceptable evidence. This experiment showed that there is less forgetting during sleep (when there is presumably little interference) than during waking hours. However, their data indicate considerable forgetting during sleep, which is usually accounted for by saying that even during sleep there is some interference (possibly from incoming stimuli, dreaming, etc.). To determine whether or not this is so, to test the theory of disuse strictly, we must have a condition in which a person has zero interference. Technically, there would seem to be only one condition which might satisfy this requirement — death. Thus, the Jenkins-Dallenbach experiment does not provide a completely general test of the theory of disuse. Therefore, we must consider the problem of whether, during a condition of zero interference, there is no forgetting, as a presently unsolvable problem, though it is potentially solvable (perhaps by advances in cryogenics wherein we can freeze, but still test, people).

The interested student should consider a number of other problems in psychology to determine whether they are solvable. For example: "Does a person perform exactly the same regardless of whether or not the individual is aware of participating in an experiment?" Can we answer the question of whether the person performs differently just because apparatus or a questionnaire or a test is used?

Vicious Circularity

Before concluding this section it may be instructive to discuss a particular kind of reasoning that, when it occurs, is outrightly disastrous for the scientific enterprise. This fallacious reasoning, called vicious circularity, occurs when an answer is based on a question and the question on the answer, with no appeal to other information outside of the vicious circle. The issue is relevant to the second criterion above as to the proper formulation of a solvable problem. A historical illustration is the development and demise of the instinct doctrine. In the early part of our century "instinct naming" was a very popular game, and it resulted in quite a lengthy list of such instincts as gregariousness, pugnacity, etc. The goal was to explain the occurrence of a certain kind of behavior, call it X, by postulating the existence of an instinct, say Instinct Y. Only eventually did it become apparent that this endeavor led exactly nowhere, at which time it was discontinued. The game, to reconstruct its vicious circularity, went thusly – Question: "Why do organisms exhibit behavior X?" Answer: "Because they have instinct Y." But the second question: "How do we know that organisms have Instinct Y?" Answer: "Because they exhibit behavior X." The reasoning goes from X to Y and from Y to X, thus explaining nothing. Problems that are approached in this manner constitute a unique class of unsolvable ones, and we must be careful to avoid the invention of new games such as "drive naming." Let us illustrate the danger from a more contemporary point of view. One question that may be important in certain contexts is why a given response did not occur. A possible answer is that an inhibitory neural impulse prevented the excitatory impulse from producing a response. That is, recent neurophysiological research has indicated the existence of efferent neural impulses that descend from the central nervous system, and they may inhibit responses. Behaviorists who rely on this concept may fall into a trap similar to that of the instinct doctrinists. That is, to the question "Why did response X fail to occur?" one could answer "Because there was an inhibitory neural impulse." Whereupon we must ask the second question again: "But how do you know that there was an inhibitory neural impulse?" and if the answer is, in effect, "Because the response failed to occur," we can immediately see that the process of vicious circularity has been invoked. To avoid this fallacious reasoning, the psychologist must rely on outside information. In this instance, one should independently record the inhibitory neural impulse, so that there is a sound, rather than a circular, basis for asserting that it occurred. Hence, the reasoning could legitimately go as follows: "Why did response X fail to occur?" "Because there was a neural impulse that inhibited it." "How do we know that there actually was such an impulse?" "Because we recorded it by a set of separate instruments," as did Hernádez-Péon, Scherrer, and Jouvet (1956).

The lesson from these considerations of vicious circularity is, to be brief, that there must be documentation of the existence of phenomena that is independent of the statement of the problem and its proposed solution. Otherwise the problem is unsolvable – there is no alternative to the hypothesis than that it be true. Guthrie's principle of learning states that "A combination of stimuli which has accompanied a movement will on its recurrence tend to be followed by that movement" (1952, p. 23). Stated more simply, when a response is once made to a stimulus pattern, the next time the stimulus pattern is presented, the organism will make the same response. To test his principle, suppose that we record a certain response to the stimulus. Then we later present the stimulus and find that a different response occurs. One might conclude that this finding disconfirms Guthrie's principle. Or, the scientist who falls victim to the vicious circularity line of reasoning might say that, while the second presentation of the stimulus appeared to be the same as the first, it must not have been. Because the response changed, the stimulus, in spite of efforts to hold it constant, must have changed in some way that was not readily apparent. A scientist who reasons thusly would never be able to falsify the principle, and hence the principle becomes untestable. To render the principle testable there must be a specification of whether or not the stimulus pattern changed from the first to the second test of it that is independent of the response finding.

Some Additional Considerations of Problems

Even after we have determined that a problem is presently testable, there are other requirements to be met before considerable effort is expended in conducting an experiment. One such requirement is that the problem be sufficiently important. Numerous problems arise for which the psychologist will furnish no answers immediately or even in the future, though they are in fact solvable problems. Some problems are just not important enough to justify research—they are either too trivial or too expensive (in terms of time, effort, and money) to answer. For instance, the problem of whether rats prefer Swiss or American cheese is likely to go unanswered for centuries; similarly, "why nations fight"—not because it is unimportant, but because its answer would require much more effort than society seems willing to expend on it.

Sometimes the psychologist is aware of a problem that is solvable, adequately phrased, and important, but an accumulation of experiments on the problem shows contradictory results. And often there seems to be no reason for such discrepancies. That is what might be called "the impasse problem." When faced with this situation, it would not seem worthwhile to conduct "just another experiment" on the problem, for little is likely to be gained, regardless of how the experiment turns out. That is, if the experiments are numerous and contradictory, little is to be gained by adding one more set of data to either side. Unless an experimenter can be extremely imaginative and develop a totally new approach that has some chance of systematizing the knowledge in the area, it is probably best to stay out of that area and one's limited energy to perform research on a problem that has a greater chance of contributing some new knowledge.

Some aspects of this general discussion may strike you as representing a "dangerous" point of view. One might ask how we can ever know that a particular problem is really unimportant. Perhaps the results of an experiment on what some regard as an unimportant problem might turn out to be very important—if not today, perhaps in the future. Unfortunately, there is no answer to such a position. Such a situation is, indeed, conceivable, and our position as stated above might "choke off" some important research. It is suggested, however, that if an experimenter can foresee that an experiment will have some significance for theory of some applied practice, the results are going to be more valuable then if such consequences cannot be foreseen. There are some psychologists who would never conduct an experiment unless it has some specific influence on a given theoretical position. This might be too rigid a position, but it does have merit.

There is no clear delineation between an important problem and an unimportant one, but it can be fairly clearly established that some problems are more likely to contribute to the advancement of psychology than are others. And it is a good idea for the experimenter to try to choose what is considered an important problem rather than a relatively unimportant problem. Within these rather general limits, no further restrictions are suggested. In any event, science is the epitome of the democratic process, and any scientist is free to work on whatever problem he or she chooses. What some scientists would judge to be "ridiculous problems" may well turn out to have revolutionary significance. Some psychologists have wished for the creation of a professional journal with a title like *The Journal of Crazy Ideas*, to encourage wild and speculative research.

Psychological Reactions to Problems

Unfortunately the existence of problems that lead to scientific advances are frequently a source of anxiety for some people. When there is a new discovery, people tend to react in one of two ways. The curious, creative person will adventurously attempt to explain it. The incurious and unimaginative person, on the other hand, may attempt to ignore the problem, hoping it will "go away." A good example of the latter type of reaction occurred around the fifteenth century when mathematicians produced a "new" number they called "zero." The thought that zero could be a number was disturbing, and a number of city legislative bodies even passed laws forbidding its use. The creation of imaginary numbers led to similar reactions; in some cases the entire arabic system of numerals was outlawed. Fortunately, this legislation was not effective.

Strangely, negative reactions to scientific discoveries have not been confined to the layman; in fact they have frequently been quite pronounced and emotional on the part of outstanding scientists. A fascinating historical consideration of factors that limit the openmindedness of scientists has been presented by Barber (1961). He discussed, for example, the reasons that it took astronomer-scientists so long to accept the Copernican theory of planetary motion. Among the reasons for rejecting the Copernican theory was that it is "simply absurd" to think that the earth moves; this is hardly a basis for scientific reasoning. Similarly, Mendel's great achievement — the development of his theory of genetic inheritance — failed to be accepted because it ran counter to the existent theories of the time, because it was "too mathematical," and so forth. And, merely because English astronomers of 1845 distrusted mathematics, Adams' discovery of a new planet (Neptune) was not published.

One major error that has been committed by scientists throughout history, even up to the present time, is judging the quality of scientific research by the status of the researcher. The most interesting problems that Mendel faced in this respect are worth exploring a bit further. Mendel, it seems, wrote deferentially to one of the distinguished botanists of the time, Carl von Nägeli of Munich. Mendel, the unimportant monk from Brüun was obviously a mere amateur expressing fantastic notions that ran, incidentally, counter to those of the master. Nevertheless, von Nägeli honored Mendel by answering him and by advising him to change from experiments on peas to hawkweed. It is ironic, indeed, that Mendel took the advice of the great man and thus labored in a blind alley for the rest of his scientific life on a plant not at all suitable for the study of inheritance of separate characteristics.

It is to be hoped that society in general, and scientists in particular, will eventually learn to assess advances in knowledge on the basis of a truth criterion alone, and that the numerous sources of resistance to discoveries will be reduced and eliminated.

Review of Chapter 2

At the end of chapter 1, we suggested some general methods that might enhance your effectiveness of studying. Perhaps at the end of each chapter you might review those suggestions and see how you can apply them to the new study unit. Remember, always try to study the *whole unit*, so that you will end up with the entire study unit defined, by this book, as experimental psychology. When preparing for your final examination you will be able to review the entire field, forming a whole unit from your entire course. For now, however, some questions from this chapter for you to consider are as follows:

1. Distinguish between a ("true") problem and a pseudoproblem—this leads you into the question of the distinction between solvable versus unsolvable problems.

2. Can you make up some examples of your own of problems that are unsolvable? Perhaps as you are walking through your everyday life you might observe and ponder events about you, and use them as stimuli for the formation of some unsolvable problems.

3. Why is it necessary in science that a problem be solvable, at least in principle?

4. What is an operational definition? Do *all* terms used in psychology need to be operationally defined? (This question should also be held for consideration throughout your more advanced study of scientific methodology.)

5. Finally, you might start the formulation of your answer to the question that is the focus of all of our academic endeavors—"What is knowledge?"

3
The Hypothesis

The Nature of a Hypothesis

We have previously noted that a scientific investigation starts with the statement of a solvable problem. Following this, a tentative solution to that problem is offered in the form of a proposition. The proposition must be testable—it must be possible to determine whether it is probably true or false. Thus, a hypothesis is a testable proposition that may be the solution to a problem. To enlarge on this definition of a hypothesis, briefly consider the relationship between the problem and the hypothesis.

If it is found after suitable experimentation that the relevant hypothesis is probably true, then we may say that the confirmed hypothesis has solved the problem to which it was addressed. If the relevant hypothesis is probably false, we may say that it does not solve the problem. Let us consider the problem: "Who makes a good bridge player?" Our hypothesis might be that "people who are intelligent and who show a strong interest in bridge make good bridge players." The collection and interpretation of sufficient data might confirm the hypothesis. In this case we say that we have solved the problem because we can answer the question.[1]

On the other hand, let us say that we fail to confirm our hypothesis. In this event it should be apparent that we have not solved the problem, i.e., we have failed to obtain definite information that these specific qualities make a good bridge player.

Frequently when we obtain a true hypothesis and thus solve a problem we may say that the hypothesis *explains* the phenomena with which the problem is concerned. Let us assume that a problem exists because we are in possession of a certain fact. As noted in the last chapter, this fact, existing in isolation, requires an explanation. The need to explain the fact presents us with our problem. If we can relate that fact to some other fact in an appropriate manner, we can say that the first fact is explained. A hypothesis is the tool by which we seek to accomplish such an explanation. That is, we use a hypothesis to state a possible relationship between one fact and another. If we find that the two facts are actually related in the manner stated by the hypothesis, then we have accomplished our immediate purpose—we have explained the first fact. (A more complete discussion of explanation is offered in chapter 15.)

To illustrate, let us return to an example used in chapter 2. Becquerel was presented with the fact that a certain photographic film had been fogged. This fact demanded an explanation. In addition to noting this fact, Becquerel also noted a second fact: that a piece of uranium was lying near the photographic film. His hypothesis was that some characteristic of the uranium produced the fogging. Becquerel's test of this hypothesis proved successful; he established that his hypothesis was true. Thus, by relating the fogging of the film to a characteristic of the uranium he explained his fact.

But what do we mean by "fact?" "Fact" is a common-sense word, and as such its meaning is rather vague. We understand something by it, such as a fact is "an event of actual occurrence." It is something that we are quite sure has happened (Becquerel was quite sure that the photographic film was fogged). But it may facilitate our task if we replace this common-sense word with a more precise term. For in-

---

[1] The only qualification is that the problem is not *completely* solved. We are using "solved" in an approximate sense and further research is required to enlarge our solution, as, for instance, finding other factors that make good bridge players. In this case we may eventually arrive at a hypothesis that has a greater probability than the earlier one; and since it is more general, containing more variables, it offers a more complete solution.

stance, instead of using the word *fact,* suppose that we conceive of the fogging of the film as a *variable,* that is, the film may be fogged in varying degrees, from a zero amount up to some large amount, such as total exposure. Similarly, the amount of radioactive energy that is given off by a piece of uranium may be conceived of as a variable; it may be an amount anywhere from zero to a large amount. Therefore, instead of saying that two *facts* are related, we may now make the more productive statement that two *variables* are related. The advantages of this procedure are sizable. For example, we may now hypothesize a *quantitative* relationship—the greater the amount of radioactive energy given off by the piece of uranium, the greater the fogging of the photographic plate. Hence, instead of making the rather crude distinction between fogged and unfogged film, we may now talk about the amount of fogging. Similarly, the uranium is not simply giving off radioactive energy, it is emitting an amount of energy. We are now in a position to make statements of wide generality. Before, we could only say that if the uranium gave off energy, the film would be fogged. Now we can say, for instance, that if the uranium gives off a little energy, the film will be fogged a small amount; if the uranium gives off a lot of energy, the film will be greatly fogged, and so on. Of course, this example is only illustrative; we can make many more statements about the relationship of these two variables with the use of numbers. But quantitative statements of hypotheses will be taken up more thoroughly shortly.

These considerations now allow us to enlarge on our preceding definition of a hypothesis. For now we may define a hypothesis as a *testable statement of a potential relationship between two (or more) variables.*

There are a number of terms used in science other than "hypothesis" to refer to statements of relationships between variables. They are such words as "theories," "laws," "principles," and "generalizations." The discussion in this and the next several chapters will be applicable to any statement of an empirical relationship between variables, without here distinguishing among them. The point that we wish to focus on is that an experiment is conducted to test an empirical relationship and, for convenience, we will usually refer to the statement of that relationship as a hypothesis. That a hypothesis is *empirical* means that it directly refers to data that we can obtain from our observation of nature. More precisely, the variables contained in an empirical hypothesis are operationally definable and hence refer to events that can be directly observed and measured. We shall talk more about statements of relationships called "theories" and "general explanations" when we talk about explanation in chapter 15.

This last point above about empirical statements is important for the understanding of the nature of a hypothesis. It will be advantageous to consider it more fully. To accomplish this let us note that all possible statements fall into one of three categories: *analytic, contradictory,* or *synthetic.* These three kinds of statements differ on the basis of their possible *truth values.* By *truth value* we mean whether a statement is true or false. Thus, we may say that a given statement has the truth value of *true* (such a statement is "true") or that it has the truth value of false (this one is "false"). Because of the nature of their construction (the way in which they are formed), however, some statements can take on only certain truth values. Some statements, for instance, can take on the truth value of *true* only. Such statements are called analytic statements (other names for them are "logically true statements" or "tautologies"). Thus, an *analytic statement is a statement that is always true—it cannot be false.* The statement "If you have a brother, then either you are older than your brother or you are not older than your brother" is an example of an analytic statement. Such a statement exhausts the possibilities, and since one of the possibilities must be true, the statement itself must be true. A *contradictory statement* (sometimes also called a "self-contradiction" or a "logically false statement"), on the other hand, *is one that always assumes a truth value of false.* That is, because of the way in which it is constructed, it is necessary that the statement be false. A negation of an analytic statement is obviously a contradictory statement. For example, the statement "It is false that you are older than your brother or you are not older than your brother" (or the logically equivalent statement "If you have a brother, then you are older than your brother and you are not older than your brother") is a contradictory statement. Such a statement includes all of the logical possibilities, but says that all of these logical possibilities are false.

The third type of statement is the synthetic statement. A synthetic statement may be defined as any statement that is neither an analytic nor a contradictory statement. In other words, a *synthetic statement is one that may be either true or false; or, more precisely, it has a probability of being true or false.* An example of a synthetic statement would be "You are older than your brother," a statement that may be either true or false. And the important point to observe in this discussion is that a *hypothesis must be a synthetic statement.* Thus any hypothesis must be capable of being proven (probably) true or false.

The differences among the three types of statements are highlighted in Table 3-1. There we may see that the symbolic statement of an analytic proposition is "*a* or not *a*." For instance, if we let *a* stand for the sentence, "I am in Chicago," then the appropriate analytic statement is "I am in Chicago or I am not in Chicago." This statement is necessarily true because no other possibilities exist. The synthetic statement is symbolized by *a*. Thus, following our example, it says "I am in Chicago," and the probability of this statement is necessarily less than (<) 1.0, but necessarily greater than (>) 0. The symbolic statement of the contradictory type of proposition is "*a* and not *a*" — "I am in Chicago and I am not in Chicago." Clearly, such a statement is absolutely false, barring such unhappy possibilities as being in a severed condition.

Table 3-1 Possible Kinds of Statements

	Analytic	*Synthetic*	*Contradictory*
Symbolic statement	*a* or not *a*	*a*	*a* and not *a*
Example of statement	"I am in Chicago, or I am not in Chicago."	"I am in Chicago."	"I am in Chicago and I am not in Chicago."
*Truth or probability value**	(Absolutely) true	$1.0 > P > 0$	(Absolutely) false

*The symbol $>$ may be read as "greater than" and $<$ as "less than." Thus, for the synthetic statement, P is less than 1.0 but zero is less than P. Or alternatively, 1.0 is greater than P; but P is greater than 0.

Now, why should we state hypotheses in the form of synthetic statements? Why not use analytic statements, in which case it would be guaranteed that our hypotheses are true? The answer to such a question appears in an understanding of the function of the various kinds of statements. The reason that a synthetic statment may be true or false is that it refers to the empirical world, i.e., it is an attempt to tell us something about nature. And as we previously saw, every statement that refers to natural events might be in error. An analytic statement, however, is empty. That is, while it is absolutely true, it tells us nothing about the empirical world. This characteristic results because an analytic statement includes all of the logical possibilities, but it does not attempt to inform us which is the true one. And this is the price that one must pay for absolute truth. If one wishes to state information about nature, one must use a synthetic statement, in which case the statement always runs the risk of being false. Thus, if someone asks me if you are older than your brother I might give my best judgment, say, "you are older than your brother," which is a synthetic statement. I may be wrong in this statement, but at least I am trying to tell the person something about the empirical world. Such is the case with our scientific hypotheses; they may be false in spite of our efforts to assert true hypotheses, but they are potentially informative in the sense that they attempt to say something about nature.

Now, if analytic statements are empty and thus do not tell us anything about nature, why do we bother with them in the first place? The answer to this question could be made quite detailed. Suffice it to say here that analytic statements are valuable in facilitating deductive reasoning (logical inferences). The statements in mathematics and logic are analytic and contradictory statements and are valuable to science because they allow us to transform synthetic statements without adding additional knowledge (recall that analytic statements are empirically empty). The essential point is that sciences use all three types of statements, but use them in different ways. Here we emphasize the point that the synthetic type of proposition is used for the statement of hypotheses; for, in stating a hypothesis, we attempt to say something informative about the natural world. This attempt carries with it the possibilities that our hypothesis is probably true or false.

The Manner of Stating Hypotheses

Granting, then, that a hypothesis is a statement of a potential empirical relationship between two or more variables, and also that it is possible to determine whether the hypothesis is probably true or false, we might well ask what form that statement should take. That is, precisely how should we state hypotheses in scientific work?

Lord Russell answered this question by proposing that the logical form of the general implication be used for expressing hypotheses (cf. Reichenbach, 1947, p. 356). Using the English language, the general implication may be expressed as: "*If* . . . , *then*. . . ." That is to say, *if* certain conditions hold, *then* certain other conditions should also hold. To better understand the "If . . . , then. . . ." relationship, let *a* stand for the first set of conditions and *b* for the second set of conditions. In this case the general implication would be "If *a*, then *b*." But in order to communicate what the conditions indicated by *a* are, we must make a statement. Therefore, we shall consider that the symbols *a* and *b* are actually statements that express these two sets of conditions. And if we join these two simple statements, as we do when we use the general implication, then we end up with a single compound statement. This compound statement *is* our hypothesis.

The statement *a*, incidentally, is referred to as the *antecedent condition* of the hypothesis (it "comes first"), and *b* is called the *consequent condition* of the hypothesis (it follows the antecedent condition). We have previously noted that a hypothesis is a statement relating two variables. Since we have said that antecedent and consequent conditions of a hypothesis are stated as propositions, it follows that the symbols *a* and *b* are *propositional variables.* A hypothesis, thus, proposes a relationship between two (propositional) variables by means of the general implication as follows: "If *a* is true, then *b* is true." The variables *a* and *b* may stand for whatever we wish. If we suspect that two particular variables are related, we might hypothesize a relationship between them. For example, we might think that industrial work groups that are in great inner conflict have decreased production levels. Here the two variables are: (1) the amount of inner conflict in an industrial work group, and (2) the amount of production that work groups turn out. We can formulate two sentences: (1) "An industrial work group is in great inner conflict," and (2) "That work group will have a decreased production level." If we let *a* stand for the first statement and *b* for the second, our hypothesis would read: "*If* an industrial work group is in great inner conflict, *then* that work group will have a decreased production level."

With this understanding of the general implication for stating hypotheses, it is well to inquire about the frequency with which Russell's suggestion has been accepted in psychology. The answer is clear: The explicit use of the general implication is almost nonexistent. Two samples of hypotheses, essentially as they are stated in professional journals, should illustrate the point:

1. The purpose of the present investigation was to study the effects of a teacher's verbal reinforcement on pupils' classroom demeanor.
2. Giving students an opportunity to serve on university academic committees results in lower grades in their classes.

Clearly these hypotheses, or implied hypotheses, fail to conform to the form specified by the general implication. Is this bad? Are we committing serious errors by not precisely heeding Russell's advice? Not really. For as Hempel and Oppenheim (1948) pointed out, it is always possible to restate such hypotheses as general implications. For example, the above hypotheses could be restated as follows.

The first hypothesis contains two variables: (1) amount of verbal reinforcement, and (2) amount of acceptable classroom behavior. The propositions concerned with these variables are: (1) A teacher verbally reinforces a student for desirable classroom performance, and (2) the student's demeanor improves. The hypothesis relating these two variables may now be expressed. "*If* a teacher verbally reinforces a student for acceptable classroom behavior, *then* the student's classroom behavior will improve."

We may similarly identify the variables contained in the second hypothesis and state the propositions as follows: (1) Students are given the opportunity to serve on university academic committees, and (2) those students achieve lower grades in their classes. The hypothesis: "*If* students are given the opportunity to serve on university academic committees, *then* those students will achieve lower grades in their classes."

It is apparent that these two hypotheses fit the "If *a*, then *b*" form, although it was necessary to modify somewhat the original statements. Even so, these modifications did not change the nature or the meaning of the hypotheses.

What we have said to this point, then, is that Russell has suggested the use of the general implication for stating hypotheses, and that psychologists do not take his advice in that they express their hypotheses in a variety of ways. However, we can restate their hypotheses as general implications. The next question, logically, is why did Russell offer this advice, and why are we making a point of it here? Briefly, the way in which we determine whether or not a hypothesis is confirmed depends on our making certain inferences from experimental findings to the hypothesis. The rules of logic tell us what kind of inferences we can legitimately make (they are called valid inferences). But in order to determine whether or not the inferences are valid, the statements involved in the inferences (e.g., the hypotheses) must be stated in certain standard forms, one of which is the *general implication*. Hence, in order to discuss the nature of experimental inferences and really to understand them, we must use standard logical forms. This area will be covered more completely in chapter 13 when we consider the nature of experimental inferences.

Another reason is that if the experimenter attempts to state a hypothesis as a general implication it may help to clarify the prime reason for conducting the experiment. That is, by succinctly and logically writing down the purpose of the experiment as a test of a general implication, the experimenter is forced to come to grips with the precise nature of the relevant variables. Any remaining vagueness in the hypothesis can then be removed when operational definitions of the variables are stated.

There is yet another form that is used in stating hypotheses. It involves certain mathematical statements that are essentially of the following nature: $Y = f(X)$, That is, a hypothesis stated in this way proposes that some variable, Y, is related to some variable, X, or, alternatively, that Y is a function of X. Such a mathematically stated hypothesis clearly fits our more general definition of a hypothesis, namely that two variables are related. Although the variables in this case are quantitative (their values can be measured with numbers), the two variables may still refer to whatever we wish. In psychology the classical paradigm for the statement of our laws has been in the form of R as a function of S: that is, $R = f(S)$ — in this instance we identify a response variable *(R)* that systematically changes as the stimulus *(S)* is varied. For instance, we might refer to hypothesis 2 above and assign numbers to the independent variable. The independent variable, which would be X in the equation $Y = f(X)$, might be stated as the extent to which students serve on committees. We might develop a scale such that zero would indicate no service, one a little service, two a medium amount of service, and so on. Course grades, the dependent variable, Y, are similarly quantified such that an A is 4.0, a B is 3.0, and so on. Thus, the hypothesis could be tested for all possible numerical values of the independent and the dependent variables.

In any event, the important point here is that even though a hypothesis is stated in a mathematical form, that form is basically of the "If *a*, then *b*" relation. Instead of saying "If *a*, then *b*" we merely say "If (and only if) X is this value, then Y is that value." For example if X is 3 (medium committee service) then Y is 2.0 (an average grade).

There are two common misconceptions about the statement of hypotheses as general implications that we should consider. First, it is erroneous to say that the antecedent conditions *cause* the consequent conditions. This may or may not be the case. The general implication merely states a potential relationship between two variables — *if* one set of conditions holds, *then* another set will be found to be the case — not that the first set causes the second. Thus, if the hypothesis is in fact highly probable, we can expect to find repeated occurrences of both sets of conditions together. But the general implication says nothing about *a* causing *b*.

Second, the general implication does not assert that the consequent conditions are true. Rather, it says that *if* the antecedent conditions are true, *then* the consequent conditions are true. For example, the statement, "*If* I go downtown today, *then* I will be robbed" does not mean that I *will be* robbed. Even if the compound statement is true, I might not go downtown today. Thus, *if* the hypothesis is highly probable, then whether or not I will be robbed depends on whether or not I satisfy the antecedent conditions.

We might emphasize that all hypotheses have a probability character in that none of them can be absolutely true or false. Yet the above forms for stating hypotheses, whether they are the logical form of "If a, then b," or the mathematical form of $R = f(S)$, they yield statements that are absolute in that we do not attach a probability value to them. The statement "If a, then b," for instance, can only be true or false. When we use these forms of statements, then, it is in a sense of approximation and we include the qualifier that the statements are "probably true" or "probably false."

There is one final matter about the logical character of scientific laws that we will consider before concluding this section. Nagel (1961) pointed out that our laws must express a stronger connection between the antecedent and consequent conditions than what he refers to as a mere "matter of fact concommitance." By this term he meant that the antecedent and consequent conditions are only accidentally occurring together. As an illustration of a statement in which there is mere accidental concommitance, Nagel considered "All screws in Smith's current car are rusty." It is apparent that there is no necessary connection in this statement between the antecedent condition of screws being in Smith's car and the consequent condition of those screws being rusty. No one is likely to maintain, for instance, that ". . . if a particular brass screw now resting on a dealer's shelf were inserted into Smith's car, that screw would be rusty" (Nagel, 1961, p. 52). It is clear that statements of this kind are universal, and since the relationship is accidental, Nagel refers to them as instances of *accidental universality*. Rather than having laws that have accidental universality, Nagel suggested that our statements should express a *nomic universality* (lawfullness), in which case there is some element of necessity between the antecedent and consequent conditions. Nagel offered the statement that "Copper always expands on heating" as an example of one that has nomic universality. By rephrasing this sentence we can note that it fits the form of the implication as follows: "If copper is heated, then it will expand." In this case heating the copper (the antecedent condition) physically necessitates expansion (the consequent condition). In contrast, merely placing a new brass screw into Smith's car does not, in any sense of the word, "necessitate" or produce another rusty screw.

Types of Hypotheses

We have suggested that the general implication is a good form for stating hypotheses. We have discussed the use of the implication, but have said nothing explicitly about the generality of the implication. In one of the previously cited examples it was said that, if an industrial work group has a specific characteristic, certain consequences follow. We did not specify what industrial work group, but it was understood that the hypothesis concerns at least *some* industrial work groups out of all possible groups. Are we now justified in asserting that the hypothesis holds for all industrial work groups? The answer to this question is ambiguous, and there are two possible courses. First, we could say that the particular work group out of all possible work groups with which we are concerned is unspecified, thus leaving the matter up in the air. Or, second, we could assume that we are asserting a universal hypothesis, i.e., that it is implicitly understood that we are talking about *all* industrial work groups that are in conflict. In this instance, if you take *any* industrial group in conflict, the consequences specified by the hypothesis should follow. In the interest of the advancement of knowledge, we lean toward the latter interpretation, for if the former interpretation is followed, no definite commitment is made on the part of the scientist, and if nothing is risked, nothing is gained. If it is found that the hypothesis is not universal in scope (that it does not apply to *all* industrial work groups), it must be further limited. This is at least a definite step forward. That this is not an idle question is made apparent by reviewing the psychological literature. Hull, for example, states a number of his empirical generalizations in this manner. His classical Postulate IV says: "If reinforcements follow each other at evenly distributed intervals . . . the resulting habit will increase in strength. . . ." (Hull, 1952, p. 6). Without worrying here about what the specific variables in Postulate IV mean, we may observe the form of the principle. Is it clear that Hull is asserting some relationship between *all* reinforcements and *all* habits? It is by no means, but the most efficient course open to us, as we have discussed it above, would be to *assume* that such a universal relationship is being asserted.

Although it is recognized that the goal of the scientist is to assert hypotheses in as universal a fashion as possible, it is also clear that we should explicitly state the degree of generality with which we are asserting our hypotheses. With this in mind, let us investigate the possible types of hypotheses that are at the disposal of the scientist.

The first type of hypothesis is what is called the *universal hypothesis,* which asserts that the relationship in question holds for all the variables that are specified, for all time, and at all places. An example of a universal hypothesis would be "For all rats, if they are rewarded for turning left, then they will turn left in a T maze." We may add that universal hypotheses in psychology typically have to be restricted in scope.

Another type of hypothesis is the *existential hypothesis,* which asserts that the relationship stated in the hypothesis holds for at least one particular case ("existential" implies that one exists); for instance,

"There is at least one rat, that if it is rewarded for turning left, then it will turn left in a T maze.[2] Examples of this type of hypothesis abound. Hull's Postulate V (D), for instance, says, "At least some drive conditions tend partially to motivate into action habits which have been set up on the basis of different drive conditions" (Hull, 1952, p. 7). Again, we need not be concerned here with what the postulate actually says, but merely note that it is in the form of the existential hypothesis. (It might be better to say, "At least one drive condition . . ." to make it clear that Hull's hypothesis assumes the existential form as previously discussed.) It may be added that this form of hypothesis can be very useful in psychological work, for many times a psychologist asserts that a given phenomenon exists, regardless of how frequently it occurs. In this connection Bugelski said, "often one subject is as useful as many, if the problem involved is of the 'is-it-possible' nature. Hermann Ebbinghaus used himself as a subject and contributed greatly to our knowledge of learning. Raymond Dodge studied his own knee jerk for several years and made notable contributions to our information about reflex action. It takes only one positive case to prove something can happen" (Bugelski, 1951, pp. 115–16). The increased frequency with which one participant is studied, as in the instance of single case methodology (frequently referred to as $n = 1$ designs as in chapter 11), provides other illustrations of the employment of existential hypotheses.

Once a scientist has confirmed an existential hypothesis, the immediate task is accomplished. But more than establishing the existence of a phenomenon we may wish to entertain the question of the phenomenon's generality. Typically, phenomena specified in existential hypotheses are difficult to observe, and one cannot easily leap from this type of highly specialized hypothesis to an unlimited, universal one. Rather, the scientist seeks to establish the conditions under which the phenomenon does and does not occur so that we can eventually assert a universal hypothesis with necessary qualifying conditions. Let us illustrate by oversimplifying a research project that sought to enhance our understanding of hallucinations (McGuigan, 1966a). In the problem formulation stage the author asked some clinical psychologists about the mechanisms of this most interesting phenomenon. The initial reply was that hallucinations were "ideational in nature." Perhaps the quizzical look on the author's face communicated the difficulty he was having about how to operationally define "ideational in nature," thus stimulating a reply that was more concrete: "Well,

---

[2] There are other types of hypotheses that could be discussed but because they are relatively rarely used they will not be considered here. (cf. Hempel, 1945; Reichenbach, 1949).

they're cortical events." Though buried deep in the central nervous system, hallucinations so conceived at least had some sort of potential reality status. Their direct study was another problem. A more feasible approach was to consider behavioral (muscular response) manifestations of hallucinations, regardless of possible central nervous system involvement. But even behavioral aspects of hallucinations would be difficult to record, since behavior would clearly not be overt and readily observable. Rather, any responses involved in the occurrence of hallucinations would have to be so small that one could not see them with the naked eye; they would, therefore, have to be electronically amplified to make them "visible." Furthermore, since there are a number of different kinds of hallucinations, the location of these small, covert responses would probably differ according to variety of the hallucination. One could hypothesize, for instance, that covert *speech* responses might be involved in the occurrence of auditory hallucinations, that covert *ocular* responses might uniquely occur in the case of visual hallucinations, and so forth. Even leaving aside the difficulties of recording covert responses, there are questions of how to obtain patients who will admit that they hallucinate (most psychotics who evidently hallucinate refuse to so admit on the grounds that their physician might think they are "crazy"), or how to know when a person actually experiences these private events, or how to adequately communicate with a schizophrenic who hallucinates. Perhaps by now you are getting a suitable universal hypothesis about covert responses in all patients who have all kinds of hallucinations. Consequently, rather than attempt to take a giant step that would be extremely difficult, if not impossible, the more modest approach in this research was to test an existential hypothesis. The notion was, so to speak, that auditory hallucinations are the product, at least in part, of a person covertly speaking (slightly "whispering") to himself. More precisely, the existential hypothesis was that "There is at least one paranoid schizophrenic who, if he auditorally hallucinates, then he emits covert oral responses." The research *did* confirm this hypothesis, i.e., it was found that slight speech responses coincided with the patient's report of "hearing voices." Once it was established that the phenomenon existed and that it could be recorded, the credibility of some sort of universal hypothesis increased; the question is just how a universal hypothesis should be qualified. To answer this question one would next attempt to record covert oral responses during the hallucinations of other patients. No doubt failure should sometimes be expected and the phenomenon might be observable, for instance, only for paranoid schizophrenics who have auditory hallucinations and not for visual, olfactory, etc., hallucinators. Furthermore, success might occur only, say, for "new" patients, and not for chronic psychotics. But, whatever the specific conditions under which the phenomenon occurs, research should eventually lead to a universal hypothesis which includes a statement that limits its domain of application. For instance, it might say that "For all paranoid schizophrenics who will admit to hallucinations of the auditory variety and who have been institutionalized for less than a year, if they auditorially hallucinate, then they emit covert oral responses."

We can thus see how research progresses in a piecemeal, step by step fashion. Our goal is to formulate propositions of a general nature, but this is accomplished by studying one specific case after another, only gradually arriving at statements of increasing generality.

One of the reasons that we seek to establish universal statements is that the more general statement has the greater predictive power. Put the other way, a specific statement has extremely limited predictive power. Consider the question, for example, of whether purple elephants exist. Certainly no one would care to assert that all elephants are purple, but it would be most interesting if one such phenomenon were observed; the appropriate hypothesis, therefore, is of the existential type. Should it be established that the existential hypothesis was confirmed, the delimiting of conditions might lead to the universal hypothesis that "For all elephants, if they are in a certain location, are 106 years old, and answer to the name 'Tony,' then they are purple." Though such a highly specific hypothesis would clearly not be very useful for predicting future occurrences—an elephant that showed up in that location at some time in the distant future would be unlikely to have the characteristics specified.

Arriving at a Hypothesis

It is difficult to specify the processes by which we arrive at a hypothesis with any finality. Although considerable research is being conducted on this problem, it is not possible at this time to specify adequately just what phases a scientist goes through in arriving at a hypothesis. We *can* say that the process of arriving at a hypothesis is a creative matter that has been the object of studies in thinking, imagination, concept formation, and the like. This leads to a distinction advanced by Reichenbach (1938), who said that the manner in which the scientist actually arrives at his hypothesis falls within the *context of discovery*, and the presentation of the proof that a hypothesis is probably true is in the *context of justification*. Thus, science in general is not interested in the context of discovery (however, psychology is interested, since this is a portion of its subject matter). Science is concerned with the context of justification, for here, instead of presenting the thought processes as they occurred in the development of the hypothesis, the scientist reconstructs those thought processes in a logical manner and systematically presents them in a scientific report; we set forth a justifiable set of inferences that lead from one statement to another. The adequate expression of our thoughts in science is through the rational reconstruction of those thoughts. When the scientist publishes a hypothesis, and material related to it, the scientist does not relate how the hypothesis was actually arrived at. We do not say, "While I was sitting in the bathtub the following hypothesis occurred to me...." Rather scientists justify their hypotheses. What they write falls within the context of justification, which is the concern of the major part of this book. In this connection we might note that some critics of the study of scientific methodology (or philosophy of science) say that such an endeavor is worthless. They say that scientific discoveries are not made in a logical, step-by-step fashion in strict conformity with the scientific method; a scientist does not sit down with a problem and rationally go through the phases of the scientific method as listed in chapter 1. To this criticism we must answer that this may be true, or in fact it may not be true, but whether it is true or false is really irrelevant. What is relevant is that when scientists communicate findings to others, they employ the context of justification. Whether or not the use of the context of justification will facilitate the discovery of hypotheses is an empirical question for studies in the psychology of thinking. It *is* possible that the making of valuable discoveries can be facilitated by studying the scientific method, for it seems reasonable that if we learn how scientific discoveries have been made in the past, and set such procedures down in a systematic way, we may be able to make new discoveries more efficiently in the future.

Dealing further with the context of discovery, we may note that when a scientist arrives at a hypothesis, a mass of data is surveyed (implicitly or explicitly), certain aspects of those data are abstracted, the scientist perceives some similarities in the abstractions, and relates those similarities in order to arrive at generalizations. For instance, the psychologist particularly observes stimulus and response events. It is noted that some stimuli are similar to other stimuli and that some responses are similar to other responses. Those stimuli that were perceived as being similar according to a certain characteristic are then defined as belonging to the same class, and similarly for the responses. Consider the following situation in a Skinner Box. A rat presses a lever and receives a pellet of food. At about the time the rat presses the lever, a click is sounded. After a number of associations between the click, pressing the lever, and eating the pellet, the rat will learn to press the lever when a click is sounded.

In this situation the experimenter judges that all of the separate instances of the lever-pressing response are sufficiently similar for them to be classified together. A class of lever-pressing responses is thus formed out of a number of similar lever-pressing response instances. In like manner a class of all of the stimulus instances of clicks is formed, judging that all of the clicks are similar enough to form a general class. It can thus be seen that the psychologist uses classification to distribute a large amount of data into a smaller number of categories that can be handled efficiently. The handling of these data is then facilitated by assigning symbols to the classes. Then attempts are made to formulate relationships between the classes. By "guessing" that a certain relationship exists between the classes, the hypothesis is formulated. For example, one might state that when a click is made, a certain response will follow. The hypothesis would be: If a click stimulus is presented a number of times to a rat in a Operant Box, and if the click is frequently associated with pressing a lever and eating a pellet, then the rat will press the lever in response to the click on future occasions. Parenthetically, this seems to be typical of the process that the scientist goes through, more or less as an ideal. Some scientists, probably the more compulsive ones, go through each step in considerable detail, while others do so in a more haphazard fashion. But whether or not scientists go through these steps in arriving at a hypothesis, explicitly, they all seem to approximate them to some extent.

We may note the view of some others on this topic. Dubs wrote: "It is a well known fact that most hypotheses are derived from analogy.... Indeed, careful investigations will very likely show that all philosophic theories are developed analogues" (Dubs, 1930, p. 131). In support he pointed out that Locke's conception of simple and complex ideas was probably suggested by the theory of chemical atoms and compounds that was becoming prominent in his day. Underwood has written: "How does one learn to theorize? It is a good guess that we learn this skill in the same manner we learn anything else—by practice" (Underwood, 1949, p. 17).

Of course, some hypotheses are more difficult to formulate than others. It seems reasonable to say that the more general a hypothesis is, the more difficult it is to conceive. The important general hypotheses must await the genius to proclaim them, at which time a science usually makes a sizable spurt forward, as happened in the cases of, say, Newton and Einstein. It appears that, to formulate useful and valuable hypotheses, a scientist needs, first, sufficient experience in the area and, second, the quality of "genius." The main problem in formulating hypotheses in complex and disorderly areas is the difficulty of establishing a new "set"—the ability to create a new solution that runs counter to, or on a different plane from, the existing knowledge. This is where scientific "genius" is required.

Consider the source of hypotheses from a somewhat different point of view, in particular with reference to the results of a scientific inquiry. The findings can be regarded, in a sense, as a stimulus to formulate new hypotheses—results may be used to test a hypothesis, but they can also suggest additional ones. For example, if the results indicate that the hypothesis is false, they can possibly be used to form a new hypothesis that is in accord with the results of the experiment. In this case the new hypothesis must be tested in a new experiment. Now, what happens to a hypothesis if it turns out to be false? If there is a new (potentially better) hypothesis to take its place, it can be readily discarded. But if there is no new hypothesis, then we are likely to maintain the false hypothesis, at least temporarily, for no hypothesis ever seems to be finally discarded in science unless it is replaced by a new one.

Criteria of Hypotheses

After we have formulated our hypothesis (or better, our hypotheses), we must determine whether or not the hypothesis is a "good" one. Of course, we eventually will test our hypothesis to determine whether the data confirm or disconfirm it, and certainly, other things being equal, a confirmed hypothesis is better than a disconfirmed hypothesis, in the sense that it offers one solution to a problem and thus provides some additional knowledge about nature. But even so, some confirmed hypotheses are better than other confirmed hypotheses. We must now inquire into what we here mean by "good" and by "better." To answer this question we offer the following criteria by which to judge hypotheses. Each criterion should be read with the understanding that the hypothesis that meets it to the greatest extent is the best hypothesis, assuming that the hypothesis satisfies the other criteria equally well and that the criteria are about equal in worth. It should also be understood that these are flexible criteria. They are offered tentatively, and as the information in this important area increases, they will no doubt be modified. The hypothesis:

    1. . . . must be testable. The hypothesis that is presently testable is superior to the hypothesis that is only potentially testable.

    2. . . . should be in general harmony with other hypotheses in the field of investigation. While this is not essential, in general the

disharmonious hypothesis has the lower degree of probability. For example, consider the hypothesis that eye color is related to intelligence. This hypothesis is at an immediate disadvantage because it conflicts with the existing body of knowledge. We know, for instance, that hair color has never been demonstrated to be related to intelligence. There is considerable additional knowledge of this sort which, taken together, would suggest that the "eye color" hypothesis is not true — it is not in harmony with what we already know.

3. . . . should be parsimonious. If two hypotheses are advanced to answer a given problem, the more parsimonious one is to be preferred. For example, if we have evidence that a person has correctly guessed the symbol (hearts, clubs, diamonds, spades) on a number of cards significantly more often than chance, we might advance several hypotheses to account for this phenomenon. One might be to postulate extrasensory perception (ESP) and another to say that the subject "peeked" in some manner. The latter would be more parsimonious because it does not require that we hypothesize new very complex mental processes. The principle of parsimony has been previously expressed in various forms. For instance, William of Occam advanced a rule (called *Occam's razor*) to the effect that entities should not be multiplied without necessity. A similar rule was expressed by G. W. Leibniz' principle of the identity of indiscernibles. Lloyd Morgan's canon is an application of the principle of parsimony to psychology: "In no case is an animal activity to be interpreted in terms of higher psychological processes, if it can be fairly interpreted in terms of processes which stand lower in the scale of psychological evolution and development" (Morgan, 1906, p. 59). It is apparent that these three principles have the same general purpose — they all seek, other things being equal, the most parsimonious explanation of a problem (cf. Newbury, 1954). Thus, we should not prefer a complex hypothesis if a simple one has equal explanatory power; we should not use a complex concept in a hypothesis (e.g., ESP) if a simpler one will serve as well (e.g., peeking at the cards); we should not ascribe higher capacities to organisms if the postulation of lower ones can equally well account for the behavior to be explained.

4. . . . should answer (be relevant to) the problem. It would seem unnecessary to state this criterion, except that examples can be found in the history of science where the right answer was given to the wrong problem. It is often important to make the obvious explicit.

5. . . . should have logical simplicity. By this we mean logical unity and comprehensiveness, not ease of comprehension. Thus, if one hypothesis can account for a problem by itself, and another hypothesis can also account for the problem but requires a number of

supporting hypotheses or ad hoc assumptions, the former is to be preferred because of its greater logical simplicity (cf., Cohen & Nagel, 1934, pp. 212–15). (The close relationship of this criterion to that of parsimony should be noted.)

6. . . . should be expressed in a quantified form, or be susceptible to convenient quantification. The hypothesis that is more highly quantified is to be preferred. The advantage of a quantified over a non-quantified hypothesis was illustrated earlier in the example from the work of Becquerel.

7. . . . should have a large number of consequences, should be general in scope. The hypothesis that yields a large number of deductions (consequences) will explain more facts that are already established, and will make more predictions about events that are as yet unstudied or unestablished (some of which may be unexpected and novel). In general it may be said that the hypothesis that leads to the larger number of deductions will be the more fruitful hypothesis.

The Guidance Function of Hypotheses

We have already discussed the ways in which hypotheses allow us to establish "truth." It is well here to ask how an inquiry gets its direction. In nature, for instance, how do we know where to start our search for "truth"? The answer is that hypotheses direct us. An inquiry cannot proceed until there is a suggested solution to a problem in the form of some kind of hypothesis.[3]

Francis Bacon proposed that the task of the scientist is to classify the entire universe. But the number of data in the universe is, if not infinite, at least indefinitely large. To make observations in such a complex world, we must have some kind of guide. Otherwise we would have little reason for not sitting down where we are and describing a handful of pebbles or whatever else happens to be near us.

---

[3] Whether or not one chooses to assume that a hypothesis guides an inquiry in a strictly exploratory type of experiment (see p. 75) is an arbitrary decision. We are taking the position here that some guidance is offered in this type of situation, that there is some reason that data of a particular kind are gathered, and whether or not there is an explicit hypothesis, it is assumed that there is at least an implicit one, no matter how vague it might be. As Underwood said ". . . probably no one undertakes an investigation without some thought as to 'what will happen.' Such thoughts may not be verbalized, but that they are almost universally there no one can doubt" (Underwood, 1957b, p. 208). And again, "Research in a relatively new area of investigation is seldom undertaken without some conceptual scheme in mind . . . without some preconception as to the nature of the phenomena and perhaps the processes lying behind them. These predilections are usually lightly held but they do afford the initial working hypotheses . . ." (Underwood, 1957b, p. 258).

We must set some priority on the kind of data that we study, and this is accomplished by the hypothesis. Hypotheses, then, serve to guide us to make observations that are pertinent to our problem; they tell us which observations are to be made and which observations are to be omitted. If, for instance, we are interested in the problem of why a person taps every third telephone pole that is passed, our hypothesis would probably guide us in the direction of a better understanding of compulsions. It would take us a long time if we started out in a random direction to solve our problem and commenced, for instance, counting the number of blades of grass in a field. That this point is not universally accepted, however, is apparent when one notes that others have asserted that hypotheses are not valuable. They ask, for instance, what there is to guide us in selecting our hypotheses. To this we must answer that we know very little, but that we must be confident that more complete answers are, at least eventually, forthcoming as our research on the thinking process accumulates. Such a question is one that clearly lies within the context of discovery, and to place it within the context of justification is a serious error.

On Accident, Serendipity, and Hypotheses

The goal of understanding that part of nature which is called "the behavior of organisms" is probably as difficult a task as the human race has ever set itself. One of the reasons for the difficulty is the great expanse of the behavioral realm; the number of response events that we *could* conceivably study staggers the imagination. Consequently, we need to assign priorities to the kinds of behavioral phenomena that we study in any detail. Hypotheses, we have said, serve this function— they help to tell us which of an indefinitely large number of responses are more likely to justify our attention. During the conduct of an experiment in which a certain hypothesis is being tested, however, one need not be blind to other events. In fact, to reach a goal psychologists must necessarily be alert to all manner of happenings other than that to which they are *primarily* directing their attention. For, on occasion, some chance observation that is irrelevant to, or different from, the hypothesis being tested may lead to the formulation of an even more important hypothesis. We have, you will note, mentioned several examples of the role of accidental observations in science. This is, in fact, a sufficiently important phenomenon that it justifies the use of a unique term. "Serendipity" is the word that has recently become quite popular. "Serendipity" was borrowed from Walpole's "Three Princes of Serendip" by the physiologist Cannon (1945). Walpole's story concerned a futile search for something, but the finding of many valuable things which were not sought. And so it is in science—the scientist may vainly seek "truth" by being guided by one hypothesis, but in the search accidentally observe an event that leads to a more fruitful hypothesis. An interesting case in point was related by Fisher (1964). This researcher was interested in setting off drives such as hunger and thirst by direct chemical stimulation of specific brain cells. It had been established that the injection of a salt solution into the hypothalamus of goats increased the thirst drive, thus resulting in their drinking large quantities of water. Analogously, Fisher sought to test the hypothesis that injection of the male sex hormone into a rat's brain would trigger male sexual behavior. As he told the story of the "The Case of the Mixed-up Rat":

> By one of those ironic twists that are so typical of scientific research, the behavioral change produced in my first successful subject was a completely unexpected one. Within seconds after the male hormone was injected into his brain he began to show signs of extreme restlessness. I then put in his cage a female rat that was not in the sexually receptive state. According to the script I had in mind, the brain injection of male hormone should have driven the male to make sexual advances, although normally he would not do so with a nonreceptive female. The rat, however, followed a script of his own. He grasped the female by the tail with his teeth and dragged her across the cage to a corner. She scurried away as soon as he let go, whereupon he dragged her back again. After several such experiences the male picked her up by the loose skin on her back, carried her to the corner and dropped her there.
>
> I was utterly perplexed and so, no doubt, was the female rat. I finally guessed that the male was carrying on a bizarre form of maternal behavior. To test this surmise I deposited some newborn rat pups and strips of paper in the middle of the cage. The male promptly used the paper to build a nest in a corner and then carried the pups to the nest. I picked up the paper and pups and scattered them around the cage; the male responded by rebuilding the nest and retrieving the young.
>
> After about 30 minutes the rat stopped behaving like a new mother; apparently the effect of the injected hormone had worn off. Given a new injection, he immediately returned to his adopted family. With successive lapses and reinjections, his behavior became disorganized; he engaged in all the same maternal activities, but in a haphazard, meaningless order. After an overnight rest, however, a new injection the next day elicited the well-patterned motherly behavior.
>
> The case of the mixed-up male rat was a most auspicious one. Although the rat had not followed the experimenter's script, the result of this first experiment was highly exciting. It was an encouraging indication that the control of behavior by specific neural systems in the brain could indeed be investigated by chemical means. We proceeded next to a long series of experiments to verify that the behavior in each case was actually attributable to a specific chemical implanted at a specific site in the brain rather than to some more general factor such as

mechanical stimulation, general excitation of the brain cells, or changes in acidity or osmotic pressure (Fisher, 1964, pp. 2–4).

Fisher's research is another good example of serendipity and well illustrates the flexibility that is characteristic of the successful scientist. What is being urged, then, is that while testing a hypothesis we continue to be alert to accidental occurrences that will stimulate other research. Almost without exception, in fact, the experimenter who patiently and flexibly observes the participants in an experiment gets many hints for the development of hypotheses other than the one presently being tested. The position has been taken, though, that scientists should not, or do not, explicitly test hypotheses (cf., Sidman, 1960). This may strike you as an extreme position, but the advocacy of it has been quite explicit. Bachrach (1965), in referring to this matter, advanced his first law that "People don't usually do research the way people who write books about research say that people do research" (Bachrach, 1965, p. ix). More specifically, he coined the phrase "*hypothesis myopia,* a common disease among researchers holding certain preconceived ideas that might get in the way of discovery" (Bachrach, 1965, p. 22). The argument, in short, is that if one seeks to test a hypothesis, one may thereby be blinded to events other than those related to the hypothesis, and these other events are potentially very important.

The primary fault with this type of argument is that it erroneously places the blame on the hypothesis, not where the blame properly belongs. We have taken the position that hypotheses are valuable, that the research of any experimenter is guided by some sort of hypothesis, even though that hypothesis consists only of vague notions. And regardless of the precision with which a hypothesis is stated, the possession of a hypothesis per se does not blind the experimenter from making observations other than those called for by the hypothesis. The hypothesis, furthermore, is not the only "set of preconceived ideas that might get in the way of scientific discovery." All manner of biases, we have seen (pp. 42–43), may operate against scientific discovery. In short the term *hypothesis myopia* should be replaced with *experimenter myopia.* To exorcise the hypothesis from scientific research is to throw the baby out with the bath. Our experimentation can, in all probability, best proceed by explicitly formulating and testing hypotheses; at the same time we should keep alert for accidental occurrences that may lead to the development of even more valuable hypotheses. It only needs to be added that overemphasis of the role of accident in scientific discovery also has its dangers. One cannot, for example, enter the laboratory with confidence that "serendipity will save the day."

The hard facts of everyday experimentation are that the extremely large majority of accidental occurrences have little significance—the investigation of each odd little rise in an extended performance curve, the inquiry into every consequence of equipment malfunction on a subject's behavior, the quest for the "why" of every unscheduled response can consume all of an experimenter's energy. We must, at least most of the time, keep our eyes on the doughnut and not on the hole.[4]

Let us now review what we have said. We first started with a consideration of a problem that a psychologist seeks to solve. The psychologist initially states the problem very clearly, and then proposes a solution to the problem in the form of a hypothesis. The psychologist should formulate both the problem and the hypothesis quite clearly and succinctly. These formulations can also be used in the later write-up of the experiment. As we shall see in the next chapter, the psychologist will formulate the problem with such phrases as "The statement of the problem is . . ." or "It was the purpose of this experiment to. . . ." The hypothesis is then expressed in such ways as "It was expected that . . ." or "It was hypothesized that. . . ." We defined the hypothesis, the tentative solution to the problem, as a testable statement of a potential relationship between two or more variables. Statements, we said, fall into one of three categories. The analytic statement is one that is formed in such a way that it is necessarily true, but empty. The contradictory statement is also empty, and it is necessarily false. Hypotheses are synthetic statements, and they are neither absolutely true nor false—rather, they have a determinable degree of probability, and they are not empty in that they are our attempts to say something about nature. The most prominent type of hypothesis is the universal one, and it is, at least ideally, stated in the form of a general implication. Existential hypotheses, those that state that there is at least one phenomenon that has a given characteristic, are useful in science too.

---

[4] In their satire on methodological errors entitled "Some Unprinciples of Psychological Research," White and Duker (1971) illustrated the ultimate in relying on serendipity with the researcher who seized the opportunity to collect data on a stalled, darkened subway. Their basic unprinciple relevant here was entitled *"Peripatetic Passim, or, Something happened to me on the way to class today."*
"Basic unprinciple. The investigator makes maximum use of the unusual, timely, or newsworthy situation with whatever is at hand.
Example. A random sample of subway riders was tested during a prolonged service blackout to identify those personality characteristics associated with panic during enforced isolation. Analysis of handwriting samples and of figure drawings indicates that uneven line pressure and poorly articulated limbs are associated with verbal and motoric indices of panic. These results are interpreted somewhat guardedly because illumination was inadequate during part of the experiment" (p. 398).

The manner in which a scientist arrives at a hypothesis falls within the context of discovery, and the manner in which we report scientific reasoning that justifies our conclusions falls within the context of justification. We also considered how we can evaluate hypotheses and, to this end, advanced seven tentative criteria for judging hypotheses. One of the functions of hypotheses is that they give direction to the experimenter. They indicate which kinds of data should be collected and which kinds should be ignored, though one should constantly be alert for hints from data that might lead to other hypotheses.

With this background, we are now ready to consider, in some detail, the methods by which the experimenter subjects hypotheses to empirical test. After formulating a problem and the proposed solution to it, the experimenter turns attention to the plan for conducting the experiment.

Review of Chapter 3

1. A hypothesis is posed as a potential solution to a problem. In formulating the problem the experimenter has to isolate potentially relevant variables for study. How do you define the hypothesis conceived in this way, and in what sense might it be a solution to the problem?

2. The term "empirical" is central to our study. What do we mean when we refer to an *empirical* hypothesis? In your later study, if you cannot now, you might consider the question of what are some "nonempirical hypotheses?"

3. Discuss the statement of hypotheses as nonquantified and as quantified statements. What is the advantage of stating hypotheses as quantified relationships?

4. Can an empirical hypothesis be strictly true or false, or must it always have a probability value attached to it that is less than certainty?

5. Distinguish between existential and universal hypotheses. In anticipation of chapter 11, you might carry on the question for yourself of the relationship between existential hypotheses and those concerned with $n = 1$ research.

6. Why is the distinction between the context of discovery and the context of justification an important one?

7. How do we determine the value of a hypothesis? Are there a limited number of criteria for this purpose?

8. Could science proceed as efficiently without hypotheses and theories as with them?

4
The Experimental Plan

The Evidence Report

We have noted that a scientific inquiry starts with a problem. The problem must be *solvable* and may be stated in the form of a question (chapter 2). The inquiry then proceeds with the formulation of one or several hypotheses as possible solutions to the problem (chapter 3). In the next phase of the scientific investigations the hypothesis (or the hypotheses if there are more than one) is tested to determine whether it is probably true or false. This amounts to conducting a study in which certain empirical results that relate to the hypothesis are obtained. The results of the study are then summarized in the form of an *evidence report*. For our immediate purposes we shall simply consider that an evidence report is a summary statement of the results of an investigation, i.e., it is a sentence which concisely states what was found in the inquiry. Once the evidence report has been formed it is related to the hypothesis. By comparing the hypothesis (the prediction of how the results of the experiment will turn out) with the evidence report (the statement of how the results *did* turn out), it is possible to determine whether the hypothesis is probably true or false. We now need to inquire into the various methods in psychology of obtaining data that may be used to arrive at an evidence report. This amounts to an inquiry into the various methods of scientific investigation in psychology.

Methods of Obtaining
the Evidence Report

The methods to be discussed have in common the fact that they allow the investigator the opportunity to collect data from which, regardless of the specific method used to obtain them, an evidence report can be formulated.

Nonexperimental Methods

Since this is a book on experimental methodology we will naturally emphasize experimental methods. But to enhance our perspective here, it is valuable to briefly consider the nonexperimental ones, too. In chapter 12 we shall enlarge on some research alternatives to experimentation.

The manner of classifying the nonexperimental methods is somewhat arbitrary and varies with the classifying authority. There are two general types of methods that can be contrasted with the experimental method. The first to consider is the *clinical method* (sometimes called the "case history method" or the "life history method"). Traditionally, the psychologist used this method in an attempt to help a client solve personal problems, be they emotional, vocational, or whatever. In a common form of the clinical method, the psychologist attempts to help by collecting all relevant information about the person. The interest is in all aspects of a person's life, from birth to the present time. Some of the techniques for collecting this information might be an intensive interview, perusing records, administering psychological tests, questioning other people about the individual, studying written works of the person, or obtaining biographical questionnaires. Then, on the basis of such information, the psychologist tries to determine the factors that led to the development of the person's problem. This leads to the formulation of an informal hypothesis as to the cause of the person's problem; and the collection of further data will help to determine whether the hypothesis is probably true or false. Once the problem and the factors that led to its development are laid bare for the person, the psychologist will try to help the individual achieve a better adjustment to the circumstances. It should be noted that the clinical method is generally used in an applied as against a basic sense, since its usual aim is to solve a practical problem, not to advance science. However observation of behavior through this method can be a source of more general hypotheses that can be subjected to stringent testing.

The second general nonexperimental method is that of *systematic observation.* In using this method the investigator takes an event as it occurs naturally and studies it, with no effort to produce or control the event, as in experimentation. The popular technique of *naturalistic observation,* as used in the study of children at free play, would be one example of the use of this method. The purpose there might be to determine what kinds of skills children of a certain age possess. A number of types of play apparatus would be made available for the children, and their behavior would be observed and recorded as they played. Another example of the use of systematic observation might be a study of panic. We do not ordinarily produce panic in groups of people for psychological study. Rather, psychologists usually wait until a panic occurs naturally, and then set out to study it. An example of how social psychologists studied a panic was after the Orson Welles' radio dramatization of H. G. Wells' *War of the Worlds.* Psychologists were interested in why the panic occurred and thus interviewed people who participated in it (Cantril, 1940). Ethologists typically use this approach in their study of animals in their natural environment. More sophisticated instances of the method of systematic observation will be presented in chapter 12 on quasi-experimental designs.

The data obtained by the method of systematic observation can be used for testing hypotheses through the construction of evidence reports, but it should be obvious that some important limitations exist. Such limitations will be briefly discussed after a consideration of the experimental method.

Experimental Methods

In the early stages of the development of a science, the nonexperimental methods tend to be more prominent. In some sciences, sociology, for example, there is little hope that anything but nonexperimental methods can be generally used. This is primarily because sociology is largely concerned with the effect of the prevailing culture and social instructions on behavior, and it is difficult to manipulate these two factors as independent variables in an experiment. In those fields that are ultimately susceptible to experimentation, however, a change in methodology eventually occurs as knowledge accumulates.

Scientific investigations become more and more searching because the "spontaneous" happenings in nature are not adequate to permit the necessary observations. This leads to the setting up of special conditions to bring about the desired events under circumstances favorable for scientific observations, and experiments then can originate. Thus, the experimenter takes an active part in producing the event. Some advantages of this approach were well expressed by Woodworth and Schlosberg:

1. The experimenter can make the event occur when he wishes. So he can be fully *prepared* for accurate observation.
2. He can *repeat* his observation under the same conditions for verification; and he can describe his conditions and enable other experimenters to duplicate them and make an independent check of his results.
3. He can *vary the conditions* systematically and note the variation in results . . . (1955, p. 2).

Since psychology is concerned with the behavior of organisms, in using nonexperimental methods the psychologist must wait until the behavior of interest occurs naturally. The researcher does not have control over the variable to be studied, but can only observe the event in its natural state. The one characteristic that all the nonexperimental methods have in common is that the variables that are being evaluated are not purposefully manipulated by the researcher.[1] It follows that when a theory is tested through the use of the experimental method, the conclusion is more highly regarded than if it is tested by a nonexperimental method. Put another way, the evidence report obtained through experimentation is more reliable than that obtained through the use of a nonexperimental method. This is true largely because the interpretation of the results is clearer in an experiment. Ambiguous

[1] To emphasize this definition of an experiment, suppose that we are interested in the way that learning speed changes with age. We might have a group of 20-year-old people and a group of 60-year-old people. Both groups would learn the same task. Although this might appear to be an "experiment" at first glance, we would not so classify it because we did not *purposely manipulate* the "independent variable" (age of the participants). Rather, we *selected* one participant to fit certain age requirements. It is apparent that "age of participants" is not a variable over which a researcher has control. We cannot say to one participant: "You will be 20 years old," and to another participant: "You will be 60 years old." Hence, this investigation typifies the use of the method of systematic observation. Put another way, as we shall see, participants in an experiment are randomly assigned to the experimental and control groups. But in the method of systematic observation, they are not. Although two groups may be used in systematic observation, what would be called a "control group" in an experiment is called a "comparison group"—one to which participants are *not* randomly assigned. This is a critical difference in design.

interpretation of results can occur in nonexperimental methods primarily because of a lack of control over extraneous variables. It is difficult, or frequently impossible, in a systematic observation study to be sure that the findings, with respect to the dependent variable, are due to the independent variable, for they may result from some uncontrolled extraneous variable that happened to be present in the study. In nonexperimental methods it is also usually more difficult to define the variables studied than where they are actually produced, as in an experiment.

All of this does not mean, however, that the experimental method is a perfect method for answering questions. Certainly it can lead to errors and in the hands of poor experimenters the errors are sometimes great. Relatively speaking, however, the experimental method is preferred *where it can be appropriately used.* But if it cannot be used, then we must do the next best thing and use the method of systematic observation. Thus when it is not possible to produce the events that we wish to study, as in the example of panic, we must rely on nonexperimental methods. But we must not forget that when events are *selected* for study, rather than being *produced* and *controlled,* caution must be exercised in accepting the result.

One frequent criticism of the experimental method is that when an event is brought into the laboratory for study (as it usually is) the nature of the event is thereby changed. For one thing, the event does not naturally occur in isolation, as it is made to occur (relatively so) in the laboratory, for in natural life there are always many other variables that influence it. Criticism of experiments on such grounds is unjustifiable, since that is what we really want to know—we intentionally seek to find out what the event is like when it is uninfluenced by other events. It is then possible to transfer the event back to its natural situation at which time we know more about how it is produced. The fact that any event may appear to be different in the natural situation, as compared to the laboratory, simply means that it being influenced by other variables, *which in their turn* need to be brought into the laboratory for investigation. Once all the relevant variables that exist under natural conditions have been studied in isolation in the laboratory, and it has been determined in what way they all influence the dependent variable and each other, then a thorough understanding of the natural event will have been accomplished. Without such a piecemeal analysis of events in the laboratory, it is likely that we would never be able to understand them adequately.

Now, although we have made some strong statements in favor of the laboratory analysis of events, we must recognize the possibility that the events actually may be changed or even "destroyed." This occurs when the experimenter has simply not been successful in transferring the event into the laboratory. A truly different event was produced than that which was to be studied. For instance, it may be that the person's behavior in an experiment may be different than it would be in "real life." That is, a participant who is aware of performing in an experiment may behave differently than usual. The best that can be said for this situation is that frequently events cannot be studied adequately unless they *are* brought into the laboratory, and at least in the laboratory suitable controls can be introduced so that even if the event is distorted by observation, the event is studied with this effect held constant.

Skinner (1953, p. 435) made a similar point when he said, in essence, that certain characteristics of behavior are too complex to understand through casual observation of everyday life — to find the relevant variables that determine a certain kind of behavior would require considerable time, and even then it may not be possible to find them through such a common sense approach. But with adequate recording devices in the laboratory, and under controlled conditions, we can determine the variables that are responsible for an event. These findings can be used to good advantage in further study of the complex world at large. To illustrate his point, Skinner suggested that casual observation has led lay people to conclude that a particularly undesirable type of behavior may be eliminated by punishing a person. Punishment gives quick results; a particular type of behavior seems to disappear immediately if its perpetrator is punished. However, actual laboratory studies have indicated that punishment is rather ineffective in eliminating a response. While it may effect a temporary suppression, it does not seem to permanently eliminate behavior, even when it is applied over a long period of time. A more effective technique for eliminating a response is the process of extinction, but the discovery of this finding could not have been made from the observation of everyday life. It required laboratory investigation.

Types of Experiments

In your reading you may run across a number of terms that refer to different types of experiments; to prevent possible confusion some of them will be discussed here. First, note well the the *general experimental method* remains the same, regardless of the type of problem to which it is applied, so that, strictly speaking, we are not talking about different types of experiments, but different types of problems and purposes for which they are used. To clarify this matter, contrast *exploratory* with *confirmatory* experiments. The type the experimenter uses depends on the state of the knowledge relevant to the problem being dealt with. If there is little knowledge about a given problem, the experimenter performs an exploratory experiment. Lacking much knowledge about the problem we are usually not in a position to formulate a possible solution — generally, we cannot postulate an explicit hypothesis that might guide us to predict such and such a happening. The experimenter is simply curious, collects some data, but doesn't really have any basis on which to guess how the experiment will turn out. Although we *may not* have an explicit hypothesis, there is an "informal" hypothesis we have decided to investigate — the effect of one specific variable on another, rather than just any variable on a host of other variables. But this type hypothesis is not sufficiently advanced to say what kind of effect the one variable will have on the other, or even to say that there *will* be an effect. It can be seen that the exploratory experiment is performed in the earlier stages of the investigation of a problem area. As we gather data relevant to the problem, we become increasingly capable of formulating hypotheses of a more clear-cut nature. We thus become able to predict, on the basis of a hypothesis, that such and such an event should occur. At this stage of knowledge development we perform the confirmatory experiment, i.e., we start with an explicit hypothesis that we wish to test. On the basis of that hypothesis we predict an outcome of the experiment; we set up the experiment to determine whether the outcome is, indeed, that predicted by our hypothesis. Put another way, the exploratory experiment is used primarily to find new independent variables that affect a given dependent variable, while the confirmatory experiment is used to confirm that a given variable is influential. In the confirmatory experiment we may also want to determine the extent and precise way in which one variable influences the other, or, more generally, to determine the functional (quantitative) relationship between the two variables. Underwood (1949, pp. 11–14) used two descriptive terms to refer to the problems which the two types of experiments are used to solve. The exploratory experiment refers to his "I-wonder-what-would-happen-if-I-did-this" type of problem, while the confirmatory experiment is analogous to his "I'll-bet-this-would-happen-if-I-did-this" type of problem.

Regardless of the state of knowledge in a given problem area, the immediate purpose of an experiment is to arrive at an evidence report. If the experiment is exploratory, the evidence report can be used as the basis for formulating a specific, precise hypothesis. In a confirmatory experiment the evidence report is used to determine whether the hypothesis is probably true or false. In the latter case, if the hypothesis is not in accord with the evidence report, it might be modified to better fit the data and then tested in a new experiment. If the hypothesis is supported by the evidence report, then its probability of being true is increased. In addition to providing such a general understanding, the distinction between these two types of experiments has direct implications for the types of experimental designs that are employed, one type of design being more efficient for the exploratory experiment, while another is more efficient for the confirmatory experiment. These designs will be discussed later.

Sometimes you may run across the term "crucial experiment" *(experimentum crucis)*. This term is used to describe an experiment that purports to test one or several "counter-hypotheses" simultaneously with the primary hypothesis of interest. For instance, it may be possible to design the experiment in such a manner that if the results come out one way, one hypothesis can be said to be confirmed and a second hypothesis disconfirmed, and if the results point another way, the first hypothesis is said to be disconfirmed and the second hypothesis is confirmed. Ideally, a crucial experiment is one whose results support one theory and disconfirm all possible alternative theories. However, we can seldom if ever be sure that we can state at any given time all of the possible alternative hypotheses. Accordingly, perhaps we can never have a crucial experiment, but only approximations of one.

The term *pilot study* or *pilot experiment* has nothing to do with the behavior of aircraft operators, as one student thought, but refers to a preliminary experiment, one conducted prior to the major experiment. It is used, usually with only a small number of subjects, to suggest what specific values should be assigned to the variables being studied, to try out certain procedures to see how well they work, and more generally to find out what mistakes might be made in conducting the actual experiment so that the experimenter can be ready for them. It is a dress rehearsal of the main performance.

Planning an Experiment

Given a problem to be solved and a hypothesis, which is a tentative solution to that problem, we must design an experiment that will determine whether that hypothesis is probably true or false. In designing an experiment the researcher uses his ingenuity to obtain data that are relevant to the hypothesis. This involves such problems of experimental technique as: What apparatus will best allow manipulation and observation of the phenomenon of interest? What extraneous variables may contaminate the phenomenon of primary interest and are therefore in need of control? Which events should be observed and which should be ignored? How can the results of the experiment best be observed and recorded? By considering these and similar problems an attempt is made to rule out the possibility of collecting irrelevant evidence. For instance, if the antecedent conditions of the hypothesis are not satisfied, the evidence report will be irrelevant to the hypothesis, and further progress in the inquiry is prohibited. Put another way, the hypothesis says that *if* such and such is the case (the antecedent conditions of the hypothesis), *then* such and such should happen (the consequent conditions of the hypothesis). The hypothesis amounts to a contract that the experimenter has signed — the experimenter has agreed to make sure that the antecedent conditions are fulfilled. If the experimenter fails to fulfill that agreement, then whatever results are collected will have nothing to do with the hypothesis; they will be irrelevant and thus cannot be used to test the truth of the hypothesis. This points up the importance of adequately planning the experiment. For if the experiment is improperly designed, then either no inferences can be made from the results, or it may only be possible to make inferences to answer questions that the experimenter has not asked. That this is not an idle warning is indicated by the frequency with which the right answer is given to the wrong question, particularly by neophyte experimenters. And if the only result of an experiment is that the experimenter learns that these same errors should not be made in the future, this is very expensive education indeed.

It is a good idea for the experimenter to draft a thorough plan of the experiment before it is conducted. Once the complete plan of the experiment is set down on paper it is desirable to obtain as much criticism of it as possible. The experimenter often overlooks many important points, or looks at them with a wrong, preconceived set, and the critical review of others may bring potential errors to the surface. No scientist is beyond criticism, and it is far better for us to accept criticism before an experiment is conducted than to make errors that might invalidate the experiment. We shall now suggest a series of steps that the experimenter can follow in the planning of an experiment. Note that while the experimental plan preceeds the eventual write-up of the experiment, sizable portions of the plan can later be used for the write-up.

Outline for an Experimental Plan

1. Label the Experiment. The title should be clearly specified, as well as the time and location of the experiment. As time passes and the experimenter accumulates a number of experiments in the same problem area, this information can be referred to without much chance of confusing one experiment with another.

2. *Survey the Literature.* All of the previous work that is relevant to the experiment should be studied. This is a particularly important phase in the experimental plan for a number of reasons. First, it helps in the formulation of the problem. The experimenter's vague notion of what problem he wants to investigate is frequently made more concrete by consulting other studies. Or, the experimenter thus may be led to modify his original problem in such a way that the experiment becomes more valuable. Another reason for this survey of pertinent knowledge is that it tells the experimenter whether or not the experiment even needs to be conducted. If essentially the same experiment has previously been conducted by somebody else, there is certainly no point in repeating the operation, unless it is specifically designed to confirm previous findings. Other studies in the same area also provide numerous suggestions about extraneous variables that need to be controlled and hints on how to control them.

The importance of the literature survey cannot be overemphasized. The experimenters who tend to slight it usually pay a penalty in the form of errors in the design or some other complication. The knowledge in psychology is growing all the time, making it more difficult for one person to comprehend the findings in any given problem area. Therefore, this step requires particularly close attention. Also, since relevant studies should be summarized and referred to in the write-up of the experiment, this might just as well be done before the experiment is conducted thus combining two steps in one.

We are very fortunate in psychology to have the *Psychological Abstracts,* which make any such survey relatively easy.[2] Every student of psychology should attempt to develop a facility in using the *Abstracts.*

3. *State the Problem.* The experiment is being conducted because there is a lack of knowledge about something. The statement of the problem expresses this lack of knowledge. Although the problem can be developed in some detail, through a series of logical steps, the actual statement of the experimental question should be concise. It should be stated succinctly and unambiguously in a single sentence, preferably as a question. The statement of the problem as a question

---

[2] The *Psychological Abstracts* is a professional journal published monthly by the American Psychological Association. It summarizes the large majority of psychological research and classifies it according to topics (and authors) so that it is fairly easy to determine what has previously been done on any given problem. The December issue of each year summarizes all the research for that year.

implies that it can be answered unambiguously in either a positive or negative manner. If the question cannot be so answered, *in general* we can say that the experiment should not be conducted. Every worthwhile experiment involves a gamble. If the problem cannot be definitely answered either positively or negatively, the experimenter has not risked anything and therefore cannot hope to gain new knowledge.

4. *State the Hypothesis.* The variables specified in the statement of the problem are explicitly stated in the hypothesis as a sentence. Natural languages (e.g., English) are usually employed for this purpose, but other languages (e.g., mathematical or logical ones) can also be used, and, in fact, are preferable. The "if . . . then . . ." relationship was suggested as the basic form for stating hypotheses.

5. *Define the Variables.* The independent and dependent variables have been specified in the statement of the problem and of the hypothesis. They must now be defined in such a manner that they are clear and unambiguous—the variables must be *operationally defined.* The importance of this phase has been previously emphasized. To repeat here, if no such operation is possible for all the experimental variables, we must conclude that the hypothesis is untestable.

6. *Apparatus.* Every experiment involves two things: (1) an independent variable must be manipulated; and (2) the resulting value of the dependent variable must be recorded. Perhaps the most frequently occurring type of independent variable in psychology is the presentation of certain values of a stimulus; and in every experiment a response is recorded. Both of these functions may be performed manually by the experimenter. However, it is frequently desirable, and in fact sometimes necessary, to resort to mechanical or electrical assistance. Thus, there are two general functions of apparatus in psychological experimentation: (1) to facilitate the administration of the experimental treatment; and (2) to aid in recording the resulting behavior. Let us consider how these two functions might be advantageously accomplished in an experiment.

Expressions from literature and from everyday language have long alluded to the emotional significance of the size of the pupils of the eyes, as, for instance, "His eyes were like saucers" or "His eyes were pinpoints of hate." Hess (1965) conducted a series of studies in which this topic was systematically investigated. His general procedure was to present a variety of stimuli and to relate these stimuli to resulting changes in pupillary size. His participants peer into a box, looking at a screen on which is projected a stimulus picture, as shown in Figure 4-1. A mirror reflects the image of the participant's eye into a motion picture camera. First a control slide is present for 10 seconds; this slide is matched in overall brightness with an experimental slide so that the participant's eyes are adapted to the appropriate light in-

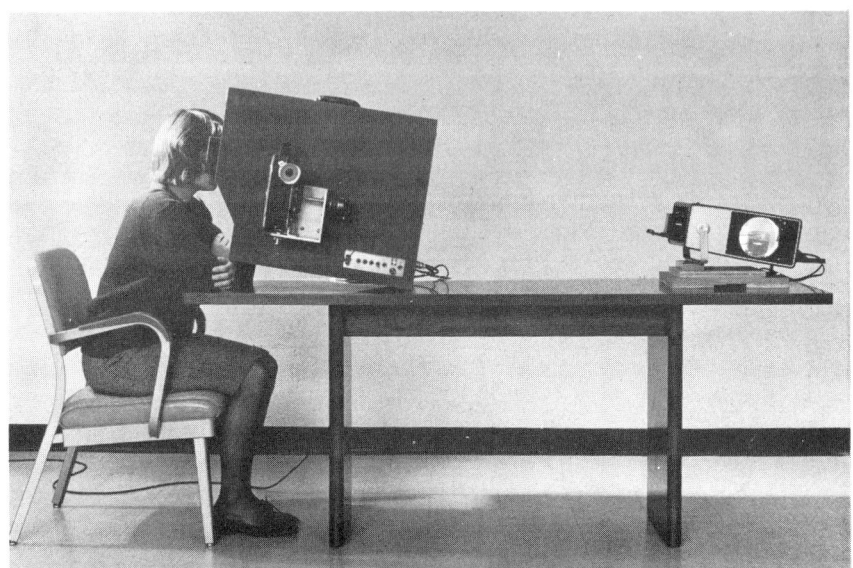

FIGURE 4-1a. Participant in pupil-response study peers into a box, looking at a rear-projection screen on which slides are flashed from the projector at right. A motor-driven camera mounted on the box makes a continuous record of pupil size at the rate of two frames a second.

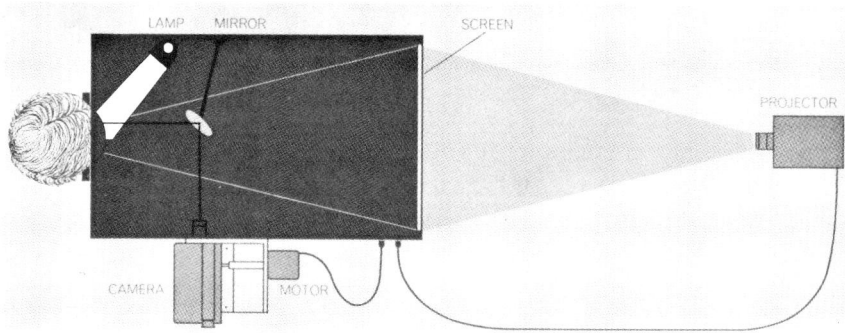

FIGURE 4-1b. Pupil-response apparatus is simple. The lamp and the camera film work in the infrared. A timer advances the projector every 10 seconds, flashing a control slide and a stimulus slide alternately. The mirror is below eye level so that view of screen is clear.

Figures 4-1, a and b, from Eckhard H. Hess, "Attitude and Pupil Size." Copyright © 1965 by Scientific American, Inc. All rights reserved. Photograph (4-1 a): Courtesy of Saul Mednick.

tensity. Then the experimental slide is presented for 10 seconds, and this sequence of alternating control and experimental slides is continued for 10 or 12 times a sitting. To quantify the dependent variable, the movie film is projected on a screen and the size of the pupil is measured either with a ruler or electronically by means of a photo cell. The results have been most intriguing: In general, interesting or pleasant pictures lead to dilation of the pupils, while unpleasant or distasteful stimuli lead to pupillary constriction. Thus the presentation of a female pinup produces greater enlargement of the pupils of men than it does of women; but women showed a greater enlargement than did men to a male pinup, or to a picture of a mother and a baby. Distasteful pictures, such as of sharks or of crippled or cross-eyed children, resulted in a decrease of pupillary size. Other studies showed that when an arithmetic problem is presented, the size of the pupil increases to a maximum at the point at which the solution is reached and returns to its base level as soon as the answer is reported.

Let us now return to the main point we wish to illustrate through these studies. We may note that apparatus was used to present the experimental treatment—a device for presenting a stimulus for whatever length of exposure the experimenter desired. In this case pictures and control slides were projected for ten seconds. Furthermore, there was an automatic timing device with a driving motor so that experimental and control slides were automatically alternated every ten seconds. The second function of apparatus in experimentation was fulfilled by the movie camera. It operated at a rate of two frames per second, so that pupillary size was photographed regularly (Figure 4-2). The use of a ruler or photo cell then allowed quantification of the dependent variable.

The types of apparatus used in behavioral experimentation are so numerous that we cannot attempt a systematic coverage of them here. You might, though, start with "How to build and use electronic circuits without frustration, panic, mountains of money, or an engineering degree (Hoenig & Payne, 1973).[3] We shall, however, briefly try to illustrate further the value of apparatus, as well as to refer to certain cautions that should be observed.

Frequently a certain stimulus, such as a light, must be presented to a participant at very short intervals. It would be difficult for an experimenter to time the intervals manually and thus make the light

---

[3] For detailed presentations of electronic instrumentation see Brown and Spence (1958), Cornsweet (1963), Jackson (1975), or Tocci (1975). A general discussion of instrumentation with specialized articles by experts in various fields of psychology is offered by Sidowski (1966). Yanof (1965), Venables and Martin (1967), and Zucker (1969) take up the area of behavioral and biomedical electronics (psychophysiology, etc.).

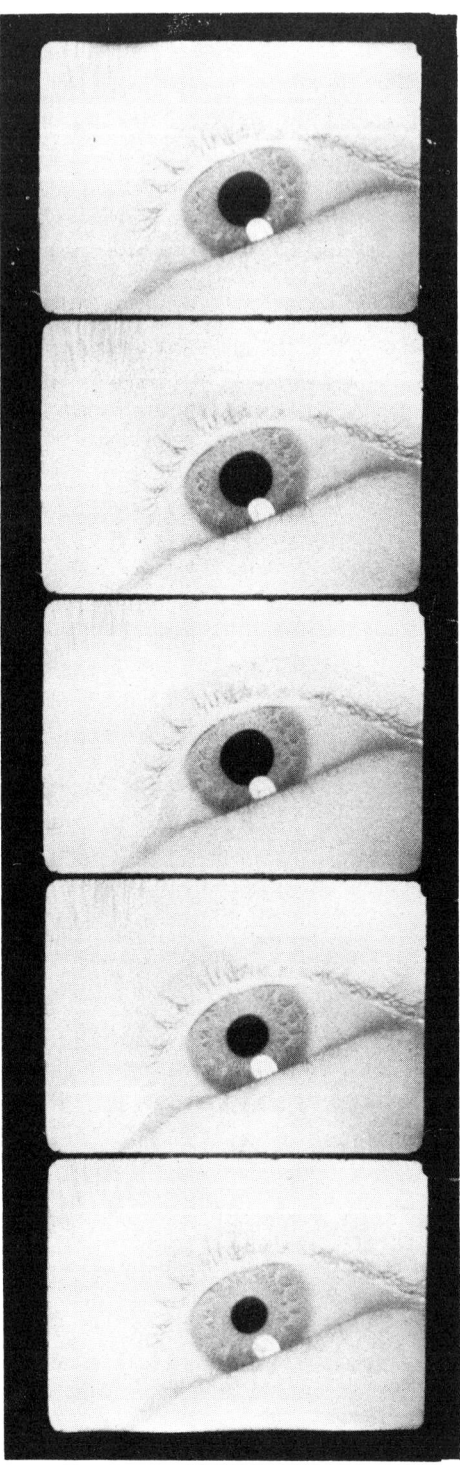

Figure 4-2. Pupil size varies with the interest value of a visual stimulus. The five frames show the eye of a male subject during the first 2 1/2 seconds after a photograph of a pinup girl appeared on the screen. His pupil diameter increased in size 30 percent.

From Eckhard H. Hess, "Attitude and Pupil Size." Copyright © 1965 by Scientific American, Inc. All rights reserved.

come on at precisely the desired moments. In addition to the considerable error that would be involved, the work required of the experimenter would be highly undesirable, and in fact might distract him from other important aspects of the experiment. This type of stimulus presentation could be handled very easily by placing a metronome in the circuit to break and complete the circuit at the proper times. As you can see, one of the main advantages of apparatus is that it reduces the "personal equation." Suppose that a reaction-time experiment is being conducted. For example, one might wish to present a series of words, one at a time, to a participant and measure the time it takes to respond with the first word that comes to mind. If the experimenter were forced to measure the participant's reaction time by starting a stop watch when the word is read and stopping it when the participant responds, considerable error would result. The reaction time of the experimenter in starting and stopping the stop watch enters into the "participant's reaction time"; this is especially bad because the experimenter's reaction time would vary and hence a constant value could not be subtracted as a correction factor. A better approach would be to use a voice key that is connected in circuit with a timing device. In this apparatus, when the word is read to the participant, the timer automatically starts, and when the participant responds, it automatically stops. The experimenter may then record the reaction time and say the next word. While the "reaction time" of the apparatus is still involved, it is at least constant and of smaller magnitude, as compared to the experimenter's reaction time. A similar example of the valuable use of apparatus would be in timing a rat as it runs a maze. Several types of apparatus that automatically record the rat's latency and running time have been developed for this problem.

We have been emphasizing the frequent *desirability* of apparatus in experimentation, but there are a large number of problems that simply cannot be studied without the use of apparatus. In some experiments apparatus is not only valuable, it is downright essential. Probably the clearest illustration would be the recording of response measures of a psychophysiological nature. If one wished to study the effect of some independent variable on electroencephalograms ("brainwaves"), an electroencephalograph is required. Similarly, for measuring pulse rate a sphygmograph is necessary, for recording galvanic skin responses a psychogalvanometer needs to be used, and so forth.

In spite of all the advantages of apparatus, there are a number of possible disadvantages. We have assumed above that, to be of value, the apparatus is suitable for the job required of it. This is not always the case, for sometimes apparatus is not accurate, not adequately calibrated, etc. Furthermore, sometimes apparatus may interfere with the event being studied. Problems of this nature are discussed in greater detail in the chapter on control (chapter 6). But there is one potential disadvantage that should be mentioned here, a disadvantage that can be illustrated by analogy with the "Law of the Hammer." This law, which is a genuinely universal one, states simply that if you give a little boy a hammer, he will find many things that need pounding. And so it often is with the scientist—we somewhat too frequently find many things that *need* to be recorded with the particular apparatus that is in our laboratories. The recent entrance into many laboratories of the small computer has often led to the belief that computers are necessary for the conduct of *all research.* In this way, then, the availability of the computer or other apparatus can determine the problem that is researched when, clearly, it should be the other way around. While no doubt a boy does find some items that really require pounding, many hammered items are satisfactory prior to a juvenile onslaught. The more fruitful approach is for the scientist to formulate a problem and then consider the equipment requirements. Otherwise one might not only needlessly expend energy on unfruitful projects, but might also be somewhat blinded to more valuable research areas.

7. *Control the Variables.* In this phase of planning the experiment the scientist must consider all of the variables that might contaminate the experiment; one should attempt to evaluate any and all variables which might affect the dependent variable. It may be decided that some of these extraneous variables might act in such a way that they will invalidate the experiment, or at least leave the conclusion of the experiment open to question. Such variables need to be controlled. To elaborate on what we mean by "control" and the techniques for achieving it will demand much attention; since it is of extreme importance in experimentation, it will be the subject of a later discussion (chapter 6). For now, let us simply note a point that will be expanded later—that we must make sure no extraneous variable will differentially affect our groups, i.e., that no such variable will affect one group differently than it will affect another group.

In the course of examining variables that may influence our dependent variable, we will have to make certain decisions. We will decide, preferably on the basis of information afforded by previous research, that a certain variable is not likely to be influential, in which case it may be reasonable to ignore it.

Other variables, however, might be considered to be relevant, but there are difficulties in controlling them. Perhaps it is reasonable to assume that such variables will exert equal effects, at random, on all conditions. If this assumption seems tenable, the experimenter might choose to proceed. But if such an assumption is rather tenuous, then the problem may well be of such serious proportions that it would be wiser to abandon the experiment.

8. Select a Design. So far we have discussed only the two-groups design, i.e., where the results for an experimental group are compared to those for a control group. We shall consider a number of other designs later, from among which the experimenter may choose the one most appropriate to the problem at hand. For example, it may be more advantageous to use several groups, instead of just two, in which case a *multigroups* design would be adopted. Another type of design, which in many cases is the most efficient and which is being used more and more in psychology, is the *factorial* design. For now, however, you should restrict your consideration to the two-groups design, at the same time realizing that a number of alternatives are possible.

9. Select and Assign Participants to Groups. The experimenter conducts an experiment in order to conclude something about behavior. To do this, of course, one must select certain participants to study. But from what group should the participants be selected? This is an important question because we want to generalize findings from the participants that we study to the larger group of participants from which they were chosen (see step 14, page 91). The larger group of participants is the *population* (or "universe") under study; those who participate in an experiment constitute a *sample*. More generally, by "population" we mean the total number of possible items of a class that might be studied — it is the entire group of items from which a sample has been taken. Thus, we may note that a population need not refer only to people, but to any type of organism: amoebae, rats, jellyfish. Furthermore, one may note that the definition is worded in such a manner that it can and does refer to inanimate objects. For example, we may have a "population" of types of therapy (directive, nondirective, etc.), of learning tasks (hitting a baseball, learning a maze, etc.), and so forth. Or an experimenter may be interested in sampling a population of stimulus conditions (high, medium, and low intensity of a light), or a population of experiments (three separate experiments that test the same hypotheses). In engineering, a person may be interested in studying a population of the bridges of the world; an industrial psychologist may be concerned with a population of whiskey products, and so on.

In designing an experiment one should specify with great precision the population (or populations) being studied. For the moment let us merely concern ourselves with participant populations, and leave other populations until later (see chapter 15). In specifying a population we must note those of its characteristics that are particularly relevant to its definition, e.g., if we are concerned with a population of people, we might wish to specify the age, sex, education, socioeconomic status, and race. If we are working with animals, we might wish to specify the species, sex, age, strain, experience, habitation procedures, and feeding schedules. Unfortunately, one can observe by reading articles in professional journals that experimenters rarely define the populations they are studying with any great precision.

Given a well-defined population that we wish to study, then, we are faced with the problem of how to actually study it. If it is a small population, it may be possible to observe each individual. And adequately studying an entire population is far preferable to studying a sample of it. We once conducted some consumer research studies in which we were supposed to obtain a sample of 18 people in a small town in the High Sierras. After considerable difficulty we were able to locate the "town" and after further difficulty were eventually able to find an eighteenth person. In this study, the entire population of the town was exhausted (as was the author), and more reliable results were obtained than if a smaller sample of the town were selected. As it turned out, however, not a single person planned on purchasing TV sets, dishwashers, or similar electrical appliances during the next year, largely because the town did not have electricity.

In any event, the population to be studied is seldom so small that it can be exhausted by the researcher. More likely, the population is so large, sometimes infinitely large, that it cannot be studied in its entirety, and the researcher must resort to studying a sample. One of the reasons that a population may be indefinitely large (or perhaps finite, but infinitely large), is that the experimenter may wish to generalize not only to people now living, but to people who are as yet unborn.

When the population is too large to be studied in its entirety, the experimenter must select a number of participants and study them. One technique that an experimenter may use in selecting a sample is that of *randomization*. In random selection of a sample of participants from a population each member of the population has an equal chance of being chosen. For instance, if we wish to draw a random sample from a college of 600 students, we might write the names of all the students on separate pieces of paper. We would then place the 600 slips of paper in a (large) hat, mix thoroughly, and, without looking, draw our sample. If our sample is to consist of 60 students, we would select 60 pieces of paper. Of course there are simpler techniques to achieve a random sample. For instance we might take a list of all 600 students and select every tenth one to form our sample. To select the first name, we would randomly select one of the first ten participants, and then count successive tens from there.

Once the experimenter randomly selects a sample, it is then assumed that the sample is typical of the entire population—that a *representative* sample has been drawn. Drawing samples at random is usually sufficient to assure that a sample is representative, but the researcher may check on this if it is desired.[4] For one, if values are available for the population, then they can be compared with the sampled values. If one is studying the population of people in the United States, there is readily available a large amount of census information about education levels, age, sex, etc.[5] We can compute certain of these statistics for a sample and compare these figures with those for the general population. If the values are close, we can assume that the sample is representative. The assumption being made, in this case, is that if a sample is similar to the population in a number of known characteristics, it is also similar with respect to characteristics for which no data are as yet available. This could be a dangerous assumption, but it is certainly better than if there were no check on representativeness.

Once the population has been specified, a sample drawn from it, and the type of design determined, it is necessary to divide the sample into the number of groups to be used. The participants must be assigned to groups by some random procedure. By using randomization we would assure ourselves that each participant has an equal opportunity to be assigned to each group. Some procedure such as coin flipping can be used for this purpose. For example, suppose that we have a sample of 20 participants, and that we have two groups. We might take the first participant and flip a coin; if it is "heads," the participant will be placed in group one, and if it is "tails," the participant will be placed in group two. We would then do likewise for the second participant, and so on until we have 10 participants in one group. The remaining subjects would then be assigned to the other group.

---

[4] A somewhat infrequent definition of a representative sample is one that has been randomly drawn from a population. Following this definition, of course, it would be foolish to check the representativeness of the sample. We are, however, following the more common procedure of assuming that a sample that is not representative (is very atypical) of a population can be randomly selected. For example let us define a population of people that consists of 90 blonds and 10 redheads. Now if we draw a random sample of 10, and find that they are all redheads, we would say that the sample is not representative of the population. See Lindquist (1953) for an excellent discussion of this topic.

[5] Of course, some census statistics are also obtained through sampling techniques, and such statistics do not guarantee that the population has those precise values. We are simply assuming here that such interval statistics are *probably* true—a safe assumption.

We now have two groups of participants who have been assigned at random. The next thing to do is to determine which group is to be the experimental group and which is to be the control group. This decision should also be determined in a random manner, such as by flipping a coin. We might make a rule that if a "head" comes up, group one is the experimental group and group two the control group; but if our coin flipping yields a "tail," group one would be the control group, group two the experimental group.

By now you no doubt have acquired a feel for the importance of randomization in experimental research—the random selection of a sample of participants from a population, the random assignment of participants to groups, and the random determination of which of the two groups will be the experimental group and which will be the control group. It is by the process of randomization that we attempt to eliminate biases (errors) in our experiment. When we want to make statements about our population of participants, we generally study a sample that is representative of that population. If our sample is not representative, then what is true of our sample may not be true of our population, and we might make an error in generalizing the results obtained from our sample to the population. Random assignment of our sample to two groups is important because we want to start our experiment with groups that are essentially equal. If we do not randomly assign participants to two groups, we may well end with two groups that are unequal in some important respect. If we assign participants to groups in a nonrandom manner, perhaps just looking at each participant and saying "I'll put you in the control group," we may have one group being more intelligent than the other; consciously or unconsciously, we may have selected the more intelligent participants for the experimental group.

Having thus emphasized the importance of these procedures, we must hasten to add that the use of randomization does not guarantee that our sample is representative of the population from whence it came, or that the groups in an experiment are equal before the administration of the experimental treatment. For, by an unfortunate quirk of fate, randomization may produce two unequal groups, e.g., one group may, in fact, turn out to be significantly more intelligent than the other. However, randomization is typically the best procedure that we can use, and we can be sure that, at least in the long run, its use is justified. For any given sample, or in any given experiment, randomization may well result in "errors," but here, as everywhere else in life, we must play the probabilities. If a very unlikely event occurs (e.g., if the procedure of randomization leads to two unequal groups) we will end up with an erroneous conclusion. Eventually, however, due to the self-checking nature of science, the error will be discovered.

We may conclude discussion of this step of the experimental plan by noting that the number of groups to be used in an experiment is determined by the number of independent variables and the number of values of them that we have selected for study, as well as by the nature of the variables to be controlled. For instance, if we have a single independent variable that we decide to vary in two ways, we would have two groups. More than likely these two groups would be called experimental and control groups. If we select three values of the independent variable for study, then obviously we would need to assign our sample of participants to three groups. It might be added that usually an equal number of participants is assigned to each group. Thus if we have 80 participants in our sample, and if we vary the independent variable in four ways, we would have four groups in the experiment, probably 20 participants in each group. It is not necessary, however, to have the same number of participants in each group, and the experimenter may determine the size of each group in accordance with criteria that we shall take up later.

10. Determine the Experimental Procedure. The procedure for conducting the data collection phase of the experiment should be set down in great detail. The experimenter should carefully plan how the participants will be treated, how the stimuli will be administered, how the response will be observed and recorded. The instructions to the participants (if humans are used) should be specified and a statement concerning the administration of the independent variable and the recording of the dependent variable should be formulated. It is very useful to make an outline of each point to be covered in the actual data collection phase. The experimenter might start his outline right from his greeting to the participant and carry through step by step to when he says "goodbye." It is also advisable to try out a few participants to see how the procedure works. More often than not such "dress rehearsals" will suggest new points to be covered and modifications of procedures already set down. And if more elaborate checking of the procedure is desired, a pilot study might be conducted.

11. Evaluate the Data. The data of the experiment are usually subjected to statistical analysis. As psychology has progressed this phase of experimentation has become increasingly important. We have witnessed the development of some very powerful statistical techniques. In some manner the reliability of the results of the experiment should be evaluated. Suppose, for instance, a two-groups design is used and the sample mean for the experimental group is found to be 14.0, the sample mean for the control group 12.1. In this case one might conclude that the population mean for the experimental group is higher than for the control group. On the basis of this limited amount of information, however, such a conclusion is not justified, since it has not been determined that the sample difference is a reliable difference. The observed difference is but a *sample* difference and may not be a "real" difference. If the difference is not reliable, the next time the experiment is conducted the outcome may be reversed. Thus, as we noted in chapter 1, we need to use a statistical technique to determine whether the sample difference between the mean scores of the two groups is reliable. The statistical analysis will tell you, in effect, the odds that the difference between the groups might have occurred by chance. If the probability that this difference could have occurred by random fluctuations is small, then we may conclude that the difference is reliable, that the experimental group is reliably superior to the control group.

The main point here is that the data are evaluated by a statistical test, a number of which are available. Some tests are appropriate to one kind of data or experimental design and some are not. The use of such statistical tests frequently requires that certain assumptions about the experimental design and the kind of data collected must be met. In taking such matters into consideration it is advisable to plan the complete procedure for statistical analysis prior to conducting the experiment. Sometimes experimenters do not do this, and find that there are serious problems in the statistical analysis which could have been prevented by a little more planning and insight. Lack of rigor in the use of statistics can invalidate the experiment. And the statistics used must be appropriate to the design selected.[6]

12. Form the Evidence Report. We have said that the evidence report is a summary statement of the findings of the experiment, but we can now add that it should tell us something more than this. It should also tell us that the antecedent conditions of the hypothesis held (were actually present) in the experiment. More completely, then, the evidence report is a statement that asserts that the antecedent conditions of the hypothesis obtained, and that the consequent conditions specified by the hypothesis were found either to occur or not to occur. If the consequent conditions were found to occur, we may refer to the evidence report as positive, and if they were not found to occur, the evidence report is negative. To illustrate, consider the hypothesis: "If a teacher praises a student for good reading performance, then the student's reading growth will increase." We will design an experiment to test this hypothesis. The participants in an experimental group will

---

[6] A "nonstatistical," though rigorous, approach will be presented later (chapter 11).

be praised each time they exhibit good reading performance. No praise will be given to the members of the control group when they show similar performance. Let us assume that the experimental group exhibits a reliably greater increase in reading growth than does the control group. Referring to the hypothesis, we may note that the antecedent condition was satisfied, and that the consequent condition was found to be the case. We may thus formulate our evidence report: "Students were praised by a teacher when they exhibited good reading performance, and they exhibited an increase in reading growth (as compared to the control group)." In short, the evidence report is a sentence that asserts that the antecedent conditions held *and* that the consequent conditions either did or did not hold—it is of the form "*a* and *b*," where *a* stands for the antecedent conditions and *b* for the consequent conditions of the hypothesis.

13. Make Inferences from the Evidence Report to the Hypothesis. In this phase the evidence report is related to the hypothesis for the purpose of determining whether the hypothesis is probably true or false. To do this we must make an inference from the evidence report to the hypothesis. Essentially the inference is the following: If the evidence report is positive, the hypothesis is confirmed (the evidence report and the hypothesis coincide—what was predicted to happen by the hypothesis actually happened, as stated by the evidence report). If, however, the evidence report is negative, we may conclude that the hypothesis is not confirmed.

14. Generalize the Findings. The extent to which the results of the experiment can be generalized depends on the extent to which the populations with which the experiment is concerned have been specified and the extent to which those populations have been represented in the experiment by random sampling. Considering only human participant populations again, let us say that the experimenter specified his population as all of (and only) the students at Ivy College. If a random sample of participants was drawn from that population, the experimental results may be generalized to that population; it may be asserted that what was true for this sample is probably true for the whole population. Of course, if the population was not adequately defined, or the sample was not randomly drawn, no such generalization can be made, and the results would apply only to the sample studied.

A Summary and Preview

We have covered the major phases of experimentation, and you should now have a good idea of the individual steps and how they fall into a logical pattern. Our first effort to present the whole picture was in chapter 1. In chapters 2 and 3 and in this section we have attempted to enlarge on some of the steps. Thus in chapter 2 we considered the nature of the problem, and in chapter 3 we discussed the hypothesis. These two initial phases of planning the experiment were summarized as steps 3 and 4 above. Next we said that the variables specified by the hypothesis should be operationally defined (step 5). The use of apparatus for presenting stimuli and for recording responses was discussed as step 6. The important topic of control was briefly considered (step 7), but will be enlarged on later (chapter 6). Following this we pointed out that several designs are possible in addition to the two-groups design that we have largely concentrated on (step 8). The ways in which several different experimental designs may be used is the subject of chapters 5, 8, 9, 10, and 11. Next we took up the selection of participants and assignment of them to groups (step 9). Step 10 consisted of a brief discussion of experimental procedure, and in step 11 we offered a preview of the techniques of statistical analysis that will be more thoroughly covered in connection with the different experimental designs. The formation of the evidence report and the way in which it is used to test the hypothesis were taken up in steps 12 and 13. A separate chapter will be devoted to the latter topic (chapter 13). Finally, we briefly considered the problem of generalization (step 14), a topic that will be more elaborately considered in chapters 14 and 15. Of course, as we continue through the book, each of the above points will continue to appear in a variety of places, even though separate chapters may not be devoted to them. As a summary of this section, as well as to facilitate your planning of experiments, we offer the following check list.

1. Label the experiment.
2. Summarize previous research.
3. State your problem.
4. State your hypothesis.
5. Define your variables.
6. Specify your apparatus.
7. State the extraneous variables that need to be controlled and the ways in which you will control them.
8. Select the design most appropriate for your problem.
9. Indicate the manner of selecting your participants, the way in which they will be assigned to groups, and the number to be in each group.
10. List the steps of your experimental procedure.
11. Specify the type of statistical analysis to be used.
12. State the possible evidence reports. Will the results tell you something about your hypothesis no matter how they come out?

13. Are you clear about the inferences that can be made from the evidence report to the hypothesis?
14. To what extent will you be able to generalize your findings?

Now that we have presented the steps of experimentation in an orderly and logical fashion, let us conclude this section with a tempering comment. An advertisement for an employment agency could well state that "there is always a future in laboratory maintenance." There is, in fact, a widely held belief among experimentalists that each has his or her personal poltergeist who capriciously intervenes in the laboratory at critical times. There are many common "laws of experimentation" that reflect the consequences: "Anything that can go wrong will go wrong"; "everything goes wrong at one time"; "things take more time to repair than to do"; "if several things can go wrong, the one that will go wrong is that which will do the most harm"; "if your lab seems to be going well, you have overlooked something"; and the like. This point is included now because, after faithfully following the prescription offered here, students typically experience difficulties in the conduct of an experiment that they summarize by such phrases as "everything's a mess." By this they apparently mean such things as: the equipment stopped working in the middle of an experimental session, some participants were uncooperative, a fatal error in control of variables was detected after the data were collected, and so forth.

Now clearly such difficulties may present themselves during the conduct of an experiment, but they are just as clearly not the sole possession of students; the sophisticated researcher also experiences his share of such grief, though he has learned to be more agile and is better able to recover when troubles appear. Adjustments nearly always are made by the professional researcher, such as replacing the data for one subject for whom procedural errors occurred by running another subject under the same condition. The list of steps presented here, in short, is meant to be a help in carefully planning your experiment. And the anticipation of potential problems will help to reduce the number of experimental errors that occur. We might only add that experienced psychologists themselves profit to the extent to which they formulate and adhere to a precise experimental plan.

Conducting an Experiment: An Example

One of the values of conducting an experiment early in your course in experimental psychology is that it affords you the opportunity to make errors that you can learn to avoid in later research. For this reason it is important for students to commence work on a problem early in the course, regardless of the simplicity of the experiment or of whether or not it will contribute new knowledge. Too many students feel that their first experiment has to be an important one. Certainly, we want to encourage the conduct of important research, but the best way to reach the point where this is possible is to practice. The following example is a realistic one at this stage of your training; the fact that it is a simple, straightforward one allows us to better illustrate the points that we have previously covered.

The problem that one class set for themselves concerned the effect of knowledge of results on performance: They wanted to know whether informing a person of how well a task is performed will facilitate the learning of that task. The title of this experiment was "The Effect of Knowledge of Results on Performance," and the students conducted a rather thorough literature survey on that topic. The problem was then stated: "What is the effect of knowledge of results on performance?" The hypothesis was: "If knowledge of results is furnished to a person, then that person's performance will be facilitated." Note that the statement of the problem and the hypothesis has *implicitly* determined the variables; they next need to be made *explicit*. The task that the participants performed was to draw, while blindfolded, a 5-inch line. The independent variable concerns the amount of knowledge of results furnished the participants, and it was varied from zero to some large amount. A large amount of knowledge of results was operationally defined as telling the participants whether the line drawn was "too long," "too short," or "right." "Too long," in turn, was defined as any line 5-1/4 inches or longer, "too short" as any line 4-3/4 inches or shorter, and "right" as any line between 5-1/4 inches and 4-3/4 inches. A zero amount of knowledge of results was defined as furnishing the participant no information about the length of his line. The dependent variable was actual length of the line drawn. Each participant was required to draw 50 lines, the proficiency being determined by the participant's total performance (the sum total of the deviations from 5-inch lines) on all 50 trials.

The apparatus consisted of a drawing board on which was affixed ruled paper, a blindfold, and a pencil. The paper was easily movable for each trial, and ruled in such a manner that the experimenter could tell immediately within which of the three intervals (long, short, or right) the participants' lines fell.

Two values of the independent variable were selected for study, a positive amount and a zero amount. Two groups were thus required, an experimental and a control. The experimental group received knowledge of results, whereas the control group did not. The subject population was defined as all of the students in the college.

From a list of the student body, 60 participants were randomly selected for study.[7] The 60 participants were then randomly divided into two groups (see p. 120 for the precise manner in which this assignment may be carried out). It was then randomly determined that one of the groups was the experimental group, and the other was the control.

Next it was determined which extraneous variables might influence the dependent variable and therefore needed to be controlled. Our general principle concerning control is that both groups should be treated alike in all respects, except as far as the independent variable is concerned (in this case different amounts of knowledge of results were administered). Hence, essentially the same instructions were read to both groups; a constant "experimental attitude" was maintained in the presence of both groups. The experimental plan specified that the experimenter should not frown at some of the subjects and be gay or jovial with others. Incidental cues were eliminated insofar as possible, e.g., the experiment might have been invalidated had the experimenter held his or her breath when the participants reached the 5-inch mark. Not only would this have furnished some knowledge of results to an alert control participant, but it would have increased the amount of knowledge of results for the experimental participants.

Is the amount of time between trials an important variable? Previous research suggested that it was. In general, the longer the time between trials, the better the performance. This variable was therefore controlled by holding it constant for all participants. It was specified that each participant should wait ten seconds after each response before the hand was returned to the starting point for the next trial. What other extraneous variables might be considered? Perhaps the time of day at which the participants are run is important; a person might perform better in the morning than in the afternoon or evening. If the experimental group were run in the morning and the control group in the afternoon, then no clear-cut conclusion about the effectiveness of knowledge of results could be drawn from the data. One control for this time variable might be to run all participants between 2 P.M. and 4 P.M. But even this might produce differences, since people might perform better at 2 P.M. than at 3 P.M. Further-

---

[7] It was correctly assumed that all 60 participants would cooperate. The fact that this assumption is not always justified leads to the widespread practice among experimenters of using students in introductory psychology classes. Such students are quite accessible to psychologists and usually "volunteer" readily. This method of selecting participants, of course, does not result in a random sample, and thus leads to the question of whether the sample is representative of some population such as that of all the students in the college.

more, it was not possible to run all the participants within this one hour on the same day, so the experiment had to be conducted over a period of two weeks. Now does it make a difference whether students participate on the first day or the last day for the two weeks? It may be that examinations are being given concurrent with the first part of the experiment, causing the participant to be nervous. Then again it may be that people who are tested on Monday perform differently than people tested on Friday.

The problem of how to control the time variable was rather complex. The following procedure was chosen (see chapter 6 for an elaboration): It was specified that all participants would be run between 2 P.M. and 4 P.M. When the first participant reported to the laboratory, a coin was flipped to determine assignment to either the experimental or control group. If it turned out that the participant was to be in the control group, the next participant was assigned to the experimental group. When the third participant reported, it was similarly determined which group he or she was assigned to, and the fourth participant was placed in the other group. The rest of the participants were similarly assigned to the groups for as many days as the experiment was conducted. By using this procedure it was rather safely assumed that whatever the effects of time differences on the participant's performance, they were balanced—that they affected both groups equally. This is so because, in the long run, we can assume that an equal number of individuals from both groups participated during any given time interval of the day, and on any particular day of the experiment.

Another control problem concerns the individual characteristics of the experimenter, a topic that we shall explore in considerable detail later (chapter 15). In this case all of the students in the experimental psychology class ran participants for this experiment. Should it have been the case that one student ran more participants than another, or that the students did not run an equal number of experimental and control participants, experimenter characteristics might have differentially affected the dependent variable measures of the two groups. The experimenter variable was, thus, adequately controlled. The illustrations given should be sufficient to illustrate the control problems involved. Think of some additional variables that the class considered, e.g., do various distracting influences exist, such as noise from radiators and people talking? These could be controlled to some extent, but not completely. In the case of those that could not be reasonably controlled, it was assumed that they affected both groups equally—that they "randomized out." For instance, there is no reason to think that the various distracting influences should affect one group to a greater extent than the other. After surveying the various extraneous variables, it was concluded that this assumption was justifiable: There were no variables that could not either be controlled or whose effect, if any, would differentially affect the dependent variable scores of the two groups.

The next step considered was the experimental procedure. The plan for this phase proceeded as follows: After the participant enters the laboratory room and is greeted, the person is seated at a table and given the following instructions: "I want you to draw some straight lines that are 5 inches long, while you are blindfolded. You are to draw them horizontally like this (experimenter demonstrates by drawing a horizontal line in the air). When you have completed your line, leave your pencil at the point where you stopped. I shall return your hand to the starting point. Also keep your arm and hand off the table while drawing your line. You are to have only the point of the pencil touching the paper. Are there any questions?" The experimenter will answer any questions by repeating pertinent parts of the instructions. When the participant is ready, the experimenter places a blindfold over the eyes ("now I am going to blindfold you"), uncovers the apparatus, and places the pencil in the participant's hand. The individual's hand is guided to the starting point and the instruction is given: "Ready? Go." The experimental participants are furnished the appropriate knowledge of results (as previously specified) immediately after their pencils stop. No information is given to the control participants. After the participant completes a trial, there is a 10-second wait, after which the hand is returned to the starting point. Then the person is told: "Now draw another line 5 inches long. Ready? Go." This same procedure is followed until the participant has drawn 50 lines. The experimenter must move the paper before each trial so that the participant's next response can be recorded. The participant's blindfold is then removed, thanks for the cooperation is given, and there is a caution to discuss the experiment with no one. Finally, the nature of the experiment is explained for the participant, and all questions are answered.

Following this, the students collected their data. It was reassuring, though hardly startling, to find that knowledge of results did, in fact, facilitate performance.

Illustration of the final steps of the planning and conduct of an experiment (statistical treatment of the data, forming the evidence report, confronting the hypothesis with the evidence report, and generalization of the findings) can best be offered when these topics are later emphasized. We will also later consider how the findings of an experiment may contribute to the formulation of empirical laws, and how they fit into some theoretical frameworks.

Writing Up an Experiment

After the data are collected, the experimenter subjects them to statistical analysis and reaches the appropriate conclusion. Then the experiment is written up. We suggest that the same general format for writing up experiments should be used regardless of whether the report is to be submitted for publication in a scientific journal or whether it is a study conducted by a beginning student in experimental psychology. This helps to maximize the transfer of learning from a course in experimental psychology to the actual conduct of experiments as professional psychologists—and the heart of this book for those students who will continue into graduate work is to help them to acquire important professional behaviors (with a minimum of "busy work").

The following is an outline that can be used for writing up the experiment. There are also a number of suggestions that should help to eliminate certain errors that students frequently make and several other suggestions that should lead to a closer approximation to scientific writing.

First, we should be aware of the fact that learning to write up research manuscripts is a most difficult (though eventually rewarding) endeavor. It is often a frustrating task for the student as well as for the professor who reads the student's write-up. Consequently, we want to concentrate especially on this section so that the end product is profitable for all.

The general goal of the research report is to facilitate the communication of scientific or technological information. If the researcher conducts an experiment but never reports it to the scientific world, the work might as well not have been undertaken. The same can be said if an article is not understandable. The scientific report is the heart of our science. We seek to learn to write well-organized reports that communicate clearly, with accuracy, and are easily understandable by the reader. How does one reach such a goal? The answer is the same as it is for achieving a high degree of proficiency for any difficult task —by practice and more practice.

Before starting your writing, you should consider a model journal, one to which you would plan submitting your experiment for publication. For instance, you might select the *Journal of Experimental*

Psychology. If so, then you should thoroughly study a number of recent issues of that journal, reading sample articles in some detail. Note precisely how the authors have handled the various points that follow. This study of the format of published articles may already have been accomplished in your literature survey, where you noticed how previous authors dealt with factors that *you* will have to consider in your write-up.

The main principle to follow in writing up an experiment is that the report must include every relevant aspect of the experiment; someone else should be able to repeat the experiment solely on the basis of the report. If this is impossible, the report is inadequate. On the other hand, the experimenter should not become excessively involved in details. Those aspects of an experiment which the experimenter judges to be *irrelevant* should not be included in his report. In general, then, the report should include every important aspect of the experiment, but should also be as concise as possible, for scientific writing is economical writing.

The writer should also strive for clarity of expression. If an idea can be expressed simply and clearly, it should not be expressed complexly and ambiguously; "big" words or "high flown" phrases should be avoided wherever possible (and don't invent words).

We shall adhere to certain standard conventions. The conventions and a number of additional matters about writing up an experiment may be found in the *Publication Manual* of the American Psychological Association.[8] The close relationship between the write-up and the outline of the experimental plan should be noted. Frequent reference should be made to that outline in the following discussion for much of the write-up has already been accomplished there.

The *Publication Manual* offers some excellent suggestions about how to write effectively, how to present your ideas with an economy and smoothness of expression, how to avoid ambiguity in your sentences, and how to generally increase the readability for your reader. Precision in the use of words is also emphasized. You can do well to study the manual in some detail.

The close relationship between the write-up and the outline of the experimental plan should be noted. Frequent reference should be made to that outline in the following discussion, for much of the write-up has already been accomplished there.

1. Title. The title should be short and indicative of the exact topic of the experiment. If you are studying the interaction of drive

---

[8] The *Publication Manual* can often be obtained in your local bookstore, from a friendly professor, or by writing: Publication Sales, American Psychological Assn., 1200 17th St. NW, Washington, D.C. 20036.

level and amount of reinforcement, include in the title a statement of the variables of drive and reinforcement. However, *every* topic included in the report need not be specified in the title. Abbreviations should not be used in the title. The recommended maximum length of the title is 12 to 15 words. The title needs to be unique—it should distinguish the experiment from all other investigations. Introductory phrases such as "A study of . . ." or "An investigation of . . ." should be avoided, since they are generally understood.

2. *Author's Name and Institutional Affiliation.* On the title page the author's name should be centered below the title, and the next line should state the university or college at which you are studying. In the case of multiple authorship where all authors are from the same institution, the affiliation should be listed last (and only once). In no case should the psychology department within the institution be specified. Frequently an entire class conducts an experiment, in which case, strictly speaking, they are multiple authors. Since the main purpose of such class experiments is to provide practice for the individual student, however, it is suggested that only the name of the student writing up the experiment be used as the author, rather than listing the entire class including the professor.

3. *Introduction.* In the literature survey portion of the experimental plan, we have already developed a basis for the introductory section of the report. The effort is to develop the problem logically, citing the most relevant studies. A summary statement of the problem should then be made, preferably as a question.

Let us emphasize that the results of the literature survey should lead smoothly into the statement of the problem. For instance, if the experiment concerns the effects of alcohol on performance of a cancellation task (e.g., striking out all letter Es in a series of letters), you might cite the results of previous experiments that show detrimental effects of alcohol on various kinds of performance. At this point you might indicate that there is no previous work on the effects of alcohol on the cancellation task and that the purpose of your experiment was to extend the previous findings to that task. Accordingly the problem is, "Does the consumption of alcohol detrimentally affect performance on a cancellation task?" The steps leading up to the statement of the hypothesis should also be logically presented, but it too should be stated in one sentence, preferably in the "If . . . then . . ." form. The statment of the hypothesis and the definition of the variables may help your reader to understand what it is you intend to do, what you expect your results to be and why you expect them. Why you expect them entails the development of a theory, if in fact you have one.

Many features of your write-up are arbitrary, such as where you define your variables—they may be precisely defined in the introduction or in the method section of the study. It is not customary to label the introductory section; rather, it is simply the first part of the article.

4. Method. The main function of this section of the report is to tell your reader precisely how the experiment was conducted. Put another way, this section serves to specify the methods of gathering data that are relevant to the hypothesis and that will serve to test the hypothesis. It is here that the main decisions need to be made as to which matters of procedure are relevant and which are irrelevant. If the author has specified every detail that is necessary for someone else to repeat the experiment, but no more, the write-up is successful. To illustrate, let us assume that a "rat" study has been conducted. The author would want to tell the reader that, say, a T maze was used, and then go on to specify the precise dimensions of the maze, the colors used to paint it, the type of doors, and the kind of covering. One would presumably not want to relate that the maze was constructed of pine, or that the wood used was one inch thick, for it is highly unlikely that these variables would influence the rat's performance. That is, it would be a strange phenomenon indeed if one could show that rats performed differently in a T maze depending on whether the walls were 3/4 inch or 1 inch thick.

While the subsections under method are not rigid and may be modified to fit any given experiment, in general the following information should be found, and usually in the following order:

a. *Participants (or Subjects).* The population should be specified in detail, as well as the method of drawing the sample studied. If any participants from the sample had to be "discarded" (students didn't show up for their appointments, they couldn't perform the experimental task, rats died, etc.), this information should be included, for the sample may not be random because of their factors. The total number of participants and the number assigned to each experimental condition should be stated. In specifying the population, such details as sex, age, general geographic location, type of institution and whence they came, any promises made to them, and the like should be specified.

b. *Apparatus.* All relevant aspects of the apparatus should be included. Where a standard type of apparatus is used (e.g., a "Hull-Type Memory Drum"), only its name need be stated. Otherwise, the apparatus has to be described in sufficient detail for another experimenter to obtain or construct it. It is good practice for the student to include a diagram of the apparatus in the write-up, although in professional journals this is only done where the apparatus is complex and novel.

c. *Design.* The type of design used should be included in a section after the apparatus has been described. The method of assigning participants to groups, and the labels attached to the groups, are both indicated (e.g., Group E may be the experimental group and Group C the control group, etc.). The variables contained in the hypothesis need to be (operationally) defined if they have not been defined in the introduction; it is also desirable (at least for your practice) to indicate which are the independent and dependent variables. The techniques of exercising experimental control may be included here. For example, if there was a particularly knotty variable that needed to be controlled, the techniques used for this purpose may be discussed. Relevant here are any unusual compromises in your experimental manipulation (if such occurred, it should be specified in detail), randomization procedures, counterbalancing, or other control procedures.

d. *Procedure.* The procedure for conducting the data collection phase of the experiment should be set down in detail. You must include or summarize instructions to the participants (if they are human), the maintenance schedule and the way in which the participants were "adapted" to the experiment (if they are infrahuman animals), how the independent variable was administered, and how the dependent variable was recorded.

5. *Results.* The results section is one of the most difficult parts of a research report for students to learn to write up. Even after the instructor has given individual comments to students on two or three experimental write-ups in a course, many students still have difficulty in effectively presenting results. Consequently, your *special* attention to this section is recommended.

The purpose of the results section is to provide sufficient information so that the reader can understand the processes by which the author reached a conclusion. This includes the systematic presentation of data and the reasoning from the data to the conclusions. Consequently, the reader is furnished the opportunity to determine whether or not the author was justified in reaching the conclusions reached, and in relating those conclusions to the empirical hypothesis.

To emphasize the importance of your attention to this section, we may note some common student errors or shortcomings. For instance, a student might propel the reader directly into a conclusion without even referring to data in the table, much less explaining how those data were arrived at. Or, some students merely include a table of means with the disarming conclusion that "There were no reliable differences." One valuable learning technique is for students in a course to select an article from a good journal and each one report to the class on the major methodological steps taken by the published author. In their brief presentations the students can pay special attention to the major components of the results section for presenting, analyzing, and reaching conclusions. Such learning experiences, incidentally, do not always have a favorable outcome for the journal articles selected, providing us with the opportunity to learn that even published articles can be sizably improved upon. Another dividend from this teaching procedure is to help the student to become familiar with the journals in our field, to acquire the habit of visiting the library and at least skimming the current journals as they come in, to build up one's storehouse of knowledge about current research, etc. The serious student who gets the "journal habit" early will benefit in many ways, including the discovery of especially interesting articles about topics that they wish to pursue (research?) in greater detail.

The heart of the results section, of course, is the presentation of the data relevant to the test of the hypothesis. These data are summarized as a precise sentence (the evidence report). If the data are in accord with the hypothesis, then it may be concluded that the hypothesis is confirmed. If they are not of the nature predicted by the hypothesis, then the hypothesis is disconfirmed.[9]

It is quite important to present a *summary* of the data under "re-

[9] Although we have not yet reached certain important matters, some advance information is summarized below for future reference. You need not worry about what this information means if it is unfamiliar, for it will become clear later. The advantages of including an outline of all relevant information for writing up a report in one place outweighs the disadvantage of prematurely presenting this information.

You should state the null hypothesis as it applies to your experiment and also the reliability level that you have adopted. Then include the results of the statistical test and indicate in detail whether or not you have rejected your null hypothesis. For example: "The null hypothesis was: There is no difference between the means of the experimental and control groups on the dependent variable." You may have found that your t test yielded a value of 2.20 which, with 16 degrees of freedom, was reliable beyond the 5 percent level. If so, you would then write this information up as follows: "t (16)... 2.20, $p < .05$," i.e., you specify that you used the t test with the number of degrees of freedom within the parentheses, that you obtained the computed value indicated, and that this value was or was not reliable at the selected probability level (here .05). You may then continue: "It is therefore possible to reject the null hypothesis. Since this finding is in accord with the empirical hypothesis, we may conclude that that hypothesis is confirmed." Of course, if the empirical hypothesis predicted that the null hypothesis would be rejected, but it was not, then it may be concluded that the hypothesis was not confirmed.

Having made a point about the null hypothesis, let us immediately direct your attention to the fact that the null hypothesis is not mentioned in journal articles. Rather, what we have above made explicit is, for professional experimenters, implicitly understood. Perhaps your understanding of the null hypothesis can be enhanced should your write-up specifically include the above mentioned steps, and once this process is clear to you, it can be dropped from later reports.

sults." This can almost always be done by using a table, but frequently figures can also be used to advantage. Whether or not tables and/or figures are used depends on the type of data and the ingenuity and motivation of the writer. Since students frequently are confused about the definitions of *table* and *figure*, as well as about their respective formats, we shall consider them in detail. Tables and figures are used for *summarizing* the data. They are not used for presenting all the data (so-called "raw data," a term that implies that the data have not been statistically treated). Nor are the steps in computing the statistical tests (the actual calculations) included under "results." In student write-ups, however, it has been found advisable to include the raw data and the steps in the computation of the statistical tests in a special appendix. The advantage of using such an appendix is that the instructor can correct any errors made in this part of the operation.

A table in the results section consists of numbers that summarize the main findings of the experiment. It should present these numbers systematically, precisely, and economically. A figure, on the other hand, is a graph, chart, photograph, or like material. It is particularly appropriate for certain kinds of data; for instance, to show the progress of learning. Information should, however, be presented only once, i.e., the same data should not be presented in a table and a figure or in the written text.

In constructing a table, one should first determine what is to be shown — the main points that should be made apparent from the table. Then should be considered the question of what is the most economical way of making these points in a meaningful fashion. Since the main point of the experiment is to determine if certain relationships exist between specific variables, the table should show whether or not these relationships exist. Of course, it is possible to present more than one table, and tables may be used for purposes other than presenting data. For example, it is frequently possible to make the overall design of the experiment more apparent by presenting the separate steps in tabular form (this use of tables is particularly recommended for students as it helps to "pull the experiment together" for them). To illustrate the format of a table, let us consider an experiment in which the effects of human environment on the cognitive ability of rhesus monkeyes were studied (Singh, 1966). This experimenter compared the problem-solving behavior of a group of *urban* monkeys with a group of *forest* monkeys; the environmental difference between the two groups was that the former had frequent and intimate interactions with human beings, while the latter (having lived in the jungles) did not. Both groups were administered a variety of tests, among which was one on visual pattern discrimination and another on object discrimination. The apparatus used was such that the animal was presented with two or more stimuli, and under one was a raisin. If the monkey reached out of the cage and displaced the correct stimulus, it received a raisin. For the visual pattern discrimination test, the animal had to respond until 45 out of 50 correct responses were made in one day. When it had thus successfully learned to visually discriminate one pattern, the monkey was presented with another, then another, until eight patterns were successfully discriminated.

To summarize his data, Singh counted the number of trials that each animal took before reaching the criterion that showed that the discrimination was learned. Then he determined the median number of trials for each group. The results are presented in Table 4-1. For example, it can be seen that the median number of trials required by the urban group to discriminate the first visual pattern was 338.0; the median number for the forest group was 491.5. Similar comparisons can be made for each of the remaining patterns. Note that the previously stated requirements of a good table are clearly satisfied in this example. Also observe the precise format used, for students have a habit of ignoring the details of the standardized conventions illustrated here.

Table 4-1 Illustration of a Good Format for a Table

Median Trials to Criterion on Successive Visual Pattern Discriminations

GROUP	PATTERN							
	I	II	III	IV	V	VI	VII	VIII
Urban	338.0	149.5	24.0	0.0	165.0	26.5	0.0	0.0
Forest	491.5	261.5	34.0	40.0	259.0	102.0	44.5	13.0

By studying Table 4-1 and other tables throughout the book, as well as those in journal articles in your library, you will begin to acquire the ability to efficiently and systematically presenting your data. In some cases you will want to include the numbers of participants in your groups; most frequently you will probably use means for your dependent variable, rather than medians; you will often want to include some measure of variability, such as standard deviations; and so forth.[10]

The same general principles stated for the construction of tables

[10] If you use abbreviations in a table in order to conserve space, be sure that the abbreviations are explained in a note to the table (same for figures). Also, the reader should not have to refer to the text in order to understand the table or figure — tables and figures should be self-contained.

holds for figures. In particular, a figure is used primarily to illustrate a relationship between the independent and dependent variables of an experiment. It is conventional to use the vertical axis (sometimes erroneously, as you can see in a dictionary, called the "ordinate") for plotting the dependent variable scores; the horizontal axis (which is *not* synonymous with "abscissa") is typically labelled "Time" or "Number of Trials." The scores for each group of participants may then be plotted and compared. As an example of a good use of a figure, let us consider how Singh presented his data on the object discrimination problem. He determined a total number of responses made by each group of monkeys for the first 48 problems that they solved. The number of correct responses out of the total number was then counted, and the percentage of correct responses was computed. In this way, then, it was ascertained that about 58 percent of the total number of responses made by the urban group to the first 48 problems were correct. During the solution of the next 48 problems, the percentage of correct responses rose; the value plotted in Figure 4-3 is approximately 69 percent. By studying Figure 4-3 we can see that the percentage of correct responses increases as the number of problems that the animals had solved increases; by the time the animals had solved over 300 problems their proficiency in solving new problems was considerably better than when they were naive. A com-

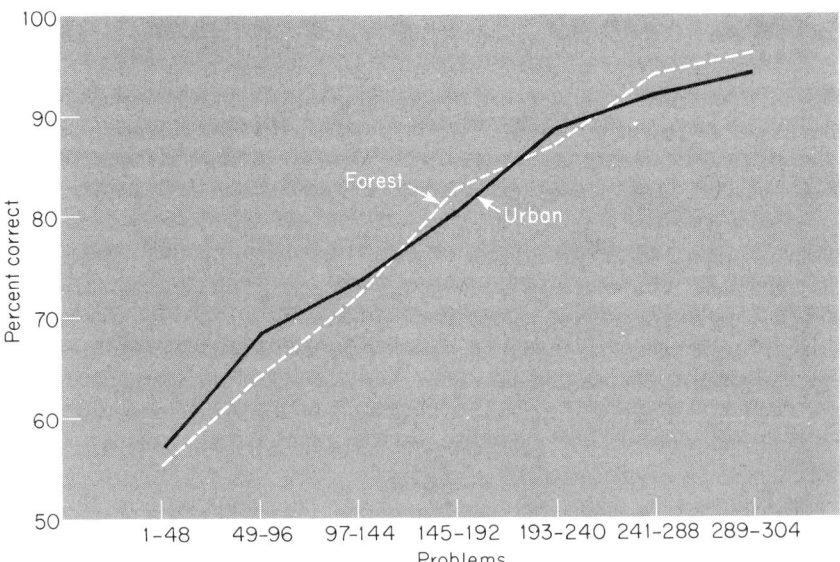

Figure 4-3. Object discrimination learning curves for the urban and forest groups (after Singh, 1966).

parison of the learning curve for the urban with that of the forest group shows that they are quite similar. In fact, Figure 4-3 is illustrative of the general conclusion of this study, viz., ". . . the results . . . do not indicate any effect of human environment on the cognitive ability of the rhesus monkeys . . ." (Singh, 1966, p. 283).

Incidentally, one question that often comes up when comparing performance curves of two groups concerns what would happen if more trials had been given. For instance, suppose that the curve for an experimental group rises in a normal manner, but that the curve for the control group is retarded. Further, suppose that the experimenter gave the participants 100 trials, and that at about trial number 90 the control group's curve markedly increases, though by trial number 100 it is still lower than that of the experimental group. What would have happened had more trials been given? Would the two curves eventually come together? If we conduct the experiment again and give the participants a larger number of test trials, we probably would find out. (Would 20 more trials be enough?) The question of what would happen to the relative position of the curves if the participants had been run on more trials seems to be a perennial one in experimentation. One might accept this as a lesson in planning an experiment: If you are going to be concerned about this question, take it into consideration before the data are collected.

The order of presenting tables and figures is important. A table of means, or a figure in which means are plotted, is used to demonstrate your major experimental effects. A table presenting your statistical analysis (such as your analysis of variance, as discussed in chapter 9) is for the purpose of stating whether or not your means are reliably different. Hence, the statistical analysis should come *after* the means. It is also important to emphasize that the source of the numbers that are presented in your tables and figures be *precisely* identified and explained. Often a reader must spend considerable time puzzling over the question of just what the numbers mean — although they may seem clear to the author, the write-up may have missed a step. For instance, rather than saying that "the mean number of errors" is plotted in Figure 2, one could be more precise and say that "The mean number of errors per ten-trial block" is plotted. Or, in another case, rather than merely referring to "the number of bar-presses," one should say "The median number of bar-presses during the 15-minute extinction period." This information may be presented in the text, in the table heading, or in the figure caption.[11]

---

[11] Thanks to Professor David A. Grant, who has evaluated many manuscripts as editor of the *Journal of Experimental Psychology*, for these and several other points.

As far as format is concerned, each table and figure goes on a separate page and is included at the end of the paper. There is a separate page which includes all figure captions, but table headings go at the top of the table. The author indicates where tables and figures should be located in the text as follows:

— —

Insert Table 1 about here

— —

The above information should be sufficient to get you started, but you are strongly advised to continue your study of techniques for constructing figures and tables. To do this, you can consult any of a number of elementary statistics books that are available in your library (e.g., Guilford and Fruchter, 1973; McNemar, 1969; Runyon and Haber, 1976). But more important, you should concentrate on figures and tables as they are presented in psychological journals.

6. *Discussion.* The main functions of this section are to interpret the results of the experiment and to relate these results to other studies. The interpretation involves essentially an attempt to explain the results. Perhaps some existing theory can be brought to bear to help understand the findings. If the hypothesis was derived from a general theory, then the confirmation of the hypothesis serves to strengthen that theory, and the findings in turn are explained by that hypothesis in conjunction with the larger theory. If the findings are contrary to the hypothesis, then some new explanation is required; to account for the results there may be advanced new hypotheses that run counter to the hypothesis tested. Or it may be that the hypothesis tested can be modified in such a way to make it consistent with the results. In this case a "patched up" hypothesis is advanced for future test.

In relating the results to other studies, the literature survey may again be brought to bear. By considering the present results along with previous ones, new insights may be obtained. The results of this particular experiment may provide the one missing piece that allows the solution of the puzzle.

Hypotheses may also be advanced about any unusual deviation in the results. For instance, one may wonder why there is a sudden rise in the terminal portion of a learning curve. Is this a reliable rise? If it is, why did it occur? In short, what additional problems were uncovered that suggest fruitful lines for further investigation? In this regard recall from chapter 3 that you, like the three princes of Serendip, may find something more valuable than that which you originally sought.

There may be certain limitations in the experiment. If so, this is the place to discuss them, e.g., what variables might have turned out to have been inadequately controlled? (If, however, you did not control a crucial extraneous variable, you probably wouldn't attempt to publish your report.) How would one modify the experiment if it were to be repeated?

You may also consider the extent to which the results may be generalized. To what populations may you safely extend your results? To what extent are the generalizations limited by uncontrolled variables, etc.?

A rather strange characteristic of some experimenters seems to be that they feel "guilty" or "embarrassed" when they have obtained negative results.[12] Whatever the reason, it is not appropriate to include long "alibis" for negative results. It is reasonable, however, to briefly speculate about why they were obtained.

"Negative results" constitute a serious problem for our science because they seldom are accepted for publication in our journals. One then wonders how biased might be the results that *are* published in our journals. To illustrate the point, if you conduct 100 experiments to test hypotheses that really are false, by our logic of testing null hypotheses, five percent of the time you will confirm those hypotheses ("by chance") when they should be disconfirmed. The possibility is alarming that many of our published experiments have, in this way, merely capitalized on chance. One answer is to publish only experiments that have been replicated, or another is to develop a *Journal of Negative Results.* The point is well illustrated by Hudson (1968) in his *A Case of Need* by the following story:

> There's a desert prison, see, with an old prisoner, resigned to his life, and a young one just arrived. The young one talks constantly of escape, and, after a few months, he makes a break. He's gone a week, and then he's brought back by the guards. He's half dead, crazy with hunger and thirst. He describes how awful it was to the old prisoner. The endless stretches of sand, no oasis, no signs of life anywhere. The old prisoner listens for a while, then says, "Yep. I know. I tried to escape myself 20 years ago." The young prisoner says, "You did? Why didn't you tell me, all those months I was planning my escape. Why didn't you let me

---

[12] "Negative results" occur when the hypothesis makes a particular prediction, but the results are contrary to that prediction. The term may also be used to indicate that the null hypothesis was not rejected; usually these two definitions amount to the same thing.

know it was impossible." And the old prisoner shrugs, and says, "So who publishes negative results?" (p. 90)

In conclusion, the introduction to the discussion may be a brief summary of the important results, followed by a clear statement as to whether or not those results supported or failed to support the hypothesis; finally, you might relate those results to other findings and theories. "In general, be guided by the following questions: What have I contributed here? How has my study helped to resolve the original problem? What conclusions and theoretical implications can I draw from my study? These questions are the core of your study, and readers have a right to clear, unambiguous, and direct answers" (*Publication Manual of the American Psychological Association,* 1974, p. 19).

To further illustrate how you can be somewhat flexible in the sections of your write-up, we may note that if the discussion is relatively brief, it might be combined with the results section, thus entitled "Results and Conclusions" or "Results and Discussion." Actually it is sometimes advantageous to the writer (and to the reader) to immediately interpret a finding after it is presented, particularly if there are a number of findings of importance that derive from a complex experiment. That is, in a separate discussion section the reader may get lost in referring a given interpretation back to the appropriate one of several findings that had been previously presented in the results section.

7. References. The main function of the reference section is to document (provide authority for) statements that you have made in your write-up. Scientists simply cannot say something like "Everybody knows that . . ." but they must be able to make reference to "the proof." Proper references enable the reader to easily locate the source in the library. References to pertinent studies throughout the write-up should be made by citing the author's (or authors') last name, the year of publication, and enclosing these in parentheses. If the name of the author occurs in the text, cite only the year of publication in the parentheses. All references should then be listed alphabetically at the end of the paper. Be sure to double-check that all references cited in the text appear in the reference list, and vice versa. Also, all quotations of more than three words must be cited with page numbers. The form and order of items for journal references is as follows: last name, initials, title of the study, the (nonabbreviated) name of the journal (underlined to indicate that it should be italicized), year of publication of the study, volume number (also underlined to indicate that it should be italicized), and pages. For example, if two references like the following are used they might be referred to in the write-up as "Many years ago Lewis (1953) showed that learning theory was applicable to human behavior. Early confirmations of this position were numerous (e.g., Calvin, Perkins, & Hoffman, 1956)." Then in the reference section of the write-up, they would be *precisely* listed as follows:

References

Calvin, A. D., Perkins, M. J., & Hoffman, F. K. The effect of non-differential reward and non-reward on discriminative learning in children. *Child Development*, 1956, *27*, 439–46.

Lewis, D. J. Rats and men. *American Journal of Sociology*, 1953, *59*, 131–35.

It is important to emphasize the word "precisely" used above — every item (number, comma, etc.) should be double-checked.[13] Instances of errors in typing references, just as in miscopying quotations or numerals representing data, display shoddy scholarship. The essence of our science is the production of scholarly works of the highest quality.

8. Abstract. The abstract should be the first page of your article, but it is listed here because you can more effectively write it after you have completed the foregoing sections. This section should be typed on a separate sheet of paper; it need not include title and author (which *is* on first page of text) and generally summarizes the article. It should quickly give the reader the essence of your research. Within 100 to 175 words in length it should present your problem, method, results, and conclusions. Statements should be made about the population of participants (number, type, age, gender), the research design, and apparatus. Results are the most important part of the abstract including a statement of levels of reliability (but no statistics) and inferences drawn therefrom.

Because the abstract is reproduced in the psychological abstracts and in other abstracting services, it should be self-contained and intelligible without the need to be rewritten by others or without making reference to the body of your write-up. Do not cite a reference in the abstract.

9. Cover Sheet. On a separate sheet of paper you should type the title in capital and lower case letters, centered on the page. Below, type the name of the author in capital and lower case letters, and below that the name of your institution. At the bottom of the cover sheet type a "running head" which is a shortened title to be used on each page of the published article.

---

[13] Refer for detail to the format for different sources (books, journals, technical reports) in the publication manual.

In typing up your report, the *entire text* should be double-spaced — merely set your typewriter on double-space and don't change it at any time from start to finish. Then collect your pages together in the following order:

1. Cover title page
2. Abstract (type the word "Abstract" at the top of the page)
3. Pages of text (first page includes title, author, and affiliation)
4. References (start on a new page)
5. Footnotes (start on a new page)
6. Tables (each on a separate page)
7. Figure captions (start on a new page)
8. Figures (each separately)

Some "Do's" and "Don't's"

Finally, here are a few suggestions and bits of information that did not fit conveniently into the previous sections. These matters are offered for students in the hope of improving their reports. We do not pretend to be very inclusive here, but shall simply point out items that continue to appear every year with new students. Some of them are minor, but if you learn these arbitrary conventions now your efficiency in writing up articles later will be increased. Furthermore, the use of many of these conventions helps to facilitate communication in scientific writing.

The first thing to do before writing up a report, and this is of *great importance,* is to consult several recent psychological journals. Since we are concerned mainly with experimentation, you might start with the *Journal of Experimental Psychology,* although experiments are certainly reported in a large number of journals (these journals may be found in college libraries). Select several articles in these journals and study them rather thoroughly, particularly as to format, e.g., study the figures, tables, sections and subsections, and their labels; note, for example, just what words and symbols are underlined. Try to note examples of the suggestions that we have offered and particularly try to get the overall idea of the continuity of the articles. But be prepared for the fact that you will probably not be able to understand every point in each article. Even if there are large sections that you do not understand, do not worry too much about it for this understanding will come with further learning. By the time you finish this book you will be able to understand most professional articles. Of course, some articles are extremely difficult, or in fact impossible, to understand even by specialists in the field. This is usually due to the poor quality of the write-ups.

In their own write-ups, students frequently make assertions such as: "*It is a proven fact* that left-handed people are steadier with their right hands than right-handed people are with their left hands," or "*Everyone knows that* sex education for children is good." Perhaps the main benefit to be derived by students in general from a course in experimental psychology is a healthy hostility for such statements. Before making such a statement you should have the data to back it up—you should cite a relevant reference. It is poor writing to use such trite phrases as "It is a proven fact that . . ." or "Everyone knows that. . . ." Such phrases probably really mean that the writer lacks data or hasn't bothered to look up the original reference. If you want to express one of these ideas, but lack data, the ideas still can have a place in the introductory section. They simply should be stated more tentatively, e.g., "It is possible that left-handed people are steadier with their right hands than right-handed people are with their left hands," or "An interesting question to ask is whether left-handed people. . . ." Our main point, then, is that if you want to assert that something is true, make sure that you have the data (a reference) to back it up; mere opinions asserted in a positive fashion are insufficient.

Sometimes authors have a tendency to "over-present" their research. One anonymous cynic offered a translation of some phrases found in technical writings and what they "really" might mean. You might ponder these when thinking about writing up a study. "Of great theoretical and practical importance . . ." ("It was interesting to me"); "While it has not been possible to provide definite answers to these questions . . ." ("The experiment was goofed up but I figured I could at least get a paper out of it"); "Data on three of the participants were chosen for detailed study . . ." ("The results on the others didn't make any sense and were ignored"); "The experimental dosage for the animals was unintentionally contaminated during the experiment . . ." ("accidentally dropped on the floor"); "The experimental dosage for the animals was handled with extreme care during the experiment . . ." ("not dropped on the floor"); "Typical results are illustrated . . ." ("The best results are shown"); "This conclusion is correct within an order of magnitude . . ." ("It's wrong").

Another point about writing up reports is that personal references should be kept to a minimum.[14] For instance, students frequently say such things as "I believe that results would have turned out differently, if I had . . ." or "It is my opinion that. . . ." Strictly speaking

---

[14] Although the standard convention in scientific writing is to avoid personal references, there are those who hold that such a practice should not be sustained. The following quotation from an article entitled "Why are medical journals so dull?" states the case for this view: ". . . avoiding 'I' by impersonality and cir-

your audience doesn't care too much about your emotional experiences, what you believe, feel, think; they are much more interested in what data you obtained and what conclusions you can draw from those data. Rather than stating what you believe, then, you should say something like "the data indicate that. . . ." The report of an experiment falls within the context of justification rather than within the context of discovery.

Harsh or emotionally loaded phrases should also be avoided. The report of scientific work should be divorced from emotional stimuli as much as possible. An example of bad writing would be: "Some psychologists believe that a few people have extrasensory perception, *while others claim it to be nonsense.*"

Misspelling occurs all too frequently in student reports. You should take the trouble to read the report over after it is written and rewrite it if necessary. If you are not sure how to spell a word, look it up in the dictionary. Unfortunately, far too few students have acquired "the dictionary habit," a habit which is the mark of an "educated person."

White (1971) asked what it is that makes engineers and scientists "such lousy writers." His answer is that it is due to "an unhealthy state of mind," and he identified several diseases such as quadraphobia (Disease Number 1) which is characterized by an overwhelming thirst for a well-established say-nothing phraseology, such as "I could care less." His Disease Number 2, statumania, is an inordinate quest for status or an overpowering desire to impress. A major symptom of statumania is the misuse of words—"If it's big it's gotta be good." An engineering executive of considerable stature, engaged in a project leading to the lunar rock hunt, wrote as follows (mutilated words in italics): "I am not the *pacific cogniss* engineer on the titanium *spear;* it's not even within my *preview.* And I want to *appraise* you right now of the fact that I'm *diabolically* opposed to the belief that the physical properties of titanium might be a *significant contributating* factor. But I'll be happy to *commencerate* with you about the problem. I'll even let you set the *tenure* of the discussion." The same quote, with word corrections italicized, follows: "I am not the *specific cognizant* engineer on the titanium *sphere;* it's not even within my *purview.* And I want to *apprise* you right now of the fact that I'm *diametrically* opposed to the belief that the physical properties of titanium might be a *significant contributing* factor.

---

cumlocution leads to dullness and I would rather be thought conceited than dull. Articles are written to interest the reader, not to make him admire the author. Overconscientious anonymity can be overdone, as in the article by two authors which had a footnote, "Since this article was written, unfortunately one of us has died" (Asher, 1958, p. 502).

But I'll be happy to *commiserate* (wrong usage) with you about the problem. I'll even let you set the *tenor* of the discussion" (p. 25).

When studying the format for writing up articles by referring to psychological journals, please note what sections are literally labeled. For instance, the introduction is not usually labeled as such, but the method section is always indicated, as are its subsections such as *Participants, Procedure,* etc.

A few final matters are:

1. Don't list minor pieces of "apparatus," like a pencil (unless it is particularly important, as in a stylus maze).

2. The word "data" is plural. "Datum" is singular. Thus it is incorrect to say "that data" or "this data." Rather one should say, "those data" or "these data." Similarly "criterion" is singular and "criteria" is plural. Thus say, "This criterion may be substituted for *those* criteria." One would not be proud of oneself by saying "This apples" or "Those apple."

3. There is a difference between a probability value (e.g., $P = .05$) and a percentage value (e.g., 5 percent). Although a percentage can be changed into a probability and vice versa, one would not say that "The probability was 5 percent," or "The percent was .05" if 5 percent is really meant.

4. When reporting the results of a statistical analysis, never say that "The data are (or are not) reliable." Data are not reliable in the technical sense. Rather, the results of your statistical analysis may indicate that there is a reliable difference between your means.

5. When you quote from an article (more than three words) put quotation marks around the quote and cite the author, year, and a page reference. Then check it for accuracy.

6. Students may systematically collaborate during the data collection phase, but they should independently write up their articles and conduct their statistical analyses by themselves. Ask your fellow student to crticize your first draft so that a later draft can be improved. If your fellow student doesn't understand what you have written, your professor probably won't either.

7. Make a distinction between "negative results" and "no results." Students sometimes say that they "didn't get any results," which communicates to an amazed listener that although they invited participants into the laboratory, for some strange reason no data were collected.

8. To help your reader understand, use abbreviations wisely. Many authors assume that the reader has the same level of knowledge about the field—the result is that the author fails to explain abbreviations. This is a disservice to the reader who is a student of psychology—and this consideration should be kept in mind by the author. On the other hand, quite often they are overdone so that the reader gets lost in a maze of abstractions. Wike (1973), in an entertaining article entitled "Water Beds and Sexual Satisfaction: Wike's Law of Low Odd Primes (WLLOP)," makes this point well: "The true scien-

tist is motivated by the quest for fame, immortality, and money (FIM) [he] . . . labored quasi-diligently for over two decades in that murky swamp (MS) called research, seeking a suitable principle, hypothesis, or phenomenon to [call his own]. Recently, in a flight from the MS [his] Wike's Law of Low Odd Primes (WLLOP) [came through] it is a superb mnemonic device (MD). Regarding MD, note that WLLOP spelled backwards is POLLW. . . . WLLOP asserts simply: *If the number of experimental treatments is a low odd prime number, then the experimental design is unbalanced and partially confounded*" (pp. 192–93). With an evaluative statement, Wike asserts that "The increasing use of abbreviated names (ANs) in the psychological literature must be regarded as a genuine breakthrough. By peppering a paper with a large number of ANs, the skilled technical writer can often approach and sometimes even achieve a nirvana-like state of total incomprehensibility in his readers" (p. 192).

9. If it is at all possible, the report should be typed, not written in longhand. Studies have shown that students who type get higher grades. We shall not consider why this is so, except to point out that one possible reason is that instructors have a "better unconscious mental set" in reading typed papers. In typing, remember to set your typewriter on *double space* and leave it there for the *entire* report, including references and tables.

In this chapter we have examined several methods for obtaining an evidence report, the most powerful being, in turn, the experimental method, the method of systematic observation, and the clinical method. The primary difference between experimental and nonexperimental methods is that in the former the researcher produces the phenomenon of interest whereas in the latter the event is studied as it naturally occurs. We noted several steps that, given close attention, can facilitate the planning and execution of the experiment, and we discussed the write-up in some detail. To emphasize and summarize points made in writing up an experiment, we shall quote from a personal communication from Professor David A. Grant who, as we noted, has had considerable experience in reviewing manuscripts for publication:

. . . a paper should tell first of all what the problem is in one sentence, then a bit on why the research was done, i.e., where one proposes to contribute to our knowledge and where the findings will be relevant, etc. Then one should clearly state what was done, what was found out, and this should be complete enough so that the reader can ascertain the bases of thinking that one has found out something. Finally, in the discussion section, it should be made clear how what one has found ties in to the current knowledge and how it advances current knowledge; in this section he can also introduce qualifying clauses, and so forth.

The write-up is important—without it, the experiment might as well never have been conducted. It should be written well. We cannot emphasize too strongly that you should study, in detail, articles as they appear in journals.

We now have a good overview of how to conduct an experiment. What remains is to elaborate on the many more specific topics of experimentation. The most immediate task will be to study the design and analysis procedures for an experiment in which participants are randomly assigned to two groups.

Review of Chapter 4

This chapter attempts to present the major aspects of experimentation as an organized unit. The effort is to help you to start with the development and formulation of a problem and to work through the important phases that allow you to arrive at a sound empirical conclusion. The formulation of your experimental plan, and later of the write-up of your research, helps you to think through each important step of your research. At selected points in your study, especially as you are planning and writing up your own research, you should find that these outlines provide valuable guidance. With your later study (of this book and in subsequent courses) you should be able to add to the skeleton presented in this chapter.

1. What is the major contrasting feature between experimental and nonexperimental methods?
2. Review (and perhaps outline) the relevant steps in planning an experiment. You might make notes about important topics mentioned but not yet covered (like statistical analysis).
3. In preparation for the first experiment to be conducted in your class, you could select a problem for yourself and develop an outline of an experimental plan. This would be especially useful to you in bringing out some questions that may not yet have occurred to you.
4. Select some psychological journals from your library and study how the various components covered here in "writing up an experiment" were handled by other authors.

5

Experimental Design

the case of
two randomized groups[1]

You have now acquired a general understanding of how to conduct experiments. In chapters 1 and 4 we covered the major phases of experimentation. In presenting an overall picture of experimentation, however, it has been necessary to cover a number of phases hastily. The remaining chapters of the book will consist of attempts to fill in these relatively neglected areas. But we should try never to lose sight of how the steps on which we momentarily concentrate fit into the general picture of designing and conducting an experiment.

We shall now focus on the phase of experimentation that concerns the selection of a design. Although there are a number of designs available to the experimenter, we have thus far limited our consideration to one that involves only two groups. In this chapter we shall discuss this type of design more thoroughly. Since the "two-groups" design is basic in psychology, an understanding of it will

[1] Logically, this chapter should follow chapter 7. But it is placed earlier because we have found it important for students to get started as soon as possible on their practical experiments, which they successfully can after study of the first five chapters.

form a sound foundation from which we can move to more complex (though not necessarily more difficult to comprehend) designs.

A General Orientation

The "two-groups" design may more completely be referred to as the "two-randomized-groups design." To summarize briefly what has been said about this design, let us recall that the experimenter defines an independent variable that is to be varied in (at least) two ways. The two values that are assigned to the independent variable may be referred to as two "conditions," "treatments," or "methods." The experimenter then seeks to determine whether these two conditions differentially affect the dependent variable. The general procedure that one follows to answer this question is to first define a certain population about which we wish to make a statement. Then we randomly select a sample of participants to study. Since that sample has been drawn randomly from the population it may be assumed to be representative of the population. Thus what is observed in the sample is used to make inferences that the same holds for the population. Let us assume that the population is defined as all students in a certain university. They may number 6,000. We decide that our sample shall be 60. One reasonable method for selecting this sample would be to obtain an alphabetical list of the 6,000 students. Then randomly select one name from the first one hundred and take every 100th student on that list after that. On the assumption that all 60 students will cooperate, we now have our sample. It has been specified that we will study two conditions in our experiment. To assign a separate group of participants to each condition, we must divide the 60 participants into two groups. Again, any method that would assure that the participants are randomly assigned to the two groups would suffice. Let us say that we write the name of each participant on a separate slip of paper and place all 60 pieces of paper in a hat. We may then decide that the first name drawn would be assigned to the first group, the second to the second group, the third to the first group, etc. In this manner we would end up with two groups, each with 30 participants. (A table of random numbers, common in statistics books, is more efficient for randomly assigning participants to groups.) A simple flip of a coin would then tell us which is to be the experimental group, and which the control group. The reason that this is called the "two-randomized-groups design" is now quite apparent: Participants are randomly assigned to two groups.

A basic and important presupposition made in any type of design is that the means (averages) of the groups do not differ significantly at the start of the experiment. In a two-groups design the two values of the independent variable are then respectively administered to the two groups. For example, some positive amount of the independent variable might be administered to the experimental group, while a zero amount is administered to the control group. Scores of all participants on the dependent variable are then recorded and subjected to statistical analysis. If the appropriate statistical test indicates that the two groups are reliably different (on the dependent variable scores), it may be concluded that this difference is due to the variation of the independent variable—assuming that the proper experimental controls have been in effect, it may be concluded that the two values of the independent variable are effective in producing the differences in the dependent variable.

"Equality" of Groups Through Randomization

Now, by randomly assigning the 60 participants to two groups, we said, it is reasonable to assume that the two groups are essentially equal; but approximately equivalent with respect to what? The answer might be that the groups as wholes are equivalent in all respects. And such an answer is easy to defend, assuming that the randomization has been properly carried out. In any given experiment, however, we are not interested in comparing the two groups in all respects. Rather, we want them to be equal on those factors that might affect our dependent variable. Suppose the dependent variable concerns the rate at which a person learns a task that involves visual abilities. In this case we would want the two groups to be equivalent at least with respect to intelligence and visual acuity. More particularly, we would want the means of intelligence and visual acuity scores to be essentially the same. For both of these factors are likely to influence scores on our dependent variable.

Students frequently criticize the randomized-groups design by pointing out that "by chance" (i.e. due to random fluctuations) we could end up with two very unequal groups. It is possible, they say, that one group would be considerably more intelligent, on the average, than the other group, that is, that one group would have a higher mean intelligence score. Even though such an event is indeed possible, it is unlikely, particularly if a large number of participants is used in both groups. For it can be demonstrated that the larger the number of subjects randomly assigned to the two groups, the closer their means approach each other. Hence, although with a small number of participants it is unlikely that the means of the two groups will differ to any great extent, it is more likely than if the number of participants is large. The lesson should be clear: If you wish to reduce the difference in the means of the two groups, use a large number of participants.[2]

Even with a comparatively large number of participants it is still possible, though unlikely, that the means of the groups will differ considerably due to random fluctuations. Suppose, for example, that we have drawn a sample of 16 participants and assigned them to two groups. Now, if we measured their intelligence, it *is* possible that we would obtain a mean intelligence quotient of 100 for one group and a mean of 116 for the second group. However, by using appropriate statistical techniques we can determine that such an event should occur by chance less than about 5 times out of 100. If we ran the experiment 100 times, and assigned participants to two groups at random in each experiment, a difference between the groups of 16 IQ points (e.g., 116–100) or more should occur by chance in only about 5 of the experiments. Differences between the two groups of less than 16 IQ points should occur more frequently. And differences between the two groups of 24 points or more should occur less than one time in 100 experiments, on the average. Most frequently, then, there should be only a small difference between the two groups.

"But," the skeptical student continues, "suppose that in the particular experiment that I am conducting (I don't care about the other 95 or 99 experiments) I *do* by chance assign my participants to two groups of widely differing ability. I would think that the group with the mean IQ of 116 would have a higher mean score on the dependent variable than does the other group, *regardless of the effect of the independent variable*. I (the experimenter) would then conclude that the independent variable is effective, when, in fact, it isn't."

One cannot help but be impressed by such a convincing attack, but retreat at this point would be premature, for there are still several weapons that can be brought into the battle. First, if one has doubts as to the equivalence of the two groups, their scores on certain variables can be computed to see how their means actually compare. Thus, in the above example, we would measure the participants' IQs and visual acuity, compute the means for both groups and compare the scores to see if there is much difference. If there is little difference, we know that our random assignment has been at least fairly successful. This

---

[2] In making this point we are ignoring the distributions of the scores. Hence, the matter is not quite as simple as we have made it, but the main point is sound.

laborious and generally unnecessary precaution actually has been taken in a number of experiments.[3]

"But," the student continues tenaciously, "suppose I find that there is a sizable difference, and furthermore, suppose that I determine this only after all data have been collected. My experiment would be invalidated." Yet, there is hope. In this case we could use a statistical technique that allows us to equate the two groups with respect to intelligence. That is, we could "correct" for the difference between the two groups and determine whether they differ on the dependent variable for a reason that cannot be attributed to the difference of intelligence. Put another way, we could statistically equate the two groups on intelligence so that differences on this extraneous variable would not differentially afffect the dependent variable scores. This statistical technique is known as the analysis of covariance and is briefly discussed later.

"Excellent," the student persists, "but suppose the two groups differ in some respect for which we have no measure, and that this difference will sizably influence scores on the dependent variable. I now understand that we can probably 'correct' for the difference between the two groups on factors such as intelligence and visual acuity, because these are easily measurable variables. But what if the groups differ on some factor that we cannot measure or do not think to measure? In this case we would be totally unaware of the difference and draw illegitimate conclusions from our data."

"You," we say to the student, secretly admiring his demanding perseverance, "have now put us in such an unlikely position that we need not worry about its occurrence. Nevertheless, it *is* possible, just as it is possible that you will be hit by a car today while crossing the street. And, if there is some factor for which we cannot make a 'correction,' the experiment might well result in erroneous conclusions." The only point we can refer to here is one of the general features of the scientific enterprise: Science is self-correcting! Thus, if any given experiment leads to a false conclusion, and if the conclusion has any importance at all for psychology, an inconsistency between the results of the invalid experiment and additional data from a later experiment will become apparent. The existence of this problem will then lead to a solution, which, in this case, will be a matter of discarding the incorrect conclusion.[4]

---

[3] In an experiment on rifle marksmanship, for instance, it was determined that four groups did not differ significantly on the following extraneous variables: previous firing experience, left or right handedness, visual acuity, intelligence, or educational level (McGuigan & MacCaslin, 1955, b).

[4] It is interesting that, of all the disciplines, science is the only self-correcting one. Scientific propositions change over time to more closely approximate truth as

Now, the matter for us to discuss here concerns a return to a question that was briefly considered in chapter 1, where we posed the following problem: After the experimenter has collected data on the dependent variable, the wish is to determine whether one group is superior to the other. The hypothesis may predict that the experimental group will have a higher mean than the control group. The first step in testing the hypothesis would be to compute the mean scores on the dependent variable for the two groups. It might be found that the experimental group has a higher mean score than the control group— say that the experimental group has a mean score of 40, while the control group has one of 35. Assuming that the higher the score the better the performance, we can conclude that this 5-point difference is reliable? Or is it merely the result of random fluctuations, of experimental error? To answer this question, we said, we must apply a statistical test. Let us now consider one statistical test that is frequently used to answer this question.

The statistical test to which we refer is known as the "t-test" (note that a lower-case "t" is used to denote this test, not a capital "T," which has another denotation in statistics).[6] The first step in computing a t-test value is the computation of the means of the dependent variable scores of the two groups concerned. The equation for computing a mean (symbolized $\bar{X}$) is

(5-1)
$$\bar{X} = \frac{\Sigma X}{n}$$

The only unusual symbol in Equation (5-1) is Σ, the capital Greek letter sigma. Σ may be interpreted as "sum of." It is a summation sign and simply instructs you to add whatever is to the right of it.[7] In this case the letter X is to the right of sigma so we must now find out what

---

we search for knowledge. The products of other disciplines (art, music, fiction) change too, but self correction is not a relevant factor in them.

[5] Before beginning this section the conscienctious student might want to read the first section of chapter 16, "Concerning the Accuracy of the Data Analysis."

[6] Before our first discussion of the statistical analysis of an experimental design, it is well to point out that the statistical tests (such as the t-test) are conducted on the supposition that certain statistical assumptions are satisfied. Since the assumptions for the statistical tests discussed in this book are similar, it is more economical to discuss them together after all of our designs have been considered (p. 418). The instructor of student who so wishes, of course, may immediately integrate this topic with the discussion of the statistical tests.

[7] More precisely, Σ instructs you to add all the values of the symbols that are to its right, values that were obtained from your sample.

values X stands for and add them. Here, X merely indicates the score that we obtained for each subject. Suppose, for instance, that we give a test to a class of five students, with these resulting scores:

	X
Joan	100
Constance	100
Richard	80
Lillian	70
Joe	60

To compute ΣX we merely need to add the X scores. In this way we find that $\Sigma X = 100 + 100 + 80 + 70 + 60 = 410$. The n in Equation (5-1) stands for the number of subjects in the group. In this example, then, $n = 5$. Thus, to compute $\bar{X}$ we simply substitute 410 for ΣX, 5 for n in Equation (5-1), and then divide n into ΣX:

$$\bar{X} = \frac{410}{5} = 82.00$$

Thus the mean score of the group of five students who took the particular test is 82.00.

Let us now turn to an equation for computing t:

$$(5\text{-}2) \qquad t = \frac{\bar{X}_1 - \bar{X}_2}{\sqrt{\left(\dfrac{SS_1 + SS_2}{(n_1 - 1) + (n_2 - 1)}\right)\left(\dfrac{1}{n_1} + \dfrac{1}{n_2}\right)}}$$

Although this equation may look forbidding to the statistically naive, such an impression should be short-lived for t is actually rather simple to compute. To illustrate, consider the fascinating (and somewhat controversial) results of one of a series of experiments on RNA in the brain. RNA (Ribonucleic acid) has been implicated in the process of memory storage, though its precise role has yet to be established. More or less grabbing the bull by the horns, Babich, Jacobson, Bubash, and Jacobson (1965) trained a group of 8 rats to approach the food cup in a Skinner Box every time a click was sounded. The animals rarely or never approached the food cup when the click was absent. On the day after this training was completed, the animals were sacrificed, their brains were removed, and RNA was extracted from a selected portion. RNA was also extracted from the brains of 9 untrained rats. Approximately eight hours after extraction, the RNA from each of the rats, trained and untrained, was injected into live, untrained rats. Hence, 17 live rats were injected with RNA: 8 of them (experimental group) received RNA from trained rats and 9 (control group) received RNA from untrained rats. Both groups were then tested in a Skinner Box by presenting a click for 25 times, and the number of times that they approached the food cup was counted (various controls were used, but they need not concern us here). The hypothesis, amazing as it might sound, was to the effect that memory storage could be passed on by means of injections of RNA or associated substances. It was therefore predicted that the experimental group would approach the food cup more often during the test trials than the control group. The number of times that each rat approached the food cup during the 25 test trials is presented in Table 5-1 (one rat from each group was discarded because it "froze" in the test situation).

Table 5-1 Number of Food Cup Approaches per Animal During 25 Test Trials

| | GROUP 1 | | GROUP 2 |
Subject Number	Experimental Rats X_1	Subject Number	Control Rats X_2
1	1	8	0
2	3	9	0
3	7	10	0
4	8	11	1
5	9	12	1
6	10	13	1
7	10	14	2
	$\Sigma X_1 = \overline{48}$	15	3
			$\Sigma X_2 = \overline{8}$

We now seek to obtain an evidence report, i.e., a summary statement of the findings of the study. This evidence report, then, will tell us whether the hypothesis is probably true or false. The first step is to compute the means of the two groups. Note that subscripts have been used in Equation (5-2) to indicate which group the various values are for. In this case $\overline{X}_1$ stands for the mean of Group 1 (the experimental group), and $\overline{X}_2$ for the mean of Group 2 (the control group). In like manner SS_1 and SS_2 stand for what is called the *sum of squares* for Groups 1 and 2 respectively; and n_1 and n_2 are the respective number of subjects in the two groups. We can now determine that $\Sigma X_1 = 48$, while $\Sigma X_2 = 8$. Since the number of subjects in Group 1 is seven, we note that $n_1 = 7$. The mean for Group 1 (i.e., $\overline{X}_1$) may now be determined by substitution in Equation (5-1):[8]

[8] In your computations you would be wise to pay attention to the significant figures, an indication of the accuracy of your measurements and computations. To determine the accuracy of a measurement you count the number of digits, e.g., 21 is correct to two significant figures, 1.2 to two significant figures, .012 to two significant figures, and 1.456 to four significant figures. The final value of statistics, like a mean or standard deviation, should be rounded off to one more significant figure than for the raw data. Intermediate calculations for the t test can be safely performed by carrying three more digits than the data.

$$\bar{X}_1 = \frac{48}{7} = 6.86$$

And similarly for Group 2 (n_2 is 8):

$$\bar{X}_2 = \frac{8}{8} = 1.00$$

We now need to compute the sum of squares (a term that will be extensively used in later chapters) for each group. The equation for the sum of squares is:

(5-3)
$$SS = \Sigma X^2 - \frac{(\Sigma X)^2}{n}$$

Equation (5-3) contains two terms with which we are already familiar, viz., n and ΣX. The other term is ΣX^2, and it instructs us to add the squares of all the values for a given group. Thus, to compute ΣX^2 for Group 1 we should square the value for the first subject, add it to the square of the score for the second subject, add both of these values to the square of the score for the third subject, etc. Squaring the scores for the subjects in both groups of Table 5-1 and summing them we obtain:

Subject Number	GROUP 1 Experimental Rats		Subject Number	GROUP 2 Control Rats	
	X_1	X_1^2		X_1	X_2^2
1	1	1	8	0	0
2	3	9	9	0	0
3	7	49	10	0	0
4	8	64	11	1	1
5	9	81	12	1	1
6	10	100	13	1	1
7	10	100	14	2	4
		$\Sigma X_1^2 = \overline{404}$	15	3	9
					$\Sigma X_2^2 = \overline{16}$

One frequent error by students should be pointed out as a precaution. That is that ΣX^2 is not the square of ΣX. That is, $(\Sigma X)^2$ is not equal to ΣX^2. For instance, the $\Sigma X_1 = 48$. The square of this value is $(\Sigma X_1)^2 = 2{,}304$, whereas $\Sigma X^2_1 = 404$.

We are now in a position to substitute the appropriate values into Equation (5-3) and compute the sum of squares for each group. We know that, for Group 1, $\Sigma X_1 = 48$, that $\Sigma X^2_1 = 404$, and that $n_1 = 7$. Hence:

$$SS_1 = 404 - \frac{(48)^2}{7} = 404 - \frac{(48 \cdot 48)}{7}$$

$$= 404 - \frac{(2304)}{7} = 404.000 - 329.143 = 74.857$$

And similarly, the values to compute the sum of squares for Group 2 are:

$$\Sigma X_2 = 8$$

$$\Sigma X_2^2 = 16$$

$$n_2 = 8$$

Therefore:

$$SS_2 = 16 - \frac{(8)^2}{8} = 16 - \frac{64}{8} = 8.000$$

We now have all of the values required by Equation (5-2), and therefore can immediately compute the value of t for this experiment. To summarize them:

$$\bar{X}_1 = 6.86 \qquad\qquad \bar{X}_2 = 1.00$$

$$n_1 = 7 \qquad\qquad n_2 = 8$$

$$SS_1 = 74.857 \qquad\qquad SS_2 = 8.000$$

And substituting these values in Equation (5-2) we obtain:

$$t = \frac{6.86 - 1.00}{\sqrt{\left(\frac{74.857 + 8.000}{(7-1) + (8-1)}\right)\left(\frac{1}{7} + \frac{1}{8}\right)}}$$

We now need to go through the following steps in computing t:

1. Obtain the difference between the means: $6.86 - 1.00 = 5.86$
2. Add $SS_1 + SS_2$: $74.857 + 8.000 = 82.857$
3. Compute $n_1 - 1$: $7 - 1 = 6$
4. Compute $n_2 - 1$: $8 - 1 = 7$
5. Add $\dfrac{1}{n_1} + \dfrac{1}{n_2} = \dfrac{1}{7} + \dfrac{1}{8} = \dfrac{8}{56} + \dfrac{7}{56} = \dfrac{15}{56}$

The results of these computations are:

$$t = \frac{5.86}{\sqrt{\left(\frac{82.857}{6+7}\right)\left(\frac{15}{56}\right)}}$$

In the next stage divide the two denominators (13 and 56) into their respective numerators (82.857 and 15):

$$t = \frac{5.86}{\sqrt{(6.374)(.2679)}}$$

then multiply the values in the denominator:

$$t = \frac{5.86}{\sqrt{1.708}}$$

The next step is to find the square root of 1.708. This may be obtained from page 447 in the Appendix, and is found to be 1.307. Dividing as indicated we find t to be:

$$t = \frac{5.86}{1.307} = 4.48$$

Although the computation of t is straightforward, the beginning student is likely to make an error in its computation. The error is generally not one of failing to follow the procedure, but one of a computational nature (dividing incorrectly, failing to square terms properly, mistakes in addition). A great deal of care must be taken in statistical work; each step of the computation should be checked in an effort to eliminate errors. As an aid to the student in learning to compute t, a number of exercises are provided at the end of this chapter. Work all of these exercises and make sure that your answers are correct.

One point in the computation of t needs to be expanded. In the numerator we have indicated that $\bar{X}_2$ should be subtracted from $\bar{X}_1$. Actually we are conducting what is known as a "two-tailed test." If you are not familiar with the distinction between one- and two-tailed tests, you need not be concerned here; the immediate point for you to observe is that we are interested in the absolute difference between the means. Hence $\bar{X}_1 - \bar{X}_2$ is appropriate if $\bar{X}_1$ is greater than $\bar{X}_2$ (i.e., if $\bar{X}_1 > \bar{X}_2$). But if in your experiment you find that $\bar{X}_2$ is greater than $\bar{X}_1$ $\bar{X}_2 > \bar{X}_1$) then you merely subtract $\bar{X}_1$ from $\bar{X}_2$, i.e., Equation (5-2) would have as its numerator $\bar{X}_2 - \bar{X}_1$.

We might also note that the value under the square root sign is always positive. If it is negative in your computation, go through your work to find the error.

The reason we want to obtain a value of t, we said, is to decide whether the difference between the means of two groups is the result of random fluctuations or whether it is a reliable difference. But several additional matters must also be discussed in relation to this difference. The first is a consideration of what is known as the "null hypothesis," a concept that it is vital to understand.[9] The null hypothesis

---

[9] The term "null hypothesis" was first used by Professor Sir Ronald A. Fisher (personal communication). He chose the term "null hypothesis" without "particular regard for its etymological justification but by analogy with a usage, formerly and perhaps still current among physicists, of speaking of a null ex-

that is generally used in psychological experimentation states that there is no difference between the population means on the dependent variable of the two groups. Note that we wish to contrast the two groups by comparing their population *means* on the dependent variable. Students often mistake the "null hypothesis" by saying that there is no difference between two groups—most assuredly there are *many* differences between any two groups, but we are only interested in a possible one of a mean difference on our dependent variable.

The null hypothesis, let us emphasize, states that there is no difference between the *population* means.[10] The reason we conduct an experiment is to make statements about populations—to determine whether the population means of our two groups differ. In a sense this may be stated otherwise as follows: we want to know whether the *true* means of our groups differ (where the true mean is taken as the population mean). Now, of course, we cannot study the population in its entirety. Rather, the way to determine whether or not the true (population) means differ is by comparing the means obtained for our two sample groups. We do this by subtracting one sample mean from the other, as specified in the numerator of Equation (5-2). This difference will almost certainly not be zero; but it will be some positive amount, the value of which may be quite small or quite large. If the difference between our sample means is quite small, we would be inclined to conclude that the difference is due to chance.

On the other hand, if the difference is large, then we might say that the difference is too large to be due to random fluctuations alone. The null hypothesis asserts that the difference between the population means is zero. In effect it says that any difference between the means

---

periment, or a null method of measurement, to refer to a case in which a proposed value is inserted experimentally in the apparatus and the value is corrected, adjusted, and finally verified, when the correct value has been found; because the set-up is such, as in the Wheatstone Bridge, that a very sensitive galvanometer shows no deflection when exactly the right value has been inserted.

"The governing consideration physically is that an instrument made for direct measurement is usually much less sensitive than one which can be made to kick one way or the other according to whether too large or too small a value has been inserted.

"Without reference to the history of this usage in physics. . . . One may put it by saying that if the hypothesis is exactly true no amount of experimentation will easily give a significant discrepancy, or, that the discrepancy is null apart from the errors of random sampling."

[10] A symbolic statement of the null hypothesis would be $\mu_1 - \mu_2 = 0$ (μ is the Greek letter mu). Here μ_1 is the population mean for Group 1 and μ_2 is the population mean for Group 2. If the difference between the sample means ($\overline{X}_1 - \overline{X}_2$) is small, then we are likely to infer that there is no difference between the population means; thus, that $\mu_1 - \mu_2 = 0$. On the other hand if $\overline{X}_1 - \overline{X}_2$ is large, then the null hypothesis that $\mu_1 - \mu_2 = 0$ is probably not true.

of the groups in your sample is merely due to random fluctuations and thus can be accounted for in terms of experimental error. If we find that the difference between the means of our groups is small, then it is likely that the difference is the result of random fluctuations and that the null hypothesis is reasonable. But if our groups differ considerably, then the difference is probably too large to be due to random fluctuations alone, and the null hypothesis is not tenable in that particular case.

In short, the null hypothesis is a statistical hypothesis and is set up for the purpose of attempting to disprove it. Our null hypothesis asserts that there is no difference between the population means of our two groups; we seek to determine that it is false, that there is a difference between the means. Hence, if it is disproven in a properly conducted experiment, we can conclude that there is a difference between our two groups and furthermore that this difference is due to variation of the independent variable. If we cannot disprove the null hypothesis, then we cannot assert that there is a difference between the two groups; variation of our independent variable is not effective.

The question now is how small the difference must be between $\bar{X}_1$ and $\bar{X}_2$ before we can say that it is due to random fluctuations of the means. Then, too, how large must the difference be before we can say that it is *not* due to random fluctuations alone? The latter question can be answered by the value of t; if t is sufficiently large, we can say that the difference between the two groups is too large to be attributed solely to random fluctuations—too large to be accounted for by experimental error. And to determine how large "sufficiently large" is we may consult the table of t. But before doing this, there is one additional value that we must compute—the degrees of freedom (df).

The degrees of freedom available for the t test are a function of the number of subjects in the experiment. More specifically, $df = N - 2.$[11] And N is the number of subjects in one group (n_1) plus the number of subjects in the other group (n_2). Hence, in our example we have:

$$N = n_1 + n_2 \quad \text{i.e.,} \quad N = 7 + 8 = 15$$

therefore:
$$df = 15 - 2 = 13$$

To determine whether our t is significant, let us now turn to a table of t (Table 5-2 on p. 131) armed with two values: $t = 4.48$ and df

[11] This equation for computing df is only for the application of the t-test to two randomized groups. We shall use other equations for df when considering additional statistical tests.

Table 5-2* Table of t

df	P	0.9	0.8	0.7	0.6	0.5	0.4	0.3	0.2	0.1	0.05	0.02	0.01
1		0.158	0.325	0.510	0.727	1.000	1.376	1.963	3.078	6.314	12.706	31.821	63.657
2		0.142	0.289	0.445	0.617	0.816	1.061	1.386	1.886	2.920	4.303	6.965	9.925
3		0.137	0.277	0.424	0.584	0.765	0.978	1.250	1.638	2.353	3.182	4.541	5.841
4		0.134	0.271	0.414	0.589	0.741	0.941	1.190	1.533	2.132	2.776	3.747	4.604
5		0.132	0.267	0.408	0.559	0.727	0.920	1.156	1.476	2.015	2.571	3.365	4.032
6		0.131	0.265	0.404	0.553	0.718	0.906	1.134	1.440	1.943	2.447	3.143	3.707
7		0.130	0.263	0.402	0.549	0.711	0.896	1.119	1.415	1.895	2.365	2.998	3.499
8		0.130	0.262	0.399	0.546	0.706	0.889	1.108	1.397	1.860	2.306	2.896	3.355
9		0.129	0.261	0.398	0.543	0.703	0.883	1.100	1.383	1.833	2.262	2.821	3.250
10		0.129	0.260	0.397	0.542	0.700	0.879	1.093	1.372	1.812	2.228	2.764	3.169
11		0.129	0.260	0.396	0.540	0.697	0.876	1.088	1.363	1.796	2.201	2.718	3.106
12		0.128	0.259	0.395	0.539	0.695	0.873	1.083	1.356	1.782	2.179	2.681	3.055
13		0.128	0.259	0.394	0.538	0.694	0.870	1.079	1.350	1.771	2.160	2.650	3.012
14		0.128	0.258	0.393	0.537	0.692	0.868	1.076	1.345	1.761	2.145	2.624	2.977
15		0.128	0.258	0.393	0.536	0.691	0.866	1.074	1.341	1.753	2.131	2.602	2.947
16		0.128	0.258	0.392	0.535	0.690	0.865	1.071	1.337	1.746	2.120	2.583	2.921
17		0.128	0.257	0.392	0.534	0.689	0.863	1.069	1.333	1.740	2.110	2.567	2.898
18		0.127	0.257	0.392	0.534	0.688	0.862	1.067	1.330	1.734	2.101	2.552	2.878
19		0.127	0.257	0.391	0.533	0.688	0.861	1.066	1.328	1.729	2.093	2.539	2.861
20		0.127	0.257	0.391	0.533	0.687	0.860	1.064	1.325	1.725	2.086	2.528	2.845
21		0.127	0.257	0.391	0.532	0.686	0.859	1.063	1.323	1.721	2.080	2.518	2.831
22		0.127	0.256	0.390	0.532	0.686	0.858	1.061	1.321	1.717	2.074	2.508	2.819
23		0.127	0.256	0.390	0.532	0.685	0.858	1.060	1.319	1.714	2.069	2.500	2.807
24		0.127	0.256	0.390	0.531	0.685	0.857	1.059	1.318	1.711	2.064	2.492	2.797
25		0.127	0.256	0.390	0.531	0.684	0.856	1.058	1.316	1.708	2.060	2.485	2.787
26		0.127	0.256	0.390	0.531	0.684	0.856	1.058	1.315	1.706	2.056	2.479	2.779
27		0.127	0.256	0.389	0.531	0.684	0.855	1.057	1.314	1.703	2.052	2.473	2.771
28		0.127	0.256	0.389	0.530	0.683	0.855	1.056	1.313	1.701	2.048	2.467	2.763
29		0.127	0.256	0.389	0.530	0.683	0.854	1.055	1.311	1.699	2.045	2.462	2.756
30		0.127	0.256	0.389	0.530	0.683	0.854	1.055	1.310	1.697	2.042	2.457	2.750
∞		0.12566	0.25335	0.38532	0.52440	0.67449	0.84162	1.03643	1.28155	1.64485	1.95996	2.32634	2.57582

*Table 5-2 is reprinted from Table IV of Fisher: *Statistical Methods for Research Workers,* published by Oliver and Boyd Ltd., Edinburgh, by permission of the author and publishers. It is also reproduced in the Appendix.

= 13. The table of t is organized around two values: a column labeled "df" and a row labeled "P" (for probability). The df column is on the extreme left, and the P row runs across the top of the table. Values of t are the numbers that complete the table. Our general purpose here is to find out what P value is associated with a specific value of t and df. To do this we must first run down the df column until we arrive at our specific value of df; in this case, 13 df. We then read across the row that is marked 13 df. This row contains a number of possible values of t: 0.128, 0.259, 0.394, etc. We must read across this row until we come to a value of t that is close to our particular value — in this case, 4.48. But the largest value of t in this row is 3.012; this value, then, is the closest match we can make to 4.48. So, we read up the column that contains 3.012 to determine what value of P is associated with it — in this case, 0.01.

Let us make a general observation; the larger the t, the smaller the P. For example, with 13 df a t of 0.128 has a P of 0.9 associated with it, while with the same df a t of 1.771 has a P of 0.1. From this observation and our study of the tabled values of t and P we can conclude that if a t of 3.012 has a P of 0.01, any t larger than 3.012 must have a smaller P than .01. It is sufficient for our purposes simply to note this fact without attempting to make it any more precise.

When we report a computed t we write an equation that indicates the numbers of df (here 13) within parentheses, e.g. t (13) = 4.48. The next step is to interpret the fact that a t of 4.48 has a P of less than 0.01 ($P < 0.01$) associated with it. This finding indicates that a difference between means of the two groups of the size that were obtained has a probability of less than 0.01; i.e., that a difference between the means of this size may be expected less than one time in a hundred by chance (.01 = 1/100). Put another way, if the experiment had been conducted 100 times, by chance we would expect a difference of this size to occur once, provided the null hypothesis is true. This, we must all agree, is a most unlikely occurrence. It is so unreasonable, in fact, to think that such a large difference could have occurred by chance on the very first of the hypothetical one hundred experiments that we prefer to reject "chance" as the only explanation. We therefore choose to reject our null hypothesis. That is, we refuse to regard it as reasonable that the real difference between the means of the two groups is zero when we have obtained such a large difference in sample means, as indicated by the respective values, in this case, of 6.86 and 1.00. But if a difference of this size cannot be attributed to chance alone, what reason can we give for it? We assume that all the proper safeguards of experimentation have been observed in obtaining these results, and that the groups therefore differed only in the respect that each was administered a different experimental treatment. It seems reasonable to assert, then, that the reason the two groups differed is that they received different values of the independent variable. This leads to the further conclusion that the independent variable is effective in influencing scores on the dependent variable; and this is precisely the purpose of the experiment.

There are still a number of questions about this procedure that need to be answered. For instance, we said that before we conduct an experiment it is unlikely that we would (by chance) obtain a P of .01 for our t. How small may P be and still be considered sufficiently likely to occur by chance alone? That is, how small must P be before we can reject the null hypothesis? For example, with 13 df, if we had obtained a value of 1.80 for t, we find that the corresponding value of P is less than .10. This would imply that such a difference between the two group means could be expected by chance less than 10 times out of 100. Now, is this sufficiently unlikely that we can reject the null hypothesis? Again, consider a t of 2.20. The corresponding P is less than 0.05. Can we reject the null hypothesis on the basis of this size of P? What if we had obtained a t of 0.90, with a corresponding P of less than 0.40; a difference of this size may be expected 40 times out of 100 by chance. Is this P sufficiently small to allow us to reject the null hypothesis? In short, the question is this: what value of P is small enough to allow us to reject the null hypothesis? Unfortunately, there is no simple answer to this question, for it depends on a number of things. The best we can say here is that the experimenter may set any value of P as the cut-off point. Thus one may say that: "If the value of t that I obtain has a P of less than 0.50, I will reject my null hypothesis." Similarly, we may set P at 0.01, 0.05, 0.30, or even 0.90 if we wish. There is only one requirement that must be satisfied in setting the value of P: we must set it *before* we conduct our experiment. The reason for this is that, for a proper test of the null hypothesis, the experimenter should not be influenced by the particular nature of the data. For example, it would be inappropriate to run a t test and determine P to be 0.06, and then decide that if P is 0.06 to reject the null hypothesis. Such an experimenter would be inclined never to fail to reject the null hypothesis, for the criterion (the value of P) for rejecting it would be determined by the value of P actually obtained. An extreme case of this would be a person who obtains a P of 0.90, and then sets 0.90 as his criterion. The sterility of such a decision is apparent, for a difference between two groups of the size that were obtained would be expected by chance 90 times out of 100. Obviously, it would be unreasonable to reject the null hypothesis with such a large P, and such an experimenter would almost surely be committing an error, i.e., the null hypothesis would be rejected when in fact it should not be rejected.

The P that an experimenter sets then is totally arbitrary. We can vary it with the particular experiment that is being conducted. For some problems it is important to have an extremely small P; for others a larger one is appropriate. Although the actual decision is arbitrary, there are a number of important considerations that will help the experimenter in arriving at a decision. The interested student will find such matters discussed in elementary statistics courses. Suffice it to say here that for general psychological experimentation a standard value of P is accepted: 0.05. Thus, unless the experimenter specifies otherwise, it is generally understood that a P of .05 is set before conducting an experiment.

Let us now apply the above considerations to our example. The hypothesis was to the effect that the sample of experimental animals should approach the food cup significantly more frequently during the test than the controls. It was found that the mean scores for the two groups were 6.86 and 1.00 respectively. Furthermore, we found that the t-test yielded a value of 4.48, which, with 13 df, had a P of less than 0.01. Assuming the conventional value of 0.05 for P as the criterion of whether or not to reject the null hypothesis, the value of less than 0.01 causes us to reject the null hypothesis. That is, we assert that there is a true difference between our two groups. Furthermore, we observe that the direction of the difference is that specified by the (empirical) hypothesis, i.e., the hypothesis predicted that the values for the experimental rats would be higher than for the control animals. Since the scores are of the nature predicted by the hypothesis (and reliably so), we may conclude that the hypothesis is confirmed.[12,13]

The following general rules may now be stated: *If the empirical hypothesis predicts that there will be a difference between two groups, and if the*

[12] One of the assumptions of parametric statistics such as the t-test, as you will see in chapter 16, is that the variances of the groups are homogeneous ("equal"). In this experiment they are not, viz., the variance for the experimental group is 12.47 and that for the control group is 1.14. One alternative to a parametric analysis is to conduct a nonparametric test, as was done in the original report of the experiment. The point that we make in chapter 16, however, is well illustrated by this example, i.e., that parametric tests are remarkably robust in that major deviations from their basic assumptions can be tolerated. Here the same conclusions follow from the t-test and from the Mann-Whitney U test, a nonparametric test.

[13] To emphasize the controversial nature of the findings reported here (and also in another experiment in chapter 9), we need only observe the results reported by other experimenters. Batkin, Woodward, Cole, and Hall (1966) report a similar effect when using carp, but Byrne (1966) cited 18 experiments on this problem that yielded negative results.

null hypothesis is rejected, and if the difference between the two groups is in the direction specified by the empirical hypothesis, then it may be concluded that the empirical hypothesis is confirmed. Thus, there are two cases in which the empirical hypothesis would not be confirmed: first, if the null hypothesis were not rejected; and second, if it were rejected, but the difference between the two groups were in the opposite direction specified by the empirical hypothesis. To illustrate these latter possibilities, let us assume that we actually obtained a t of 1.40 (which you can see has a P value greater than .05). We fail to reject the null hypothesis, and accordingly fail to confirm the empirical hypothesis. On the other hand, assume that we obtain a t of 2.40 ($P < 0.05$), but that the mean score for the controls is higher than that for the experimental rats. In this case we reject the null hypothesis, but fail to confirm the empirical hypothesis.

For further practice, and for an experiment in which the values are more typical, let us consider the following study. In this experiment the role of "contact comfort" in the development of "mother love" in dogs was studied (Igel & Calvin, 1960). The suggestion that the sense of touch is important in this regard is an old one. In 1868 Bain, for instance, said that ". . . touch is the fundamental and generic sense. . . . The soft, warm touch, if not a first-class influence, is at least an approach to that. The combined power of soft contact and warmth amounts to considerable pitch of massive pleasure. . . . In a word, our love pleasures begin and end in sensual contact. . . . It seems to me that there must be at the (parental instinct's) foundation that intense pleasure in the embrace of the young which we find characterize the parental feeling throughout" (James, 1952, p. 809).

In a series of studies Harlow and his associates have found that ". . . contact comfort is a variable of overwhelming importance in the development of affectional responses, whereas lactation is a variable of negligible importance" (Harlow & Zimmerman, 1958, p. 503). The purpose of the Igel and Calvin study was to determine whether Harlow's findings were species-specific, or whether they could be generalized to species other than the monkey. The hypothesis, thus, was to the effect that "puppies will spend more time with cloth than with wire mothers, even though they are fed by both types." Among the experimental conditions was one in which one group of puppies was raised and fed on a wire surrogate mother (Figure 5-1), while the other group was similarly raised with a terry cloth mother (Figure 5-2). The latter obviously provides more contact comfort; but would a puppy learn to love even a wire mother if, as others have held, she feeds him? The index of affection was the amount of time that each puppy spent with his mother surrogate. Hence, the more a puppy

Figure 5-1. Participant nursing from a wire surrogate mother.

"loves" his mother, the longer the amount of time he would spend with it. Amount of contact time was automatically recorded for each day from the time the pups had their eyes open (11 days old) to 31 days old. In the condition that we are considering, two puppies were randomly assigned to a wire mother and two to a cloth mother. The dependent variable scores are presented in Table 5-3.

Table 5-3 Mean Number of Minutes per Day Spent with Surrogate Mothers

Subject Number	GROUP 1 Wire Mother	Subject Number	GROUP 2 Cloth Mother
1	61.2	3	836.5
2	82.7	4	722.3
	$\Sigma X_1 = \overline{143.9}$		$\Sigma X_2 = \overline{1558.8}$
	$\bar{X}_1 = 71.95$		$\bar{X}_2 = 779.40$

Computing the necessary values for the sums of squares Equation (5-3) we find:

Group 1	Group 2
$\Sigma X_1 = 143.9$	$\Sigma X_2 = 1558.8$
$\Sigma X_1^2 = 10,584.73$	$\Sigma X_2^2 = 1,221,449.54$
$n_1 = 2$	$n_2 = 2$

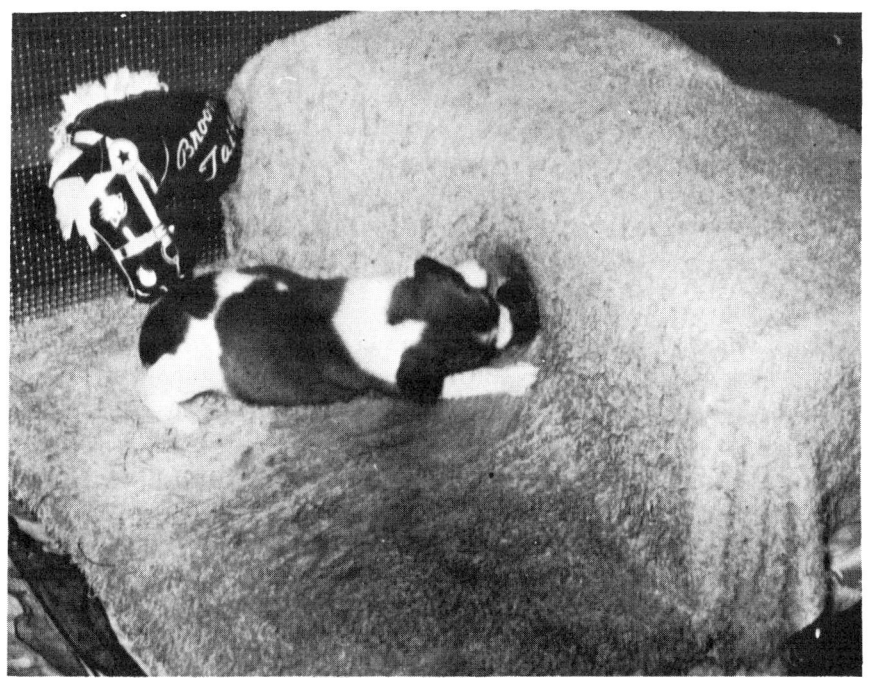

Figure 5-2. Another participant nursing from a cloth surrogate mother.

The sums of squares for Group 1 is:

$$SS_1 = 10{,}584.73 - \frac{(143.9)^2}{2} = 10{,}584.73 - \frac{(20707.21)}{2}$$

$$= 10{,}584.73 - 10{,}353.61 = 231.12$$

And for Group 2:

$$SS_2 = 1{,}221{,}449.54 - \frac{(1558.8)^2}{2}$$

$$= 1{,}221{,}449.54 - 1{,}214{,}928.72 = 6520.82$$

We now have all the necessary values for Equation (5-2) and substitute them as follows:

$$t = \frac{779.40 - 71.95}{\sqrt{\left(\dfrac{231.12 + 6520.82}{(2-1)+(2-1)}\right)\left(\dfrac{1}{2} + \dfrac{1}{2}\right)}}$$

Performing the indicated operations:

$$t = \frac{707.45}{\sqrt{\left(\frac{6751.94}{2}\right)\left(1\right)}} = \frac{707.45}{\sqrt{3375.97}}$$

$$= \frac{707.45}{58.103} = 12.18$$

We thus find that $t = 12.18$. With two degrees of freedom ($N - 2 = 4 - 2 = 2$) we refer to our table of t and find that the P associated with it is less than 0.01. Since this P is less than our conventional criterion of 0.05, we reject the null hypothesis and conclude that there is a difference between the groups. We report our t as $t(2) = 12.18$. On the assumption that necessary controls were satisfactorily effected, we may conclude that the independent variable was effective. That is, that the reason the puppies spent more time with the cloth mother was because of the difference in texture. These findings thus supported Harlow's contention that the strength of the affectional response is a function of the texture of the mother's "skin."

Summary

Let us now summarize each major step that we have gone through in testing an empirical hypothesis. For this purpose you might design a study to compare the amount of anxiety of majors in different college departments.

1. State the hypothesis, e.g., "If the anxiety scores of English and psychology students are measured, the psychology students will have the higher scores."
2. The experiment is designed according to the procedures outlined in chapter 4, e.g., "anxiety" is operationally defined (such as scores on the Manifest Anxiety Scale, Taylor, 1953), samples from each population are drawn, etc.
3. The null hypothesis is stated—"there is no difference between the population means of the two groups."
4. A probability value for determining whether or not to reject the null hypothesis is established, e.g., if $P < .05$, then the null hypothesis will be rejected; if $P > .05$, the null hypothesis will not be rejected.
5. The data are collected and statistically analyzed. For this design a t-test is conducted whereby the means of the two groups are determined. The value of t is computed and the corresponding P ascertained.
6. If the means are in the direction specified by the hypothesis (if the psychology students have a higher mean score than the English students) and if the null hypothesis is rejected, it may be concluded

that the hypothesis is confirmed. If the null hypothesis is not rejected, it may be concluded that the hypothesis is not confirmed. Or, if the null hypothesis is rejected, but the means are in the direction opposite to that predicted by the hypothesis, then the hypothesis is not confirmed.

"Borderline" Reliability

One frequently occurring problem in experimentation is that of borderline reliability. An experimenter who sets a P of 0.05 as the criterion for rejecting the null hypothesis fails to reject the null hypothesis if a P of 0.30 is obtained. But suppose that a P is 0.06. One might argue that, "Well, this isn't quite .05 but it is so close that I'm going to reject the null hypothesis anyway. This seems reasonable; after all, this means that a difference between groups of the size that I obtained can be expected only 6 times out of 100 by chance when the null hypothesis is true. Surely this is not much different than a probability of 5 times out of 100." To this there is only one answer: the t-test is decisive — a P of 0.06 is not a P of 0.05. In this case, therefore, there is no alternative but to fail to reject the null hypothesis. If the experimenter has set a criterion of a P of 0.06 before the experiment is conducted, then we would have no quarrel — the experimenter could, in this event, reject the null hypothesis. But since a criterion of a P of 0.05 was established, one cannot modify the criterion after the data are collected. A gambling analogy might be pursued: If one bets at the horse races, the bet must be placed prior to the start of the race, and the selection of a horse that "almost won" will evoke little sympathy from the cashier's window — seldom would you be able to make a bet after the conclusion of the race or would an argument that your horse lost only by a nose ("borderline reliability") be financially rewarded.

At the same time, however, we must agree with this experimenter that a P of 0.06 is an unlikely event by chance. Our advice is: "Yes. It looks like you *might* have something. It's a good hint for further experimentation. Conduct a new experiment and see what happens. If, in this replication, you come out with a reliable difference, you are quite safe in rejecting the null hypothesis. But if the value of t obtained is quite far from a computed value of 0.05 in this new, independent test, then you have saved yourself from making an error."

Systematic Observation versus Experimentation

We have previously contrasted two types of investigations: "experiments" and "systematic observation studies." The alert reader probably noticed that the two examples used to illustrate the computation of t were experiments. An example of a systematic observation study would concern anxiety of students of psychology and English. In this case *no variable is produced and purposively manipulated by the investigator*. Rather the study concerns observations of a phenomenon that was *already present in the population*. To meet the requirements for an experiment in this example, we would have to assign subjects randomly to two groups and then decree that everyone in one group would major in psychology and all those in the other, in English. If we were able to do this, we could say that our independent variable was "major of the student" and that we had varied it in two ways: English and psychology majors. Since we did not vary it in this manner, it was not purposively manipulated, and hence the study cannot be said to be an experiment. In the previous two cases, however, the requirements of experimentation *were* fulfilled, since subjects were randomly assigned to two groups and then it was determined which groups would receive which experimental treatment. Hence, the independent variable in the first experiment (p. 124) was the type of RNA, varied in two ways: (1) RNA from a trained group; and (2) RNA from an untrained group. Since the independent variable was under the control of the experimenter and since different conditions were *induced* in the two groups it may be said that the experimenter *purposively manipulated* the independent variable and thus conducted an experiment. The importance of the difference between the two types of investigations will become clear when we consider "control" (see chapter 6). It will be shown that the differences in two types of investigations lead to greater confidence in the conclusions derived from experimentation. The main point to observe here is that the t-test may be an appropriate method of statistical analysis for both types of investigations.

A Statistical Error to Avoid

You will note that in the previous examples one and only one dependent variable value was employed for each participant. The degrees of freedom for the t-test is then determined by the number of participants ($df = N - 2$) with one representative value for each participant. In some research, a number of dependent variable values may be collected for each individual who is studied, perhaps as in a learning experiment. Consider, for instance, an experiment (Table 5-4) in which there are three participants under each of two conditions (A and B) with five dependent variable scores for each participant.

Table 5-4 Illustration of the Use of Several Dependent Variable Values for Each of the Participants

Participant number	Condition A					Participant number	Condition B				
	Trial						Trial				
	1	2	3	4	5		1	2	3	4	5
1	3	2	3	5	6	1	9	4	6	8	8
2	4	1	1	4	4	2	8	9	3	7	9
3	6	9	4	9	9	3	7	8	2	8	6

The error that is being illustrated is that of separately entering *all* the data of Table 5-4 directly in the computation of the value of t. For instance, for participant 1, a student might sum and sum the squares of 3, 2, 3, 5, and 6, and similarly employ five separate dependent variable values for the other participants. Then all 30 dependent variable values might erroneously be employed to compute $N = 30$, so that $df = 30 - 2 = 28$. This is a grossly inflated value for the degrees of freedom—recall that the larger the number of degrees of freedom in Table 5-2, the smaller the value of t required for the rejection of the null hypothesis.

As we have noted, we prevent this error by *employing one and only one dependent variable value for each participant*. One could argue that in a learning experiment the last dependent variable value should be employed, so that t could be computed in this study for the values on Trial 5, viz. 6, 4, and 9, versus 8, 9, and 6. Another common method of avoiding the error of inflated degrees of freedom is to compute a representative value for each participant, e.g., to compute a mean for each row of dependent variable values, as in Table 5-5.

Table 5-5 Employing the Mean of the Trial Values for Each Participant of Table 5-4

Participant number	Condition Trial A					$\overline{X}$	Participant number	Condition Trial B					$\overline{X}$
	1	2	3	4	5			1	2	3	4	5	
1	3	2	3	5	6	3.8	1	1	4	6	8	8	7.0
2	4	1	1	4	4	2.8	2	8	9	3	7	9	7.2
3	6	9	4	9	9	7.4	3	7	8	2	8	6	6.2

In this instance, then, the t between the two groups would be based on the mean values for condition A (3.8, 2.8, 7.4) versus those for condition B (7.0, 7.2, 6.2). Did condition A differ reliably from condition B?

Summary of the Computation of t for a
Two-Randomized-Groups Design

Assume that we have obtained the following dependent variable scores for the two groups of participants:

Group 1	Group 2
10	8
11	9
11	12
12	12
15	12
16	13
16	14
17	15
	16
	17

1. We start with Equation (5-2), the equation for computing t:

$$t = \frac{\bar{X}_1 - \bar{X}_2}{\sqrt{\left(\dfrac{SS_1 + SS_2}{(n_1 - 1) + (n_2 - 1)}\right)\left(\dfrac{1}{n_1} + \dfrac{1}{n_2}\right)}}$$

2. Compute the sum of X (i.e., ΣX), the sum of X^2 (i.e., ΣX^2), and n for each group.

Group 1	Group 2
$\Sigma X = 108$	$\Sigma X = 128$
$\Sigma X^2 = 1512$	$\Sigma X^2 = 1712$
$n = 8$	$n = 10$

3. Using Equation (5-1), compute the means for each group.

$$\bar{X}_1 = \frac{108}{8} = 13.50 \qquad \bar{X}_2 = \frac{128}{10} = 12.80$$

4. Using Equation (5-3), compute the sums of squares for each group.

$$SS_1 = \Sigma X_1^2 - \frac{(\Sigma X_1)^2}{n_1} = 1512 - \frac{(108)^2}{8} = 54.000$$

$$SS_2 = 1712 - \frac{(128)^2}{10} = 73.600$$

5. Substitute the above values in Equation (5-2).

$$t = \frac{13.50 - 12.80}{\sqrt{\left(\frac{54.000 + 73.600}{(8 - 1) + (10 - 1)}\right)\left(\frac{1}{8} + \frac{1}{10}\right)}}$$

6. Perform the operations as indicated and determine that the value of t is:

$$t = \frac{0.70}{\sqrt{(7.975)(.2250)}} = \frac{0.70}{\sqrt{1.7944}} = \frac{0.70}{1.3395} = .523$$

7. Determine the number of degrees of freedom associated with the above value of t.

$$df = N - 2 = 18 - 2 = 16$$

8. Enter the table of t, and determine the probability associated with this value of t. In this example $0.70 > P > 0.60$. Therefore, assuming a required significance level of 0.05, the null hypothesis is not rejected.

Problems[14]

1. An experimenter runs a well-designed experiment wherein $n_1 = 16$ and $n_2 = 12$. A t of 2.14 is obtained. With a criterion of $P = 0.05$, can the null hypothesis be rejected?

2. An experimenter obtains a computed t of 2.20 with 30 df. The means of the two groups are in the direction indicated by the empirical hypothesis. Assuming that the experiment was well designed and that the experimenter has set a P of 0.05, did the independent variable influence the dependent variable?

3. It is advertised that a certain tranquilizer has a curative effect on psychotics. A clinical psychologist seeks to determine whether or not this is the case. A well-designed experiment is conducted with the following results on a measure of psychotic tendencies. Assuming that the criterion for rejecting the null hypothesis is $P = 0.01$, and assuming that the lower the score the greater the psychotic tendency, determine whether the tranquilizer has the advertised effect.

Values for the group that received the tranquilizer	*Values for the group that did not receive the tranquilizer*
2, 3, 5, 7, 7, 8, 8, 8	1, 1, 1, 2, 2, 3, 3

[14] Answers are on pages 456–57.

4. A psychologist hypothesizes that people who are of similar body build work better together. Accordingly, two groups are formed: Group 1 is composed of individuals who are of similar body build, and Group 2 consists of individuals with a different body build. Both groups perform a task that requires a high degree of cooperation. The performance of each participant is measured where the higher the score, the better the performance on the task. The criterion for rejecting the null hypothesis is $P = 0.02$. Was the empirical hypothesis confirmed or disconfirmed?

Group 1	Group 2
10, 12, 13, 13, 15, 15, 15, 17, 18	8, 9, 9, 11, 15, 16, 16, 16, 19, 20, 21
22, 24, 25, 25, 25, 27, 28, 30, 30	25, 25, 26, 28, 29, 30, 30, 32, 33, 33

5. On the basis of personal experience, a marriage counselor suspects that when one spouse is from the north and the other is from the south the marriage has a likelihood of being unsuccessful. Two groups of participants are selected: Group 1 is composed of marriage partners both of whom are from the same section of the country (either north or south), and Group 2 consists of marriage partners from the north and the south respectively. A criterion for rejecting the null hypothesis is not set, so that a $P = 0.05$ is assumed. Ratings of the success of the marriage (the higher the rating, the better the marriage) are obtained. Assume that adequate controls have been effected. Is the suspicion confirmed?

Group 1	Group 2
1, 1, 1, 2, 2, 3, 3, 4, 4, 5, 6, 6, 7, 7	1, 1, 2, 3, 4, 4, 5, 5, 6, 7

6
Experimental Control

The Nature of Experimental Control

The strength of civilization is based, at rock bottom, on the amount and kinds of reliable knowledge that have been accumulated. But the progress of civilization has been slow and painfully achieved; we have, in a sense, taken two steps backward for each three steps forward. Histories of western civilization typically emphasize the backward steps, as for instance in their accounts of the great wars. At least as fascinating, though, is the record of our achievements—the stories of the acquisition of knowledge and of the development of sound methods for acquiring that knowledge. Among the most striking advances in methodology was the recognition of the necessity for control conditions—so called "normal" conditions against which to evaluate experimental treatments. Unfortunately, cultural lag being what it is, the importance of control conditions has not yet pervaded the "logic" of everyday thinking; common-sense reasoning is most often wrong because it is based on observations of only one group (and frequently with an n of one, at that).

In order to reach the relatively advanced stage in which a proper appreciation of control conditions was recognized, we can suppose that methodologists had to first engage in considerable trial and error. There were, no doubt, a number of improperly controlled investigations conducted before methodologists became more sophisticated. And one must admire even these "semi-experiments," for they were imaginative indeed. An example is brought to our attention by Jones which ". . . is Herodotus' quaint account of the experiment in linguistics made by Psammetichos, King of Egypt (*Historiae* II, 2). To determine which language was the oldest, Psammetichos arranged to have two infants brought up without hearing human speech and to have their first utterances recorded. When a clear record of the children's speech had been obtained, ambassadors were sent around the world to find out where this language was spoken (specifically, where the word for "bread" was *bekos*). As a result of his experiment, Psammetichos pronounced Phrygian to be the oldest language, though he had assumed it was Egyptian before making the test" (1964, p. 419).

An account of a more sophisticated, but still ancient, investigation *did* include a control condition: "Athenaeus, in his *Feasting Philosophers* (*Deipnosophistae*, III, 84–85), describes how it was discovered that citron was an antidote for poison. It seems that a magistrate in Egypt had sentenced a group of convicted criminals to be executed by exposing them to poisonous snakes in the theater. It was reported back to him that, though the sentence had been duly carried out and all the criminals were bitten, none of them had died. The magistrate at once commenced an inquiry. He learned that when the criminals were being conducted into the theater, a market woman out of pity had given them some citron to eat. The next day, on the hypothesis that it was the citron that had saved them, the magistrate had the group divided into pairs and ordered citron fed to one of a pair but not to the other. When the two were exposed to the snakes a second time, the one who had eaten the citron suffered no harm, the other died instantly. The experiment was repeated many times and in this way (says Athenaeus) the efficacy of citron as an antidote for poison was firmly established" (Jones, 1964, p. 419). In such ways the logic of experimental control developed, slowly leading to our present level of methodological sophistication.

We have emphasized, by our repeated references to the topic of control, that it is one of the most important phases in the planning and conduct of experiments. The problem of controlling variables, therefore, requires particular vigilance on the part of the experimenter. To start, we may note that the word "control" implies that the experimenter has a certain power over the conditions of an experiment; that power is to systematically manipulate variables in an effort to arrive at a sound empirical conclusion. Let us illustrate by using the above pharmacological example.

First, the magistrate exercised control over his independent variable by producing the event that he wished to study. This is the first sense in which we shall use the word "control." We shall say that an experimenter exercises *independent variable control* when the independent variable is varied in a known and specified manner. In this example, the independent variable was the amount of citron administered, and it was purposively varied in two ways: zero and some positive amount. (Recall that independent variable control through purposive manipulation is the essential defining feature of an *experiment*, as distinguished from the *method of systematic observation*).

The second sense in which we shall use "control" may be made by restating the purpose of the experiment: the magistrate sought to determine whether variation of amount of citron administered to men who were poisoned would affect their impending state of inanimation (certainly a clear-cut dependent variable measure, if ever there was one). He was interested in finding out whether these two variables were related. There were also present, however, a number of other (extraneous) variables that might have affected the subjects' degree of viability. If there was, in fact, a relationship of the type that he sought, it might have been hidden from him by these other variables. Some substance in the subjects' breakfast, for instance, might have been an antidote; the subjects might have been members of a snake cult and thereby developed an immunity; and so forth. In the absence of knowledge of such extraneous variables, it was necessary to assume that they might have affected the dependent variable. Hence, their possible effects were controlled, i.e., the magistrate formed two equivalent groups and administered citron to only one. In this way, the two groups presumably were equated with regard to all extraneous variables so that their only difference was that one received the hypothesized antidote. The fact that only members of the group that received citron survived ruled out further consideration of the extraneous variables. With this control effected, the magistrate obtained the relationship that he sought, and our second sense of "control" is illustrated: *Extraneous variable control* refers to the regulation of extraneous variables.

In order to be clear we shall say that an extraneous variable is one that is operating in the experimental situation in addition to the independent variable. Since the extraneous variable might affect the dependent variable, and since we are not immediately interested in ascertaining whether or not it does affect the dependent variable, it must be regulated so that it will not mask the possible effect of the independent variable.

Failing to control extraneous variables adequately results in a *confounded* experiment, a disastrous consequence for the experimenter, i.e., if an extraneous variable is allowed to operate systematically in an uncontrolled manner, it and the independent variable are confounded (the dependent variable is not free from irrelevant influences). Suppose, for example, that the subjects who received citron had been served a different breakfast than the control subjects. In this case the magistrate would not know whether it was citron or something in the breakfast of the experimental subjects that was the antidote—type of breakfast would thus have been an extraneous variable that was confounded with the independent variable. We can thus see that confounding occurs when there is an extraneous variable that is systematically related to the independent variable, and it *may* act on the dependent variable; hence, the extraneous variable may affect the dependent variable scores of one group, but not the other. If confounding is present, then, the reason that any change occurs in the dependent variable cannot be ascribed to the independent variable. In summary, *confounding occurs when an extraneous variable is systematically related to the independent variable, and it might differentially affect the dependent variable values of the two or more groups in the investigation.* This is quite an important definition. Although we are cautious in recommending items for the student to commit to memory, this definition *is* worth memorizing exactly. Especially note the word "differentially." If variation of the intensity of the independent variable is systematically accompanied by variation of the intensity of an extraneous variable, and if the dependent variable value for one of the groups differs from that of another of the groups, the dependent variable is thus differentially affected; consequently that extraneous variable is confounded with the independent variable.

To illustrate further these two senses of "control," and also to get closer to home, consider a psychological example. Suppose that an experimenter is interested in determining the effect of Vitamin A on certain visual abilities. The independent variable might be operationally defined as the amount of Vitamin A administered according to a certain schedule. The dependent variable might be similarly defined as the number of letters that a participant can see on a chart placed some distance away. Since the independent variable is under the control of the experimenter it may be varied as desired. For instance, one group of participants may receive a placebo but no Vitamin A; a second group may receive a total of three units of the vitamin; while a third group is administered a total of five units. In this way we would be exercising control of the independent variable.

To illustrate extraneous variable control we might note that the lighting conditions under which the test is taken are relevant to the number of letters that the participants can correctly report. Suppose, for example, that the vision test is taken in a room in which the amount of light varies considerably during the day, and further that Group 1 is run mainly in the morning, Group 2 around noon, and Group 3 in the afternoon. In this case some participants would take the test when there is good light, others when it is poor. The test scores might then primarily reflect the lighting conditions rather than the amount of Vitamin A administered, in which case the possible effects of Vitamin A would be masked out. Put another way, the amount of lighting and amount of Vitamin A would be confounded. Lack of control over this extraneous variable would leave us in a situation where we do not know which variable or combination of variables is responsible for influencing our dependent variable.

Just to develop this point briefly, let us consider some of the possibilities when only the single extraneous variable of light is uncontrolled. Assume that the obtained value of the dependent variable increases as the amount of Vitamin A increases, i.e., that the group receiving the five-unit dose of Vitamin A has the highest dependent variable score, the three-unit group is next, and that the zero Vitamin A group has the lowest test score. What may we conclude about the effect of Vitamin A on the dependent variable? Since light is uncontrolled we do not know what effect it has. Hence, the light may actually be the factor that causes the dependent variable scores to increase. Or, it is possible that lighting has a detrimental effect such that if it were not operating in an uncontrolled fashion the apparent effects of Vitamin A would be even more pronounced, e.g., if the five-unit group received a score of 10, it might have received a score of 20, if light had been controlled. Another possibility is that the light has no effect, in which case our results could be accepted as valid. But since we do not know this, we cannot reach such a conclusion. The ambiguity in interpreting the effects of an independent variable that is confounded with a single extraneous variable should thus be apparent. But where there is more than one confounced extraneous variable, the situation is much nearer total chaos.

Experimental control, then, is the regulation of experimental variables. And we may consider two classes of experimental variables: independent and extraneous. The independent variables, we have said, are those whose effects the experimenter is attempting to determine. The wish is to know if a given independent variable affects the dependent variable. The extraneous variables are all other variables operating on the participants at the time of the experiment. By exercising independent variable control the experimenter varies the independent variable as desired. By exercising extraneous variable control one regulates the extraneous variables so that confounding is eliminated. If adequate extraneous variable control is exercised, an unambiguous statement on the relationship between the independent and dependent variables can be made. If extraneous variable control is inadequate, however, the conclusion must be tempered. The extent to which it must be tempered depends on a number of factors, but, generally, inadequate extraneous variable control leads to no conclusion whatsoever concerning the relationship.

Determining Extraneous Variables

We know that at any given moment a fantastically large number of stimuli are impinging on an organism. And we must assume that all of these stimuli are affecting the organism's behavior, if only in some subtle way. But in any given experiment we are usually interested in only one aspect of behavior—a single class of responses. Furthermore, we usually seek to determine whether a certain class of stimuli affect that response; that is, the potential independent-dependent variable relationship. Hence, for this immediate purpose we want to eliminate from consideration all other variables. If this were possible we could conclude that any change in our dependent variable is due only to the variation of our independent variable.

If these other (extraneous) variables are allowed to influence our dependent variable, however, any change in our dependent variable could not be ascribed to variation of our independent variable. *We would not know which of the numerous variables caused the change.*

We must, then, control the experimental situation so that these other extraneous variables can be dismissed from further consideration. The first step in this process is to identify them: What extraneous variables may be present in the experimental situation? Since it would be an almost endless task to list all of the variables that *might conceivably* affect the behavior of an organism, our question must be more limited: Of all the variables present, which are reasonably likely to affect our dependent variable? Even though this is still a difficult question, we can immediately eliminate from consideration a large number of unlikely influences on the organism. For example, if we are studying a learning process, we would not even consider such variables as color of the chair in which the participant sits, brand of pencil used, etc. As a first step in determining those extraneous variables that should be considered, we might refer to our literature survey (chapter 4). We can study previous experiments concerned with our dependent variable to find out which variables have been demonstrated to effect that dependent variable. We should also note what other variables previous experimenters have considered it necessary to control. Discussion sections of earlier articles may also yield information about variables that had not previously been controlled, but were recommended for consideration in the future. Together with the results of our literature survey, our general knowledge of potentially relevant variables, and considerable reflection concerning other variables, we may arrive at a list of extraneous variables that should be considered.

Specifying Extraneous Variables To Be Controlled

Once our list of potentially relevant, important extraneous variables is constructed, we must decide which should be controlled. This would include those variables that are reasonably likely to affect our dependent variable. It is to these variables that the various techniques of control will be applied. A discussion of these techniques is presented on pages 152–66. Suffice it to state here the end result—the changes in the dependent variable will be ascribable to the independent variable rather than to the controlled extraneous variables.

Specifying Extraneous Variables that Cannot Reasonably Be Controlled

A simple answer to the question of which extraneous variables should be controlled is that we should control all of them. Although it *might* be possible to control all of them, such a feat would be too expensive in terms of time, effort, and money. For example, suppose that the variation in temperature during experimental sessions is three degrees. Even though it is possible to control this variable, it is highly unlikely (in most experiments) that it would reliably (if at all) affect the dependent variable. And the experimenter's effort would be great enough to make the game "not worth the candle." This is particularly so when one considers the large number of other variables in the same category. With the limited amount of energy and resources available, the experimenter should seek to control only those variables that are potentially relevant.

Now, all of these probably minor variables might accumulate to have a rather major effect on the dependent variable, thus invalidating the experiment. And even if the effect is not so extreme, should even a minor extraneous variable be allowed to influence the dependent variable? If the experimenter is not going to control them, what can be done about them? In thinking about these points, we must remember that there always will be a large number of variables in this category. The question is, will they affect one of our experimental conditions (one of our groups) to a greater extent than another? For, if by chance such variables do not differentially affect our groups, then our worries are considerably lessened. We can make the assumption that such variables will "randomize out," that, in the long run, they will affect both groups equally. If it is reasonable to make this assumption, then this type of variable should not delay us further. A further discussion of randomization as a technique of control will allow us to consider this and similar problems later.

When To Abandon the Experiment

Up to this point we have been rather optimistic. We have assumed that we are capable of controlling all of the relevant variables that affect the dependent variable, that the effects of these variables will be essentially equal on all groups in the experiment. If it is unreasonable to make this assumption, then the experimenter must consider whether these variables are of sufficient importance to necessitate the abandonment of the experiment. Even if one is not sure on this point, perhaps it would be best if the experiment were not conducted. Sometime after considering the various control problems, the experimenter must ask what will be gained by conducting the experiment. In these cases of inadequate extraneous variable control, the answer need not be that nothing will be gained; for instance, by conducting such an experiment it may be that further insight or beneficial information will be acquired concerning the control problem. But the experimenter must realize that this situation exists and be realistic in understanding that it may be better to discontinue the project.

Techniques of Control

We have previously emphasized the importance of exercising adequate experimental control, but this phase of the experimental planning is sufficiently important that it does not seem possible to *over*-emphasize it. Although experimenters try to exercise considerable vigilance in this regard, it is frequently the case that a crucial, uncontrolled extraneous variable is discovered only after the data are compiled. Shortcomings in control are found even in published experiments. Certainly, if such variables have been discovered neither by the experimenter nor by the editors of the journals and their consultants, they are elusive and subtle. Furthermore, errors of control are not the sole property of young experimenters; they may be found in the work of some of the most respected and established psychologists.

The experimenter should give as much thought to potential errors as possible. After one has checked and rechecked it may be possible to obtain critiques from colleagues. An "outsider" can sometimes approach the experiment with a totally different set, thus seeing something that the experimenter might have missed. This is an important point. A scientist calls on colleagues to check steps of an experiment from beginning to end, including reading successive drafts of the write-up. Early in their careers students should learn to help each other in such ways, too. It is amazing to note students who will write up a paper and not even bother to read it over themselves for corrections before "handing it in."

Our main consideration in this section follows from the point at which an important extraneous variable is spotted and the experimenter asks how it is to be controlled. One must ascertain what techniques are available for regulating it in such a manner that the effects of the independent variable on the dependent variable can be clearly isolated. There are a number of such techniques; we shall attempt to classify them into several categories. This classification will necessarily be incomplete and overlapping in part, particularly as to the variations of each class. But a general understanding of the major principles should facilitate their application to a wide variety of specific control problems.

1. Elimination. The most desirable way to control extraneous variables is to eliminate them from the experimental situation. Examples of the elimination of extraneous variables in psychological laboratories would be the use of sound-deadened rooms or the Skinner box, which is sound-deadened and opaque. Unfortunately, though, most extraneous variables cannot be eliminated. The previous example concerning Vitamin A and the participant's ability to read letters from a chart is a case in point. In that example our extraneous variable was the amount of lighting. Obviously, if the method of elimination were applied, the participant would not have the light needed for them to see the chart. Other extraneous variables that one would have a hard time eliminating are participants' previous experience, sex, level of motivation, age, weight, intelligence, and so on.

2. *Constancy of Conditions.* When certain extraneous variables cannot be eliminated, we can attempt to hold them constant throughout the experiment. Control by this technique means essentially that whatever the extraneous variable, the same value of it is present for all participants. Perhaps, for instance, the time of day is an important variable. Maybe people perform better on the dependent variable early in the morning than late in the afternoon. In order to hold time of day constant, we might introduce all participants into the experimental situation at approximately the same hour on successive days. Of course this procedure would not really hold the amount of fatigue constant for all participants on all days. But it would certainly help.

Another example of effecting constancy of conditions would be to hold the lighting conditions constant in our Vitamin A chart-reading experiment. Thus, we might pull down the blinds in our experimental room and have the same light turned on for all participants. In experiments where light intensity is extremely important, we could actually measure the amount of light present for each participant. The placing of a rheostat in the lighting circuit would allow us to modify fluctuations in the electrical flow in such a manner as to hold light intensity at almost precisely the same value for all.

One of the standard applications of the technique of holding conditions constant is to conduct experimental sessions in the same room. Thus whatever might be the influence of the particular characteristics of the room (gayness, odors, color of the walls and furniture, location), that influence would be the same for all participants. In like manner, to hold various personal variables constant (educational level, sex, age), we need merely select participants with the characteristics that we want. For example we might specify that all participants must have completed the eighth grade and no more; that all participants are male; or that all participants are 50 years old.

Numerous characteristics of our experimental procedure must be subjected to this technique of control. Instructions to participants, for instance, are extremely important. For this reason experimenters read precisely the same written set of instructions to all participants (except where they must be modified for different experimental conditions). But even if the same words are read to all participants, they might be read in different ways, with different intonations and emphases, regardless of the experimenter's efforts to avoid such differences. To exercise more precise control, then, many experimenters have participants listen to the same standardized instructions from a tape recorder.

Procedurally, all participants should go through the same steps in the same order. For instance, if the steps are greet participant, seat the participant, read instructions, attach blindfold, give instructions to start the task, and so on, then one would not want to blindfold some participants *before* the instructions were read and blindfold others *after* the instructions. The attitude of the experimenter should also be held as constant as possible for all participants. If one is jovial with one participant and gruff with another, confounding of experimenter attitude with the independent variable would occur. Now, acting the same toward all participants is extremely difficult, but a strong effort should be made in this direction. The experimenter can practice the experimental procedure a number of times until it becomes so routine that each participant can be treated in mechanical fashion. It is common for the same experimenter to collect data from all the participants. If different experimenters are used unsystematically, then a rather serious error may result. In one experiment, for instance, an experimenter ran a group of rats for 14 days, but had to be absent on the 15th day. The rats' performance for a different experimenter on that day was sufficiently different from other groups who had not suffered a change of experimenters that it is reasonable to conclude that the mere handling of them by a new person (who undoubtedly used somewhat different methods of picking them up, etc.) was responsible for the change.

The apparatus that is used both in administering the experimental treatment and in recording the results should be the same for all participants. Suppose, for example, that two memory drums are used in an experiment, one of which moves more slowly than the other. If one group uses the faster drum and another the slower, confounding will result. Application of the technique of constancy of conditions dictates that all participants use the same drum. Similar precautions should be taken with regard to recording apparatus.

3. Balancing. When it is not convenient or possible to hold constant conditions in the experiment, the experimenter may resort to the technique of balancing out the effect of extraneous variables. There are two general situations in which balancing may be used: (1) where the experimenter is either unable or uninterested in identifying the extraneous variables; (2) where they can be identified and the desire is to take special steps to control them.

Consider the first situation. One group of experimenters was interested in the effect of rifle training on rifle steadiness; whether a prolonged period of training in rifle firing increased the steadiness with which soldiers held their weapons (McGuigan and MacCaslin, 1955a). Previous research had indicated that the steadier a man held a rifle, the more accurately he could shoot. Thus, if you could increase steadiness through rifle training, you *might* thereby increase rifle accuracy. The design of the experiment was a test of rifle steadiness before and after subjects received their rifle training. If the soldiers were steadier on the second test, it might be concluded that training increases steadiness. The first group of data that were analyzed are presented in Table 6-1, where the lower the score, the greater the steadiness.

Table 6-1 Mean Steadiness Scores of Soldiers before and after Rifle Training

Before Training		After Training
235.39	Training Period	194.26

From Table 6-1 we can see that the scores actually did decrease. The first thought is to conclude that training increases steadiness. When the experimenters analyzed another set of data from a control group which did not receive rifle training, the picture changes (see Table 6-2).

Table 6-2 Mean Steadiness Score of Trained and Untrained Groups

	Before Training		After Training
Trained (Experimental) Groups	235.39	Training period	194.26
Untrained (Control) Group	226.61	No training period	170.33

From Table 6-2 we can see that not only did the steadiness scores of the untrained group also decrease, but that they decreased more than those of the trained group. In order to reach the conclusion that training is the variable responsible for the decrease in scores, the experimental group had to show a significantly greater decrease in scores than did the control (no training) group. Thus, we may say that rifle training was not the reason that the steadiness scores of the trained group decreased. There must have been other variables operating to produce that change, variables that operated on both the experimental and the control groups. Whatever the variables, they were controlled by the technique of balancing (i.e., their effects on the trained group were balanced out or equalized by the use of the control group). But we may speculate about these extraneous variables. For example, the rifle training was given during the first two weeks of the soldiers' army life. It may be that the drop in scores merely reflected a general adjustment to the emotional impacts of army life. Or the soldiers could have learned enough about the steadiness test in the first session to improve their performance in the second (a practice effect).

But whatever the extraneous variables, they were controlled by the use of the control group. The logic of using a control group should now be apparent. If the experimental and control groups are treated in the same way except with regard to the independent variable, then any difference between the two groups on the dependent variable is ascribable to the independent variable (at least in the long run). Thus, we need not specify all the extraneous variables that influence the two groups during the experiment. For instance, suppose that only three extraneous variables operate on the experimental group in addition to some positive amount of the independent variable. By administering a zero amount of the independent variable to the control group and by balancing out the effects of the three extraneous variables by allowing them to operate also on the control group, we can see from Figure 6-1 that the independent variable is the only one that can differentially influence the two groups.

An additional important use of the control group as a technique of control may now be profitably discussed. Granting that: (1) a large number of extraneous variables are operating on a participant in any given situation; and (2) we cannot remove all of these variables, *then* we can use additional control groups to evaluate the influence of these variables, to analyze the total situation into its parts. Referring to Figure 6-1 we may be interested in the effect of extraneous variable 1. To evaluate that extraneous variable we need only add an additional control group which is not influenced by it (i.e., receives a zero value of it). The plan is illustrated in Figure 6-2.

For both the experimental group and control group 1, extraneous variable 1 is possibly influencing the dependent variable. Since this variable is not operating for control group 2, a comparison of the two control groups should tell us the effect of extraneous variable 1. Consider one of the extraneous variables that was operating in the

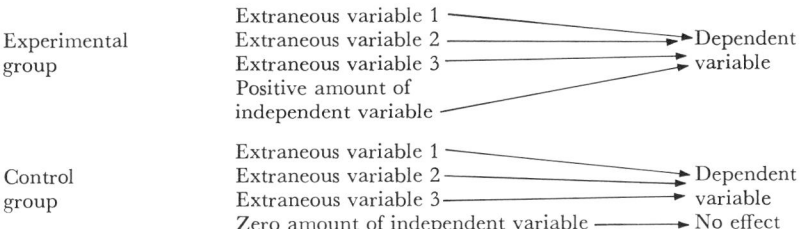

Figure 6-1. Diagrammatic representation of the use of the control group as a technique of balancing.

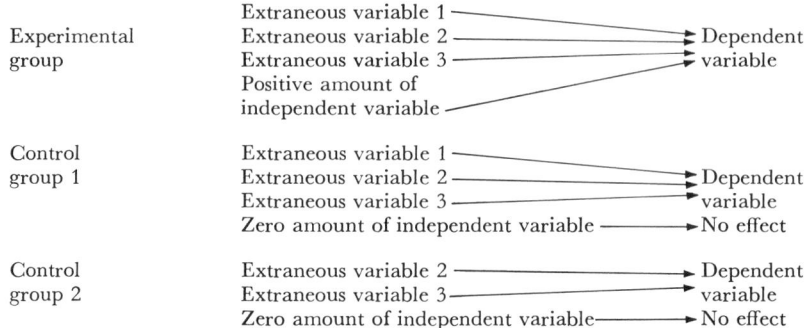

Figure 6-2. The use of a second control group to evaluate the effect of an extraneous variable.

rifle steadiness experiment: practice in the test situation. If an acquaintance with testing procedure and the specific learning of how to hold the rifle while tested led to lower scores, the addition of control group 2 should provide us with this information (see Table 6-3).

This is the same design as in Table 6-2 except for the addition of a second control group that does not take the initial test. A comparison of the steadiness scores of the two untrained groups on the test after training should tell us what the effect of the initial test is. If, for instance, untrained group 2 is less steady on the second (post-training) test than untrained group 1, we could say that the effect of taking the first (pretraining) test is to increase steadiness.

The use of additional control groups to evaluate the effect of various extraneous variables is a very important experimental technique. Frequently, the independent variable itself is of such a nature that it can be further fractionated by the addition of control groups. Some additional examples of this technique are taken up in the last section of this chapter. For additional study of this topic see the excellent (now classical) article by Solomon (1949).

Table 6-3 Possible Experimental Design for Studying the Effect of Practice on the First Steadiness Test

	Receive Test Before Training?	*Receive Training?*	*Receive Test After Training?*
Trained (Experimental) Group	yes	yes	yes
Untrained (Control) Group 1	yes	no	yes
Untrained (Control) Group 2	no	no	yes

The second situation in which balancing may be used is where there is a specific and known extraneous variable to be controlled. For instance, if an experimenter wishes to control the sex (gender) variable, participants of only one gender could be used. If both male and female participants are available, however, and not enough participants of one sex to use them exclusively, one may be forced to use both sexes. In this event we could control the effect of gender on the dependent variable by making sure that it balances out in the two groups. This would be accomplished by assigning an equal number of participants of each sex to each group. If there are 40 males and 30 females, one would make sure that 20 males and 15 females go in each group.[1] Thus, if gender is relevant to the dependent variable, its effects would be the same for each experimental condition. In a similar manner one could control the age of the participants by making sure that an equal number of each age classification is assigned to each group.

The same holds true for apparatus problems. Suppose that an experimenter has two memory drums available and wants to use both to save time. They may or may not have slight differences, but to make sure that this variable is controlled one would have half of the participants in each group use each memory drum. If there are 30 participants in each of two groups, 15 participants in each group would use drum A, while the other 15 would use drum B. Thus, whatever the differences in the memory drums, if any, their respective effects would be the same for both groups.

A major assumption in balancing the effects of extraneous variables by means of a control group is that participants in each group are equivalent on the average. They are only treated differently. However, usable data will often not be obtained for some of the participants, as called for by the experimental plan. The reasons may be many! Human participants who have been assigned to some specific experimental condition may fail to show up for an appointment; mechanical failures necessitate the discarding of some data; animals die; participants fail to follow instructions, etc.[2] Such losses of data constitute only minor problems, with one important exception: If data for noticeably more participants are lost from one group, then a com-

---

[1] Of course, in this example, even though the effect of sex is balanced out, the males will have a greater effect on the dependent variable scores than the females. The matter can be handled with appropriate statistical procedures, but it is preferable to have an equal number of each gender assigned to each group, in this example.

[2] Such would be legitimate reasons for discarding participants from an experiment. However, one must be *very* careful that legitimate reasons *are* employed — there must be some good reason that is totally independent of the participant's dependent variable value, reasons like those cited above. One should not, for instance, merely discard a dependent variable value for a participant because it was extreme or "out of line" with the rest, though amazingly this has been done.

parison of the data for the experimental and control groups may be biased. For instance, if values for 30 percent of the participants were lost from the experimental group but only 5 percent from the control group, the remaining 70 percent of the experimental group may not be representative of the population sampled—there could be some systematic reason that more were lost from one group than from another, and those who were lost may be different from those who remained (the less motivated, less intelligent, or whatever may have dropped out). Such loss of a representative sample could well invalidate the experiment.

The typical solution to loss of participants is to randomly replace them from the same population that was originally defined. It would be reasonable, in fact, to have previously selected several extra participants who would later be assigned to groups only in case of loss.

Balancing may also be applied where there is more than one experimenter. In this case we merely need to have each experimenter run an equal number of the participants in each group. To consider a situation that is a bit more complicated than any we have previously discussed let us say that we wish to balance two effects: sex and experimenter. We have two groups: 60 participants per group, two sexes, and two experimenters. In this case the balancing arrangement would look something like that presented in Table 6-4.

Table 6-4   Illustration of a Design Where Experimenters and Sex Are Balanced

Group I	*Group II*
15 Males — Experimenter 1	15 Males — Experimenter 1
15 Males — Experimenter 2	15 Males — Experimenter 2
15 Females — Experimenter 1	15 Females — Experimenter 1
15 Females — Experimenter 2	15 Females — Experimenter 2

In a final example of the balancing technique, let us say that we want to see how well rats retain a certain habit. To do this we might use a T maze which has a white and a black goal box. From the start position, the rats may run to either box. But we train them to run to one box, say the black. Hence, we always feed then when they run to the black box; they receive no food in the white box. After a number of trials they run rather consistently to the black box, avoiding the white box. After this initial training, we do not allow them to run the maze for, say, three months, at the end of which we place them in the start box of the maze for their test trials. If most of their runs are to the black box, may we assume that they "remembered" well? Our conclusion should not be so hasty. For we know that rats are nocturnal animals and that they tend to prefer dark places over light places. In particular, they have a preference for black goal boxes over white ones *before any training*. Hence it is possible that they would go more frequently to the black box on the test trials *regardless of the previous training*, and thus may not have "remembered" anything. For this reason we need to exercise control over the color of the "reward" box — we need to balance the colors. To do this we train half of our animals as above. The other half are trained in the opposite manner; they receive food when they run to the white box but no food when they run to the black box.[3] If, on the test trials, the animals trained to go to white show a preference for white, then we would have considerably more confidence in our conclusion. For regardless of the color that they were trained to, we find that they retain the habit over the three-month period. The effect of color cannot possibly be the variable that is influencing their behavior.

4. Counterbalancing. Some experiments are designed so that the same participant must serve under two or more different experimental conditions. If an experimenter were interested in whether a stop sign should be painted yellow or red, the problem would be to determine to which colored sign a participant responds faster. To answer this question one might measure a participant's reaction time, first to the yellow sign and then the red sign. By repeating this procedure with a number of participants one could reach a conclusion — perhaps that reaction time to the red sign is the smaller. Since the participants were first exposed to the yellow sign, however, their reaction time to that sign would be partially dependent on their learning to operate the experimental apparatus and on their adaptation to the experimental situation. After they have learned how to operate the apparatus and adapted to the situation, they are exposed to the red sign. Hence, their lower reaction time to the red might merely reflect practice and adaptation effects rather than effect of color — color of sign and amount of practice are confounded. The answer frequently given to the problem of how to control the extraneous variable of amount of practice is to use the method of counterbalancing. The application of this method would be to have half the participants react to the yellow sign first and the red sign second, while the other half would experience the red sign first and the yellow sign second (see Table 6-5).

The general principle of the technique of counterbalancing may be stated as: Each condition (e.g., color of sign) must be presented to

---

[3] Also alternate the positions (right and left) of the white and black boxes in order to assure that the animals are not learning mere position habits.

Table 6-5 Demonstrating Counterbalancing to Control
an Extraneous Variable

	EXPERIMENTAL SESSION	
	1	*2*
1/2 Participants	Yellow Sign	Red Sign
1/2 Participants	Red Sign	Yellow Sign

each participant an equal number of times, and each condition must occur an equal number of times at each practice session. Furthermore, each condition must precede and follow all other conditions an equal number of times. It can be seen that this principle is applicable to any number of conditions. For example, the principle of counterbalancing could be applied where we have three colors of signs, in which case one-sixth of the participants would be presented with each order specified in Table 6-6. Color of sign presented at each session is indicated by R (Red), Y (Yellow), or G (Green).

By studying Table 6-6 we can observe that the requirements for counterbalancing the effects of three variables are satisfied. In particular, if 6 subjects are used (any multiple of 6 such as 12 or 18 would suffice for this design), we can observe that each color of sign is presented twice at each session, that each participant receives each color once, and that each color precedes and follows each other color two times.

Table 6-6 A Counterbalanced Design for Three Independent Variables

	EXPERIMENTAL SESSION		
	1	*2*	*3*
1/6 Participants	R	Y	G
1/6 Participants	R	G	Y
1/6 Participants	Y	R	G
1/6 Participants	Y	G	R
1/6 Participants	G	R	Y
1/6 Participants	G	Y	R

If a number of experimental sessions is involved, not only would a certain amount of improvement in the participants' performance due to practice be expected, but also a certain decrement in performance due to fatigue. The method of counterbalancing attempts to distribute these effects equally to all conditions. Hence, whatever the practice and fatigue effects (they are called "order effects"), they presumably influence each condition equally since each condition occurs equally often at each stage of practice.

In extending the general principle of counterbalancing to a large number of variables, the number of orders (and therefore the numbers of participants required) soon becomes unrealistic. For example, to counterbalance four variables one would require at least 24 participants; for five variables, there are 120 orders; and for six variables, a minimum of 720 participants would be run. To solve this problem, one may resort to *incomplete* (as against *complete*) counterbalancing. An incomplete counterbalancing design still requires that each participant receive each treatment once, and only once, and that each treatment occur an equal number of times (once) during each session; but it does not require all possible orders of the variables to be presented. Incomplete counterbalancing still adequately controls practice of fatigue effects, and it is generally recommended by Underwood (1966b). Another presentation of designs that may be used in incomplete counterbalancing is given in Edwards (1960, pp. 275–76).

For a more extensive discussion of this topic, read the excellent presentation by Underwood (1966a, pp. 459–66). Let us here complete our discussion by pointing out a difficulty that one might encounter in counterbalancing: In using the technique, one assumes that the effect of presenting one variable before a second is the same as presenting the second before the first, e.g., that the practice effects of responding to the red sign first are the same as for responding to the yellow sign first. This might not be the case, so that seeing the red sign first might induce a greater practice (or fatigue) effect, possibly leading to erroneous conclusions. More generally, counterbalanced designs entail the assumption that there is no differential (asymmetrical) transfer between conditions. By *differential* or *asymetrical transfer* we mean that the transfer from condition one (when it occurs first) to condition two is different than the transfer from condition two (when it occurs first) to condition one. If this assumption is not justified, there will be interactions among the (order and treatment) variables that will lead to difficulties in the statistical analysis. That differential transfer in counterbalanced designs is not a completely unlikely possibility has been pointed out by Poulton and Freeman (1966). In one example, the purpose was to study the effects of variation of air pressure on card sorting behavior. One group of men first sorted cards when the pressure surrounding them was 3.5 atmospheres absolute (call this condition A); they then sorted cards at a normal pressure (condition B). A second group of men experienced condition B first, followed by condition A. Many slow responses occurred for the first group of men under condition A, as one might expect. But when they sorted cards under normal pressure, these men made almost as many slow responses as they did under condition A. The second group, on the other hand, made a fewer number of slow responses under normal pressure (condition B) and made almost the same number of slow responses when they shifted to condition A. In other words, card sorting behavior (the dependent variable) was influenced by the *order* of presenting the experimental conditions. As a result, when the results for the first and second sessions were combined, the effect of variation of pressure was obscured and the statistical test indicated (erroneously) that it was not a reliable effect. In general, these authors conclude, asymmetrical transfer reduces the recorded difference between two conditions to less than what it really should be, but it may also exaggerate the difference.

The lesson, then, is that if you use counterbalancing as a technique of control, you should examine your data for asymmetrical transfer effects. If you find yourself in possession of such, you might study your findings as interactions and consult the more detailed advice given by Poulton and Freeman (1966). To end on a happy note, however, let us observe that Underwood (1966b) emphasizes the many advantages of counterbalancing and recommends its use "for many experiments."

Balancing and *counterbalancing* can be contrasted to emphasize their differences. Students sometimes confuse balancing and counterbalancing, perhaps because they are both techniques of control with "balancing" in common for their names. A little reflection should eliminate such confusion: *Counterbalancing* is used when each participant receives more than one treatment (AB or BA) and the effort is to distribute order effects (fatigue, practice) equally over all experimental conditions (as in Table 6-5). In *Balancing* each participant receives only one experimental treatment — extraneous variables are balanced out by having them affect members of the experimental group equally with those of the control group. A participant thus serves only under an experimental *or* a control condition such that any extraneous variables (practice, fatigue) exert equal influence on members of the two groups (are "balanced out").

5. *Randomization.*[4] This technique has two general applications: (1) where it is known that certain extraneous variables operate in the experimental situation, but it is not feasible to apply one of the above

[4] Randomization is included as a technique of control because the experimenter takes certain steps to insure its operation and thus equalizes the effects of extraneous variables.

techniques of control; (2) where we assume that some extraneous variables will operate, but cannot specify them and therefore cannot apply the other techniques.[5] In either case we take precautions that enhance the likelihood of our assumption that the extraneous variables will "randomize out," i.e., that whatever their effects, they will influence both groups to approximately the same extent.

Consider some examples. Participants' characteristics are important in any psychological experiment. Such variables as previous learning experiences, level of motivation, amount of food eaten on the experimental day, relations with boy or girl friends, and money problems may affect our dependent variable. Of course, the experimenter cannot control such variables by any of the previous techniques. If, however, there is an experimental and a control group, say, and if participants are randomly assigned to the two groups, we may assume that the effect of such variables is about the same on both groups. We may expect the two groups to differ on such variables only within the limits of random sampling. Hence, the extraneous variables should not differentially affect the dependent variable. And whatever the differences (small, we expect) between the groups on such variables, they are taken into account by our method of statistical analysis. For instance, the t-test is designed so that it will tell us whether the groups differ on other than a basis of random fluctuations.

One of the most incredible examples of confounding that the author has ever encountered occurred because the experimenter failed to randomly assign his participants to groups. It would not have been incredible, perhaps, had it been committed by a high school student, but it happens that it occurred for a master's thesis. Without going into the details of the experiment, the student used speed of running a maze as an index of the strength of a theoretical variable. His reasoning was to the effect that Group 1 should have a larger amount of the theoretical variable present (due to various training conditions) and should thus have the greater speed. But in retrospect, the experiment never got off the ground. In assigning rats to groups, the "experimenter" merely reached into the cages and the first rats that came into his hands were assigned to Group 1, the remainder to Group 2. The more active animals no doubt popped their heads out of the cage to be grasped by the experimenter, while the less active ones cowered

---

[5] An additional reason that randomization is important (such as random assignment of subjects to groups) is to insure the validity of the statistical test. But this point is covered later under the topic of assumptions of statistical tests.

in the rear of the cage. Regardless, therefore, of the training administered to the groups, Group 1 had, in all likelihood, the speedier rats. Experimental treatments and initial (native) running ability were thus confounded. This example, incidentally, serves to justify what otherwise might seem as an arbitrary decision to include randomization as a technique of control. For the experimenter who does not take specific steps to assure randomization (such as randomly assigning participants to groups) can become the victim of a confounded experiment.

The potential extraneous variables that might appear in the experimental situation are considerable. Various events might occur in an unsystematic way, such as the ringing of campus bells, the clanging of radiator pipes, peculiar momentary behavior of the experimenter (such as a tic, sneezing, or scratching), an outsider intruding, odors from the chemistry laboratory, and the dripping of water from an overhead pipe. Now it might be possible to anticipate many of these variables and control them with one of our techniques, but even if it is possible, it might not be feasible. Signs may be placed on the door of the laboratory to head off intrusions, but signs are not always read. A sound-deadened room is the answer to many of the problems, but such facilities are not always available in psychological laboratories. It is unlikely that *all* such variables will be controlled by means of the previous techniques. Accordingly we can do the next best thing—we take steps to ensure that their effects will randomize out so that they will not differentially affect our groups. To facilitate the credibility of this assumption we might make sure that the order in which we run our participants is approximately that of alternation. Thus, if we randomly assign the first participant we run to the experimental group, the next would be in the control group; the third participant would be randomly assigned to either the control or experimental group, whereupon the fourth would be in the alternative group; and so forth. In this way we could expect, for example, that if a building construction operation is going on that is particularly bothersome, it will affect several participants in each group and both groups approximately equally.

An Example of Exercising Extraneous
Variable Control

To illustrate some of our major points, and to try to unify our thinking about control procedures, consider an experiment that has as its purpose the determination of whether the amount of stress present on members of a group influences the amount of hostility they will verbally express toward their parents while talking in that group situation. To answer this question we would first plan on collecting a number of individuals. Since we need to vary the amount of stress present on the members, we form two groups. A fairly heightened amount of stress is exerted on the experimental group (by some means that need not detain us here), while the control group experiences only the normal stress present in such a social situation. Our independent variable is thus amount of stress (which is varied in two ways), and the dependent variable is amount of hostility verbally expressed toward parents. Referring to Figure 6-3 we note that, as far as control is concerned, our first step is to determine the extraneous variables that are present. Through the procedures previously specified we might arrive at the following list: gender and age of participants, whether their parents are living or dead, place of the experiment, time of day, characteristics of experimenter, lighting conditions, various noises, number in the groups, family background and ethnic origin of participants, their educational level, recent experiences with parents, general aggressive tendencies, frustrations, previous feelings towards parents, and eye color.

From Figure 6-3 we note that the next step is to determine those extraneous variables that might reasonably influence the dependent variable. Merely for illustrative purposes we have included one that

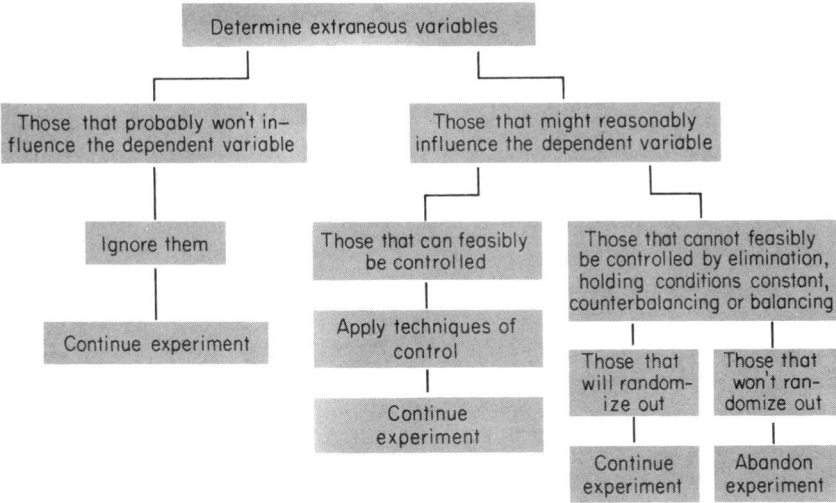

Figure 6.3. An overall diagram of steps to be followed in planning an experiment.

probably will not influence the dependent variable and thus will be ignored: eye color of subjects.[6] All the rest must be dealt with. Those that might feasibly be controlled by elimination, holding conditions constant, or balancing, we might decide, are sex, age, place, time, lighting, group number, education, whether parents are living or dead, and experimenter characteristics. Of these, we could control the following by holding conditions constant: place of experiment—by holding sessions for both groups in the same room; time of day—by holding sessions at the same time for both groups (on different days); lighting conditions—by having the same lights on for both groups with no external light present; number in the groups—by having the same number in each group; and experimenter characteristics—by having the same experimenter, with the same experimenter attitude, appear before both groups.

The variables of sex, age, educational level, and parents living or dead could be controlled by balancing. We could assign an equal number of each sex to each group, make sure that the average age of each group is about the same, distribute educational level equally between both groups, and assign an equal number of participants whose parents are living to each group. Now, it is obvious that simultaneous balancing of all these variables would be rather difficult (if not impossible) with a small number of participants. In fact, two variables would probably be as many as we could feasibly handle by this technique. We might select sex and parents living or dead as the most important and balance them out. For, if we are using college students (as we probably would), educational level and age would be about the same. Hence, we shall lump these two variables with the following as variables that we do not consider it feasible to control by the above techniques: various noises, family background, ethnic origin, recent experiences with parents, general aggressive tendencies, extent of frustration, and previous feelings toward parents. Some of these might be considered important, and it would certainly be desirable to control them. Most of them are difficult to measure, however, and thus are hard to balance out in a specific manner.

Now, is it reasonable to assume that such variables will randomly affect both groups to approximately the same extent? If participants are randomly assigned to groups (except insofar as we have restricted the random process through balancing), the assumption should be valid. And as previously noted we can always check the validity of this

---

[6] Even though there are variables that we ignore we have several times made the point that they are, in actuality, controlled through the techniques of balancing and randomization.

assumption by comparing the groups on any given variable. Since this assumption seems reasonable to make in the present example we shall conclude that the conduct of the experiment is feasible as far as control problems are concerned; we have not been able to specify any confounding effects that would suggest that the experiment should be abandoned.

The Experimenter as an Extraneous Variable

Several times we have mentioned the matter of controlling experimenter influences on the dependent variable, but the topic is sufficiently important that we shall concentrate on it here, and in a later section, too. Even though we have long been aware that experimenter characteristics may have a substantial effect on participants, researchers have essentially ignored this variable in the design of their experiments. To document this point, a sample of articles from the *Journal of Experimental Psychology* was studied (McGuigan, 1963).[7] The conclusion was that although more than one person collected data in a large number of the experiments reported, in no case was any mention made of techniques of controlling the experimenter variable, and in only one of the sample of articles was the number of data collectors actually specified. The possibility is alarming that in these studies adequate control of the experimenter variable was not exercised. Perhaps the most apparent violation of sound control procedures occurs where one experimenter collects data for a while, after which he is relieved by another experimenter, with no plan for assigning an equal number of participants in each group to each experimenter. But far more subtle effects are also possible. Rosenthal (1966), for instance, has shown that wishes and expectations of the experimenters can actually influence the nature of the data they collect. In one experiment Rosenthal informed one group of experimenters that their rats had been bred so that they were "maze-bright" ("intelligent"). A second group of experimenters were told that they had "maze-dull" rats. The expectation implanted in the experimenters, then, was that the former would be able to quite readily learn a simple discrimination problem, while the learning of the latter animals would be slow. The results indicated that experimenters who believed that their animals were bright obtained performance from them that was significantly superior to that obtained by experimenters who be-

[7] Ten years later the problem still existed, as shown by a sample of studies in five journals in which the experimenter variable was typically handled in an inadequate fashion (Page and Yates, 1973).

lieved that their animals were dull, even though both groups of rats were of equal intelligence.

The problem of experimenter bias is not simple. Though many researchers believe it exists as an important variable to be controlled (90 percent of a sample of 120 reported by Page and Yates, 1973), Barber and his colleagues (e.g., Barber and Silver, 1968, in "Fact, Fiction, and the Experimenter Bias Effect") concluded that it is relatively difficult to demonstrate, is not very persuasive, and can often be explained more readily by various experimental errors.

Even though much remains to be learned about the experimenter as a stimulus variable, we can at least attempt to take precautions to control it when feasible. Clearly, balancing of participants across experimenters (p. 155) is called for when more than one data collector is involved. The technique of elimination could possibly be resorted to, in which case participants could be run entirely by means of automated equipment such as tape recorders. In some cases it might be possible to keep knowledge of the nature of the hypothesis, or of the data actually collected, away from the experimenter. In an article "How Blind Is Blind?" Beatty (1972) reported that observers who judge dependent variable scores for different treatments often are biased in that they can use various cues to detect which scores are from which treatment condition. He suggested that where observer judgments represent the primary data (as they do in much research), they be tested for bias—the observer is asked to guess the treatment combinations each participant has received and, by comparing the frequency of correct identifications obtained to that expected by chance, one can estimate whether or not the observers were actually naive. If the observers successfully identify (beyond chance) the conditions received by the participants, they are biased.

Some Control Problems

In the following experiments you should attempt to determine what the control problems are and to specify how you would apply the appropriate techniques to solve the problems. After considering the various experiments you should then reach a conclusion as to whether or not they should have been conducted. Should you like additional practice on this important topic, study the problems offered by Underwood (1966b).

To set the tone for this section we would like to discuss an experiment in which the control, if such existed, was outlandish. One day, a general called the author to say that he was repeating an experiment that the author had conducted on rifle markmanship, and asked if the author could visit him for the purpose of discussing the experiment. The trip was made and the general immediately drove out to the rifle range where the experiment was in progress. We visited the experimental group to observe their progress. During the visit it was more enjoyable watching the general than the soldiers, who were newly "enrolled" army trainees. For while the trainees were practicing firing, the general would walk along the line, kicking some into the proper position, lying down beside others to help them fire, etc. After awhile the general suggested that we leave. That was fine, except for one thing, namely a desire to observe the control group. (By this time the author was beginning to wonder if there was a control group, but this concern was unfounded.) The general suggested a walk over the next hill, for that was where the control group was located. On his way the author stopped to talk privately with the sergeant, particularly commenting on how lively and enthusiastic the experimental participants were. The sergeant explained that that was what the general wanted—that the general expected the experimental group to fire better than the control group and they "darn" well knew that was what had better happen. When the other side of the hill was reached the author was amazed at the contrast. The control participants were the most morose, depressed, laconic group of participants he had ever seen. In talking to the sergeant in charge of this group he was informed that the general had never been to visit them. What is more, this group knew that they should not perform as well as the experimental group, for nobody wanted the general to be disappointed (their motivations are too numerous to cite here). Needless to say, when the general reported the results of the experiment they were highly significant in favor of the experimental group. Let us now see if you can spot the errors, if any, in the following experiments.

1. The problem of whether children should be taught to read by the word method or by the phonics method has been a point of controversy for many years. Briefly, the word method teaches the child to perceive the word as a whole unit, whereas the phonics method requires that he break the word into parts. To attempt to decide this issue an experimenter plans to teach reading to two groups, one by each method. The local school system teaches only the word method. "This is fine for one group," the experimenter says. "Now I must find a school system that uses the phonics method." Accordingly a visit is made to another town that uses the phonics method.

A sample of third-grade children in town is tested to see how well they can read. After administering a long battery of reading tests it is found that the children who used the phonics method are reliably

superior to the children who used the word method. It is then concluded that the phonics method is superior to the word method.

2. A military psychologist is interested in whether training to fire
a machine gun from a tank facilitates accuracy in firing the main tank
gun. A company of soldiers with no previous firing experience, are
randomly divided into two groups. One groups receives .30 caliber
machine gun training, the other does not. Both groups are then
tested on their ability to fire the larger tank gun. To do this two tanks
are set up so that they can fire on targets in a field. The machine gun
trained group is assigned one tank and a corresponding set of targets,
while the control group fires on another set of targets from the second tank. The tests show that the group previously trained on the
machine gun is reliably more accurate than the control group. The
conclusion is that .30 caliber machine gun training facilitates accuracy
on the main tank gun.

3. A psychologist seeks to test the hypothesis that early toilet
training leads to a type of personality where children are excessively
compulsive about cleanliness; conversely, late toilet training leads to
sloppiness. The psychologist notes that previous studies have shown
that middle-class children receive their toilet training earlier than do
lower-class children. Accordingly, two groups are formed, one of
middle-class children and another of lower-class children. Both
groups are provided with a finger painting task, and a number of data
about their procedures are recorded, e.g., the extent to which they
smear their hands and arms with paints, whether or not they clean up
after the session, and how many times they wash the paints from their
hands. Comparisons of the two groups on these criteria indicate that
the middle-class children are reliably more concerned about cleanliness than are those of the lower-class. It is thus concluded that early
toilet training leads to compulsive cleanliness whereas later toilet training results in less concern about personal cleanliness.

4. A physiological psychologist seeks to determine a function of
the internal part of the brain known as the hypothalamus. A sample
of cats is obtained and they are randomly assigned to two groups. An
operation removes the hypothalamus from all the cats in one group.
The second group is not operated on. On a certain behavior test it is
found that the operated group is reliably deficient, as compared to
the control group. The psychologist concludes that the hypothalamus
is responsible for the type of behavior that is "missing" in the group
that was operated on.

5. The following hypothesis is subjected to test: Emotionally
loaded words (e.g., "sex," "prostitute") must be exposed for a longer
time to be perceived than words that are neutral in tone. To test this

hypothesis various words are exposed to participants for extremely short intervals. In fact, the initial exposure time is so short that no participant can report any of the words. The length of exposure is then gradually increased until each word is correctly reported. The length of exposure necessary for each word to be reported is recorded. It is found that the length of time necessary for participants to report the emotionally loaded words is longer than for the neutral words. It is concluded that the hypothesis is confirmed.

6. A physician conducted an experiment to study the effect of acupuncture on pain. Half of 42 participants were treated for painful shoulders through classical acupuncture, while the other half received no special treatment. It was found that the participants who received classic acupuncture treatment reported a reliable improvement in shoulder discomfort to a blind evaluator after treatment. However, no statistically reliable improvement was reported by the control group. The physician concluded that acupuncture is an effective treatment for chronic shoulder pain.

7. Two classes who were taking the same course in educational psychology were used to study the effect of reward or punishment of grades. The same instructor taught both classes. In one class students were given A, B, C, D, or F class grades, while in the other class students either passed or failed without any effect on their cumulative grade record. In the dependent variable comparison, tests before and after the course indicated that there were no reliable differences between the two classes in terms of achievement, attitudes, or values. The conclusion was that students learn just as well without the reward or punishment of grades. In addition, the researcher observed a difference in classroom atmosphere in which the pass-fail class was more relaxed and free of grade oriented tensions with better rapport between the instructor and students.

7

The Independent
and Dependent Variables

From one point of view, the primary purpose of an experiment is to test a hypothesis. And a hypothesis, we said, is a statement to the effect that two (or more) variables are related. We have referred to the two variables as the independent and the dependent variables. In this chapter we will discuss these variables in greater detail and also the types of relationships that may obtain between them.

Types of Relationships Studied
in Psychology

In general approach, we may develop an analogy between the way an engineer looks at a "machine" such as a computer and the way that a psychologist looks at an organism. For the electronic computer, the engineer first has to put some type of energy into it (they call this the "input"). The input then activates the computer in such a way that the energy is "carried" through it (this is the "throughput"). And, finally, the computer accomplishes the task for which it is built and certain actions result (the "output"). There are certain relationships among the input, throughput, and output so that for certain types or amounts of input, certain types and amounts of throughput occur, and certain types and amounts of output result. Furthermore, the characteristics of the computer limit the nature of the throughput and thus of the output. For example, only certain types or amounts of input are capable of being transmitted by the computer. And the characteristics of the computer determine what kinds of output may occur.

The psychologist's approach to behavior is analogous.[1] For we may consider that the organism corresponds to the computer; the stimuli that excite the organism's receptors are the input, and the organism's responses are the output. The analogy may be pursued in the following manner: The type of stimuli that enter the organism (the input) determine what will happen within the organism (the throughput); and what happens within the organism influences the nature of the responses (the output). Furthermore, the specific characteristics of the organism, in particular the neural connections, the past experience, the genetic makeup, also determine the nature of the organism's responses (see Figure 7-1). For example, if a light is flashed in an or-

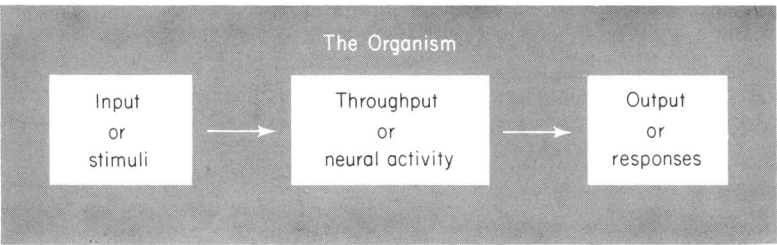

Figure 7-1. An analogy between the approach of an engineer to the study of a computer and the psychologist's approach to behavior. Feedback circuits, so common in "machines" and living organisms, are excluded.

[1] Actually, it would be more interesting to consider the psychologist's terms as a special case of, rather than an analogy to, the input-throughput-output schematization. Or, an even more radical suggestion would be to replace the psychologist's terms with those of the engineer. The implication is an important one for the generality of psychological laws and for the hierarchical status of psychology relative to other sciences. For, assuming the classical definitions of *stimulus* ("energy that excites an organism's receptors") and *response* ("the contraction of muscles or secretion of glands"), it can be noted that the psychologist's terms are relatively limited, particularly as it is often implicitly assumed that they refer to the apparatus of mammals. The use of the more general terms ("input," etc.) might facilitate our search for laws of greater generality, laws that might well apply to lower organisms with unique receptors and effectors, to all manner of "machines" such as electronic computers, and even to extraterrestrial organisms which, in all probability, have very strange receptors and effectors indeed.

ganism's eye (input), various neural pathways are excited which go to and from the visual areas of the brain (throughput), and thence to specific effectors which result in a given response (output). However, in working with an organism whose visual areas in the brain have been destroyed, different characteristics will be encountered, and a different (or perhaps no) response would occur.

There are, then, three general classes of variables with which the psychologist deals: stimulus variables (the input), organismic or mediating variables (the throughput), and response variables (the output).[2] The psychologist attempts to determine relationships between these three. The possible relationships that may be studied are shown in Figure 7-2.

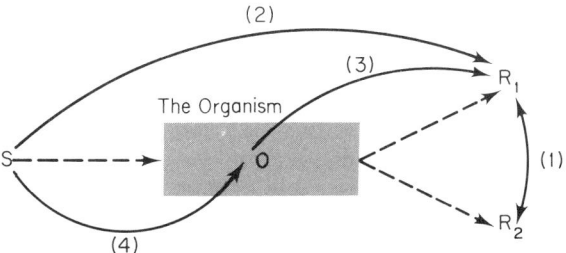

Figure 7-2. Showing the possible relationships among three classes of variables studied in psychology. *S* denotes the stimulus variables; *O*, organismic or mediating variables; *R*, response variables. The relationships indicated by numerals 1, 2, 3, and 4 are discussed in the text. (Modified from K. W. Spence, 1948.)

These possible relationships may be indicated symbolically as follows:

1. $R_1 = f(R_2)$ — response number one (any given response) is a function of response number two (any other response).[3] When determining that two classes of responses are related, you are determining the first type of relationship. It is, however, difficult to experimentally establish a relationship between two responses, and the application of correlational techniques is a more appropriate approach to this prob-

---

[2] Numerous classifications of variables with which psychologists deal are available elsewhere; e.g., Spence (1948) considers stimulus, organic, response, and hypothetical state variables; Underwood (1957) discusses environmental, task, instructional, and subject variables; Edwards (1968) uses the present classification as do Woodworth and Schlosberg (1955), though the latter add antecedent variables to the list.

[3] When we refer to a certain response we actually mean a certain response *class*, a number of quite similar instances of responses. For example, a response instance would be one hit at a baseball, whereas a response class would be made up of all of the times that we hit a baseball (a large number of response instances). Similarly, we may refer to stimulus instances and stimulus classes.

lem. For example, an experimenter may want to know whether two dependent variables are related. In running rats in a maze, for instance, one wants to find out whether the number of errors that rats make is related to the total time it takes the rats to run the maze. These are two response measures, and the correlation between them may be computed.

2. $R = f(S)$ — a certain response class is a function of a certain stimulus (class). In this case, one may vary values of the stimulus to see if values of the response change. The stimulus may thus be seen to be the independent variable, while the response is the dependent variable. This second type of relationship is that with which we are most often concerned in experimentation among the clearest examples of the areas of psychology in which this type of relationship is sought are those of perception and learning. In studies of perception we vary stimulus conditions and determine whether the perceptual responses of the organism also vary. For instance, we might vary the lighting conditions on a given object (varying the stimulus variable) and see if a person's verbal report of its size changes (a response measure).

3. $R = f(O)$ — a response class is a function of (a class of) organismic or mediating variables. The primary purpose of research aimed at this type of relationship is to determine whether certain characteristics of the organism lead to certain types of responses. We might wonder, for instance, if people who are short and stout behave differently than people who are tall and thin. More specifically, do these two types of people differ as far as happiness, general emotionality, or degree of verbosity is concerned?

4. $O = f(S)$ — a class of organismic or mediating variables is a function of a class of stimulus variables. In this case, we are primarily asking what environmental events influence the characteristics or functions of organisms. For example, we might be interested in whether the degree to which a child has been isolated influences his intelligence or motivation.

This is a brief survey of the basic types of relationships sought by psychologists.[4] We should add that more complex relationships may also be sought, as for instance those that would occur if you investigate the relationship among three responses $[R_1 = f(R_2, R_3)]$, among two stimuli and a given response $[R = f(S_1, S_2)]$, or among a stimulus, response, and organismic (mediating) variable $[R = f(O, S)]$.

---

[4] You should note that these relationships conform to our discussion in chapter 3 of the nature of hypotheses. There we showed how a hypothesis stated as a mathematical function [e.g., $R = f(S)$] is a special case of the more general "if a then b" relationship.

We would also note that the statement of the types of relationships sought depends on the way that you classify variables. Hence, other systems of classification would lead to different statements of the possible relationships.

The Independent Variable

In the first type of relationship we would vary one response to see if another is thereby affected. The response that we may vary would be our independent variable, the other response our dependent variable. However, as we said in presenting the first type of relationship, response-response relationships are not often sought with the use of standard experimental designs. In the second type of relationship we vary a stimulus and determine its effect on a response. Hence, the stimulus is the independent variable, the response the dependent variable. In the third type of relationship we vary an organismic variable as the independent variable and determine its relationship to a response, the dependent variable. And in the fourth type of relationship a stimulus is the independent variable, and an organismic variable is the dependent variable. Thus, we have three independent variables to consider: responses, stimuli, and organismic variables. And in general, the symbol to the right of each equation is the independent variable, that to the left the dependent variable.

Response Variables

The response as an independent variable is so infrequent in experimentation because it is typically difficult to purposively manipulate a behavioral variable in order to determine its influence on some other behavioral variable. For instance, suppose that you are interested in whether people who read on the subway are steadier than people who do not read on the subway. It may have occurred to some of you who are both people watchers and subway riders that a reason that some people don't read on the subway is that they can't hold newspapers steady enough. To experimentally test this proposition we would have to randomly determine that some people would be subway readers while others would be subway nonreaders, and then administer a suitable steadiness test to these two groups of participants. Clearly such a procedure would be difficult if not impossible.

One important line of research in which responses have been studied as independent variables, however, has been Premack's (1965) principle of reinforcement. The basic notion here has been that a high probability response can serve as a reinforcer for a low probability response. For instance, suppose you wished to increase the frequency with which a rat performs some infrequently made (low probability) response, like pressing a lever. The procedure would be to provide some animals with the opportunity to press the lever. Then, on successful lever pressing, they would be allowed to engage in some frequently made response, like running (perhaps in an activity wheel). The frequency of the low probability response (lever pressing) should then increase when reinforced by the high probability response (running). A control groups in which lever pressing cannot be followed by running could be used for comparison.

Stimulus Variables

Most independent variables with which we are concerned are stimulus variables, where the word "stimulus" is used in a broad sense to refer to any aspect of the environment (physical, social, etc.) that excites the receptors. The following are examples of stimulus variables as they might affect a particular kind of behavior: the effect of different sizes of type on reading speed; the effect of different styles of type on reading speed; the effect of intensity of light on the rate of conditioning; the effect of number of people present at dinner on the amount of food eaten; the effect of social atmosphere on problem-solving ability. You will note that the variables differ considerably in their complexity.

Organismic and Mediating Variables

This category refers to variables that are intermediate between stimuli and responses in an S-O-R paradigm. We mean to be broad here in our interpretation of "O" variables. By organismic variables we mean any relatively stable characteristic of the organism, including such characteristics as sex, eye color, height, weight, and body build, as well as such so-called psychological characteristics as intelligence, educational level, anxiety, neuroticism, and prejudice. Mediating variables refer to hypothetical constructs, intervening variables, logical constructs, etc. like learning, drive, semantic reactions. Perhaps we should also include here the administration of various drugs, for psychopharmacological research is increasing in its importance.

At this point, let us note that this system of classification, like any other, has its disadvantages, since it is difficult to force some variables into categories. For example, we might question the placement of intelligence in the category of organismic variables. For while it can well be considered to be a characteristic of the organism, let us observe the way in which intelligence is frequently measured: A person takes a pencil and makes a number of marks on a piece of paper—the person makes responses. Hence, it is also possible to classify intelligence as a response variable. We must, therefore, be quite arbitrary in some of our decisions, the justification being that we are trying to consider independent variables in an orderly fashion that will allow some systematic insight into the various kinds of variables used in psychological research.

A Further Consideration of
Control of the Independent Variable

In chapter 6 we said that control of the independent variable occurs when the researcher varies the independent variable in a known and specified manner. There are essentially two ways in which an investigator may exercise independent variable control: (1) purposive manipulation of the variable; and (2) selection of the desired values of the variable from a number of values that already exist. When purposive manipulation is used, we say that an experiment is being conducted; but when selection is used, we say that the method of systematic observation is being used. If an experimenter is interested in whether the intensity of a stimulus affects the rate of conditioning, he might vary intensity in two ways, high and low. If the stimulus is a light, such values as 2 and 20 candle power might be chosen. One would then, at random: (1) assign the sample of participants to two groups; and (2) randomly determine which group would receive the low intensity stimulus, which the high. In this case, we are *purposely manipulating* the independent variable (this is an experiment), for the decision as to what values of the independent variable to study and, more important, which group receives which value is *entirely up to us*. And, what is perhaps equally important, the experimenter himself "creates" the values of the independent variable.

Now, let us illustrate *independent variable control by selection of values* as they already exist (this is the method of systematic observation). Suppose that an investigator is interested in the effect of intelligence on problem solving. Further, assume that the researcher is not interested in studying the effects of minor differences of intelligence, but wants to study widely differing values of this variable. For example, the following three values might be studied: an IQ of 135, a second of

100, and a third of 65. Up to this point, the procedures in exercising the two types of independent variable control are the same; the investigator determines what values of the variables are to be studied.

However, in this case, the investigator must find certain groups that yield the desired values of intelligence. To do this one might administer a number of intelligence tests at three different institutions. First, one might select a number of bright college students, providing a group that has an average IQ of 135. Second, a visit might be made to a rather nonselective group (high school students or army personnel) to obtain an average value of 100. And third, one might find a special institution that would yield a group with an average IQ of 65. With these three groups constructed, a test of problem-solving ability would be administered and the appropriate conclusion reached. Observe that the values of the independent variable have been selected from a large population. *The IQs of the people tested determined who would be the participants. The researcher has not, as in the preceding example, determined which participants would receive which value of the independent variable.* Rather, in selection it is the other way around; the value of the independent variable determines which participants will be used. It is thus apparent that in independent variable control by selection of values as they already exist in participants, *the participants are not randomly assigned to groups* — this is a critical shortcoming.

It is not really practical, however, to settle on precise IQ values, as above. What the researcher is more likely to do is to say that "I want a very high intelligence group, a medium group, and a very low intelligence group." The investigator might then make three visits and settle for whatever IQs are obtained — in this case the averages might be 127, 109, and 72, which would probably still accomplish this particular purpose.

In short, *purposive manipulation* occurs when the investigator determines the values of the independent variable, "creates" those values, and determines which group of participants will receive which value. *Selection* occurs when the investigator selects participants who already possess the desired values of the independent variable.

The distinction between both ways of controlling the independent variable is important. To understand this, focus on the intelligence problem-solving example. What would be the investigator's appropriate conclusion? (Before hypothesizing possibilities, be sure to recall the chapter on control.) Consider the *confounded* nature of this investigation. We have three groups of participants whom we know differ in intelligence. But in what other respects might they differ? The possibilities are so numerous that we shall only list three: socioeconomic status, the degree of stimulation of their environments, and motivation to solve problems. Hence, whatever the results on the problem-solving tests, the confounding of our independent variable with extraneous variables would be atrocious. We would not know to which variable, or combination of variables, to attribute possible differences in dependent variable scores. This is not so with an experiment like the light-conditioning example above. For in that case whatever the extraneous variables might be, they would be randomized out — distributed equally — over all groups.

When a stimulus variable is the independent variable, *purposive manipulation* is generally used. If the independent variable is either a response or organismic variable, however, *selection* is the more likely independent-variable control procedure. For example, with intelligence (or number of years of schooling, or chronic anxiety, etc.) as the independent variable, we have no practical alternative but to select participants with the desired values. It might be possible to manipulate purposively some of these variables, but the idea is impractical. It is admittedly difficult, say, to raise a person in such a way (manipulating the environment, administering various drugs, etc.) that the person will have an IQ of the desired value; we doubt that you would try to do this.

A number of studies have been conducted to determine whether or not there is a relationship between cigarette smoking and lung cancer. The paradigm, essentially, has been to compare people who do not smoke with those who do. The independent variable is thus the degree of smoking. Measures on the dependent variable are then taken — frequency of occurrence of lung cancer. The results have been generally decisive in that smokers more frequently acquire lung cancer than do nonsmokers. Nobody can argue with this statement. However, the additional statement is frequently made: *Therefore,* we may conclude that smoking causes lung cancer. On the basis of the type of evidence presented above, such a statement is unfounded, for the type of control of the independent variable that has been used is that of selection of values. Numerous additional variables may be confounded with the independent variable.

The only behavioral approach to determine the cause-effect relationship is to exercise control through purposive manipulation. That is, to select at random a group of subjects who have never been exposed to the smoking habit (e.g., children or isolated cultural groups), randomly divide them into two groups, and randomly determine which group will be smokers, which the abstainers. Of course, the experimenter must make sure that they adhere to these instructions over a long period. As members of the two groups acquire lung cancer, the accumulation of this evidence would decide the question. Unfortunately this experiment will probably never be conducted.[5] But the main point of this discussion should now be quite apparent: Confounding is very likely to occur when *selection* of independent variable values is used (the method of systematic observation) but can be prevented when purposive manipulation through experimentation is properly employed. To highlight this difference, we refer to "normal" groups of participants as *control* groups if we conduct an experiment; but in the method of systematic observation they are *comparison* groups.

In studies involving more than one independent variable, the values of one variable might be purposively manipulated and the values of the other selected as they naturally occur. In the first edition of this book (1960) such an investigation was referred to as a *quasi-experiment*, but since such "compromises" with true experiments (see chapter 12) has since been more fully developed by Campbell and Stanley (1963), another term should probably be used for these half-experiments, half-nonexperiments ("semi-experiment"?).

The Dependent Variable

Measures of
the Dependent Variable

Since in psychology we study behavior, and since the components of behavior are responses, response measures are the dependent variables in psychological experimentation. The class of response measures is extremely broad (comparable to "stimulus," as discussed earlier in this chapter). By "response measures" we mean to include such diverse phenomena as number of drops of saliva a dog secretes, number of errors a rat makes in a maze, time it takes a person to solve a problem, amplitude of *electromyograms* (electrical signals given off by muscles when they contract), number of words spoken in a given period of time, accuracy of throwing a baseball, and judgments of people about certain traits. But whatever the response, it is best to measure it as precisely as possible. In some experiments great precision can be achieved, and in others the characteristics of the events

[5] Two points might be added here: (1) The research cited does not say that there is no causal relationship, so the wise person, considering the mathematical expectancy, probably would not want to bet that smoking does *not* cause cancer; and (2) at times one can move from the findings gained from an R-R approach with humans like this one to an S-R approach with animals. Hence, one of the values of animal experimentation is that S-R and O-R experiments can be conducted that are not possible with humans, though there remain problems of generalizing from the animal to the human level (chapter 15).

dictate cruder measures. To enhance our understanding, we shall briefly list some standard ways of measuring responses.

1. Accuracy of the Response. Several ways of measuring accuracy are possible. For example, we might have a metrical system, such as when we fire a rifle at a target. Thus, a hit in the bullseye might be scored a five, in the next outer circle a three, and in the next circle a one. Another type of response measure of accuracy is to count the number of errors the participant makes. For example, the number of erroneous movements a person makes in putting a puzzle together, or the number of blind alleys a rat enters in running a maze.

2. Latency of the Response. This is a measure of the time that it takes the organism to begin the response, as in the case of reaction time studies. The experimenter may provide a signal to which the participant must respond. Then, measures are made of the time interval between the onset of the stimulus and the onset of the response. Or, in the case of a rat running a maze, the latency might be the time interval between the raising of the start box door and the time the rat's hind feet leave the box.

3. Speed (Duration) of the Response. This is a measure of how long it takes the organism to complete its response, once it has started. If the response is a simple one like pressing one of two telegraph keys, the time measure would be quite short. But if it is a complex response, such as solving a difficult problem or assembling a complicated device, the time measure would be long. A measure of the speed of the response in the case of a rat running a maze would be the length of time between leaving the start box, until the goal box is reached. To emphasize the distinction between latency and speed measures—latency is the time between the onset of the stimulus and the onset of the response, and speed or duration is the time between the onset and termination of the response.

4. Frequency of the Response. One might also measure the number of times a response occurs, as in the case of how many responses an organism makes before extinction sets in. If the frequency of responding is counted for a given period of time, the rate of responding can be computed. If a response is made ten times in one minute, the rate of responding is ten responses per minute. The rate gives an indication of the probability of the response—the higher the rate, the greater the probability that it will occur in the situation at some future time. Rate of responding is most often used in experiments involving operant conditioning, e.g., the organism is placed in a Skinner Box and each depression of a lever is automatically recorded on a moving strip of paper.

Additional measures of the response might be level of ability that a person can manifest (e.g., how many problems of increasing difficulty a participant solves with an *un*limited amount of time), or the intensity of a response (e.g., the amplitude of the galvanic skin response in a conditioning study). Frequently it is impossible to obtain an adequate measure of the dependent variable with any of these techniques. In this event, it might be possible to devise a rating scale. For example, a rating scale for anxiety might have five gradations: 5 meaning "extremely anxious," 4 "moderately anxious," and so on. Competent judges would then mark the appropriate position on the scale for each participant. Or the participants could even rate themselves.

Objective tests frequently serve as dependent variable measures. For example, you might wish to know whether or not psychotherapy decreases a person's neurotic tendencies, in which case you might administer a standard test for this purpose. If a suitable standard test is not available, it might be that you could construct your own, such as one student did in developing the "Dollenmayer Happiness Scale."

These are some of the more commonly used measures of dependent variables. By combining some of the above ideas and with your own ingenuity, you should be able to arrive at an appropriate measure of a dependent variable for the independent variable that you wish to study.

Selecting a Dependent Variable

The experimenter seeks to determine if an independent variable affects a dependent variable. But how does the experimenter determine what dependent variable to measure and record? Behavior is exceedingly complex, and at any given time an organism makes a fantastically large number of responses. Take Pavlov's simple conditioning experiment with dogs. His dependent variable (as it is most frequently cited) was amount of salivation. However, that is not the dog's only response when a conditional and an unconditional stimulus are presented. For in addition to salivating, the dog also breathes at a certain rate, wags its tail, moves its legs, pricks up its ears, and so on. Now, out of this mass of behavior, Pavlov had to select a particular response to measure and record. He might have studied some response other than salivation, a response that might or might not have also been related to his independent variable. Why did he choose salivation?[6]

Presumably every stimulus-independent variable leads to certain responses. The problem of selecting a dependent variable, then,

---

[6] See Skinner (1953, pp. 52–54) for a brief, interesting answer.

would seem simply to find all the responses that are influenced by a given stimulus-independent variable. But the problem is not quite that simple, and even this answer is not a simple one to follow in practice. Look at the matter in terms of our previous distinction between exploratory and confirmatory experiments. In the exploratory experiment, the experimenter asks: "I wonder what would happen if I did this?" Some response measures are selected to see if they are affected by the independent variable. It is impossible in any practical sense to study all of them; you simply pick and hope. An interesting example of this procedure is offered in the following quotation:

> ... The discovery that serotonin is present in the brain was perhaps the most curious turn of the wheel of fate. ... Several years ago Albert Hofman, a Swiss chemist, had an alarming experience. He had synthesized a new substance, and one afternoon he snuffed some of it up his nose. Thereupon he was assailed with increasingly bizarre feelings and finally with hallucinations. It was six hours before these sensations disappeared. As a reward for thus putting his nose into things, Hofman is credited with the discovery of lysergic acid diethylamide (LSD), which has proved a boon to psychiatrists because with it they can induce schizophrenic-like states at will. ... (Page, 1957, p. 55).

In this "pick and hope" procedure you can reach two possible conclusions from your data: (1) the independent variable did not affect the particular variable; or (2) it did. In the confirmatory experiment, on the other hand, you have a precise hypothesis that indicates the dependent variable in which you are interested; it specifies that a certain independent variable will influence a certain dependent variable. Your procedure is straightforward, at least in principle. You merely select a measure of the dependent variable in question and subject your hypothesis to empirical test.

Validity of the Dependent Variable

Since the dependent variable has already been specified by the hypothesis, you must be careful to obtain a proper measure of it. That is, you must be sure that the data you record are actually measures of the dependent variable *in which you are interested*. Suppose, for instance, that instead of measuring amount of salivation Pavlov measured the change in color of his dog's hair; the whole concept of conditioning would then have been delayed until the appearance of a shrewder investigator. This is a grotesque example, but more subtle errors of the same type are frequently made in psychological research. One experimenter wishing to study the effect of a certain independent variable on emotionality might select several judges to rate the apparent emotionality of the participants after the independent variable has been introduced. Whatever the results, one should ask the question: Did the judges actually rate on the basis of emotionality, or did they unknowingly rate them on some other characteristic? It may have been that the participants were actually rated on "general disposition to life," "intelligence," "personal attractiveness," or whatever. If this actually happened, we could not say that emotionality was really the dependent variable that was measured. This brings us to the first requirement that a dependent variable in a confirmatory experiment must meet; it must be *valid*. The data recorded must actually be measures of the characteristics that the experimenter seeks to measure. Hypotheses specify the dependent variable so we attempt to obtain measurements of the dependent variables as they are operationally defined. The question of validity is: Is the operationally defined dependent variable actually a measure of that which is specified in the consequent condition of the hypothesis?

"Now," you might say, recalling our discussion on operational definitions, "if the experimenter defined emotionality as what the judges reported, then that is by definition emotionality—you can't quarrel with that." And so we can't, at least on the grounds that you offered. We recognize that anyone can define anything any way that one wants. You can, if you wish, define the typical four-legged object with a flat surface on top from which we eat "a chair" if you like. Nobody would say that you can't. However, at the same time, we must ask you to look at a social criterion: Is that the name usually used by other people? Obviously the object is usually referred to as a "table." And if you insist on referring to what the rest of us call a "table" as a "chair," nobody should call you wrong, for this is your privilege. However, you will be at a distinct disadvantage when you try to communicate with other people. When you invite your dinner guests to be seated on a table and to eat their food from a chair some very quizzical responses will be evoked, at least.

So the lesson is this: Although you may define your dependent variable as you wish, it is wise to define your dependent variable as it is customarily used—at least, if it is customarily used. And if you are lucky enough to investigate a problem that has a certain widely accepted definition for the dependent variable involved, you should either use that dependent variable or one that correlates highly with it. Perhaps the clearest example of this point is when a psychometric test is used, one that has demonstrated validity of the appropriate kind and degree; such a standardized test would obviously be a satisfactory measure of the dependent variable operationally defined. For instance, one's dependent variable of anxiety may be defined operationally as scores on the manifest anxiety scale (Taylor, 1953). The only additional comment is that that might be but one kind of anxiety so that other operational definitions of anxiety might lead to different results.

Consider dependent variable validity by some additional examples. Suppose you are interested in determining the influence of a given independent variable on problem solving. You might define your dependent variable as the number of problems of a certain nature solved within a given period of time. At first glance, this *seems* a fine example of a valid dependent variable. And, if the test has a large number of problems and if these problems are arranged in ascending order of difficulty, then it probably *is* valid. But if the test is lengthy and the problems are all easy, you probably would not be measuring problem-solving ability but, rather, reading speed. That is, regardless of the fact that "problems" are contained in the test, those who read fast would get a high score and those who read slowly would get a low score. Clearly this would not be a valid measurement of problem-solving ability (unless, of course, problem-solving ability and reading speed are reliably correlated). Or, to make the matter even simpler, if you construct a very short test composed of extremely easy problems, all the participants will get a perfect score, unless you are working with feeble-minded individuals or the like. This test is not a valid measure of the dependent variable.

In the example where we were interested in whether rats could learn to run to a white or black goal box, the training procedure consisted of feeding them in a certain goal box (say a white one). The test consists of running the rats for a number of trials in a two-choice maze that contains one black and one white goal box. Let us say, for the purposes of making our point, that the white box is always on the right and the black box is always on the left. Assume that the preponderance of runs we record are to the white box. We conclude that the running response was successfully reinforced because the rats ran to the box of the color in which they were previously fed. Now, are we really measuring the extent to which they run to the *white* box? Rats have position habits; they frequently are either "left turners" or "right turners" (or they may alternate in certain patterns). If we have selected a group of rats that are all right turners, our measure may be simply of position habits, rather than of the dependent variable we are interested in. Hence, in this example, we are measuring "frequency of turning right" rather than "frequency of running to white." This example also clearly illustrates the relationship of validity to the major problem of experimental control, viz., avoiding confounding. For it is clear that frequency of running to white is confounded with the extraneous variable of position habit for turning to the right. Through the proper application of principles of experimentation such confounds are eliminated.

To determine the validity of the dependent variable, the experimenter might correlate dependent variable values with values obtained by the same participants on some other measure that is known to be valid. If the correlation is high, the measure is valid; if low or not significant, it is not valid. For example, suppose we have a valid measure of anxiety available, but that, for various reasons, we wish to use a different measure of anxiety as our dependent variable. To determine the validity of the measure that we wish to use, we could take both measures on a number of participants. By computing the correlation between the two sets of scores, we could reach a conclusion as to the validity of our measure. Unfortunately such a procedure is seldom feasible because we often do not have available a sufficiently valid measure of the desired dependent variable, and because different potential measures of the dependent variables often have low intercorrelations (see our later discussion of multiple measures dependent variables). Such a procedure is seldom practical in an experimental situation, primarily because we typically do not have a measure of the desired dependent variable that has known validity.

Experts in the field of testing have reexamined the classical concept of validity and recognize several different kinds of validity for their tests. Although problems of validation are still vexing, these specialists have made considerable progress, and some of their concepts and procedures (such as that of construct validity) may well be beneficially applied in experimental psychology. A thorough consideration of the details of validation should be undertaken in a course on testing. We have here merely to bring the topic of validation of the dependent variable to your attention, to point out some of the types of problems present, and to illustrate some of the kinds of pitfalls that await unwary experimenters. As a minimum, you are now aware of the existence of these problems and potential errors. On the basis of considerable reflection and on the basis of results of previous research in any given area, your chances of selecting a valid dependent variable measure should be increased.

Reliability of the Dependent Variable

The second requirement that a dependent variable should satisfy is that of *reliability*, in part the extent to which participants receive about the same scores when repeated measurements are taken. For example, an intelligence test may be considered sufficiently reliable if participants make approximately the same scores every time they take the test. Suppose a person received an IQ of 105 the first time a test is taken, 109 the second, and 102 the third. These scores are approximately the same. If most people behaved similarly, it may be said that the test is reliable for the population sampled. However, suppose that a typical individual scored 109, 138, and 82. Such a test could not be considered reliable, for the repeated measurements vary too much.

Considering an experimental situation, reliability of a dependent variable could be measured in the following manner. First, the experimenter could obtain measures on the dependent variable, preferably from individuals not involved in the experiment. After a period of time, the same participants could be retested on that same measure. The correlation between the two sets of measures would be computed. If the correlation is high, the dependent variable measure is reliable; otherwise it is not. Another approach would be to compute a split-half reliability coefficient. Suppose that the dependent variable is a measure that could be divided into two halves. For instance, each participant might be required to solve 20 problems, and the experimenter could therefore obtain a total score for the odd numbered and for the even numbered problems. It would then be a simple matter of computing a correlation coefficient between these two resulting scores for all the participants, a value we would hope would be quite high.

We said above that reliability is in part the extent to which people receive the same scores when repeated measurements are taken. To elaborate on this, let us note that experimenters frequently study participants' behavioral characteristics that change with the passage of time. They may study a learning process or the growth of situational anxiety. In such a case we need merely note that the correlation of successive scores may be quite high providing that the participants maintain the same relative scores: that is, providing that the rank order of scores is similar on each testing. For example, if three people made scores of 10, 9, and 6 on the first testing, but 15, 12, and 10 respectively on the second testing, the correlation (reliability) would be high since they maintained the same relative ranks. If, however, the first person's score changed from 10 to 11, the second from 9 to 12, and the third from 6 to 12, the reliability would be lower. In short, then, whether or not the measures of the dependent variable change with time, a correlation coefficient can be computed to determine the extent to which the dependent variable is reliable.

Unfortunately, most experimenters do not bother to even consider the reliability of their dependent variables. Our knowledge about reliability in experimentation would be increased if they did. At the same time we must point out that the determination of reliability is sometimes unrealistic. There are situations in which the dependent variable is more reliable than a computed correlation coefficient would indicate. For one thing, the people studied are frequently too homogeneous to allow the computed correlation value to approach the true value. For instance, if all participants in a learning study had precisely the same ability to learn the task presented them, then (ideally) they would all receive exactly the same dependent variable scores on successive testings; the computed correlation would not be indicative of the true reliability of the dependent variable. Another reason for a different computed correlation than the true one is that the scale used to measure the dependent variable may have insufficient range. To illustrate again by taking an extreme case, suppose that only one value of the dependent variable were possible. In this event all participants would receive that score, say 5.0, and the computed correlation would again be untrue. More realistically an experimenter might use a five point scale as a measure of a dependent variable, but the only difference between this and our absurd example is one of degree. The five-point scale might still be too restrictive in that it does not allow one to sufficiently differentiate among true scores; two individuals for instance might have true scores of 3.6 and 3.9, but on a scale of five values they would both receive scores of 4.0. Finally, dependent variable measures sometimes cannot be administered more than once because novelty is one of their critical features, as in studies of problem solving.

Recognizing that it is desirable for the experimenter to determine the reliability of the dependent variable and that it is frequently not feasible to approximate the true value by correlating successive scores, we may ask what the experimenter does. The answer is that we plan experiments and collect our data. If we find that the groups differ reliably, we may look back and reach some tentative conclusions about the reliability of the dependent variable. For if the groups truly differ reliably, this means that they differ to a greater extent than can be accounted for by experimental error. And if the means on the dependent variable have differed more than can be expected by random fluctuations, it must possess sufficient reliability, for lack of reliability makes for only random variation in values. On the other hand, if groups do not truly differ reliably, this means that the dependent variable values are probably due to random variation, to experimental error. The conclusion that would most frequently be reached in such a situation is that variation of the independent variable does not affect the dependent variable. But other reasons are also possible. It may be that the dependent variable is unreliable. So this approach to determining reliability is a one-way affair: If there are statistically reliable differences among groups, the dependent variable is probably reliable; if there are no significant differences, then no conclusion about its reliability is possible (at least on this information only). The repetition of the experiment a number of times with consistent reliable results would increase our belief in the reliability of the dependent variable, for if the same results are continually obtained, certainly the dependent variable is reliable.

The concepts of validity and reliability have been extensively used by test constructors. They have been almost totally ignored by experimenters, and yet their great importance to experimentation should be apparent. If one selects a totally unreliable dependent variable, then values regardless of experimental conditions would vary at random. With all dependent variable values varying in a chaotic manner, it is impossible to determine the effectiveness of the independent variables. If the dependent variable is reliable but not valid, an erroneous conclusion may be reached. If one performs a learning experiment and the dependent variable actually measures drive, then obviously any conclusions with regard to learning are baseless. Confounding here is a critical deficiency—if an uncontrolled extraneous variable affects the dependent variable, the measures obtained may be valid for the extraneous variable, but not for the independent variable. In conclusion, if the experimenter has not selected a valid and reliable criterion (dependent variable) the experiment may be regarded as worthless.

Multiple Dependent Variable Measures

A given independent variable may affect a number of measures of behavior, and in many experiments a number of measures actually are recorded. For example, an experiment in which rats are run through a maze might use all of the following measures of behavior: time that it takes to leave the starting box (latency), time that it takes to enter the goal box (running time or speed), number of errors made on each trial, and number of trials required to learn to run the maze with no errors. Such an experiment could be looked upon as one with four dependent variable measures, in which case the experimenter would merely conduct four statistical analyses; one might run, say, four separate t-tests to see if the groups differ on any of the four dependent variables.

Should this procedure be used, it would be valuable to obtain the correlations among the several dependent variables. You might find that two dependent variable measures correlate quite highly, e.g., .95.[7] In this case, you would know that they are measuring largely the same thing, and there would be little point in recording both measures in future experimentation. Hence, in later work you would select one or the other, probably the easiest to record or otherwise work with. We would emphasize, however, that the correlations between your dependent variables should be quite high before you eliminate any of them. The author once conducted an experiment in which three dependent variables were used. It was found that the correlation between the first and the second was .70, between the first and the third .60, and between the second and the third .80. Yet, in spite of these high and reliable correlations, the statistical analysis indicated that there was no difference between experimental conditions on two of the dependent variables, but that there was a reliable difference (P < .01) for the conditions on the third criterion.

In short, then, it is desirable to measure every dependent variable that might reasonably be affected by your independent variable. This statement is offered as an ideal that can only be approached, for in any actual experiment, it would not be feasible. But it may be valuable to find differences on one criterion though no differences were obtained on the other. In practical research, there is a special danger in using only one dependent variable measure. For instance, if you are testing the effectiveness of a method of learning for school children, all of the important ramifications of the method should be considered. If we only measured amount of time to acquire a skill, other perhaps more important considerations might be neglected, e.g., effectiveness of a method might become clear only after some considerable passage of time, such as by measuring amount of transfer from the learning situation to employing the skill later in everyday life. If you have definite information that indicates a high correlation among some of your dependent variables, you might choose among them.

We have previously considered the question of operational definitions within the context of multiple dependent variable measures and have also noted a criticism of the use of operational definitions—that they are often specific to the particular empirical investigation in which they are used. In experimentation the *precise* definitions of an independent and a dependent variable are frequently unique to any given experiment, so that there is lack of commonality in many of our variables

---

[7] See p. 207 for a discussion of correlation and an interpretation of this number. Incidentally, such a correlation should be computed separately for each group in the experiment, i.e., one should not combine all participants from all groups and compute a general correlation coefficient. For two groups, one would compute two correlation coefficients. In your future study be alerted to the difference between intra- and inter-class correlations.

among different experimenters. "Response strength," "anxiety," and other frequently used terms are typically operationally defined, though the operational definitions differ widely. Anxiety is measured by the Manifest Anxiety Scale (Taylor, 1953), by the Palmar Perspiration Index, etc., and these different measures correlate poorly with each other. (The MAS and the Palmar Perspiration Index correlate $-.01$ according to Calvin, McGuigan, Tyrrell, and Soyars, 1956). Such specificity no doubt impedes our progress, and it would be beneficial for psychology if our important terms were more general or, as the physicists put it, were more "fundamental." Fundamental definitions are general in two senses: they are universally accepted, and they encompass a wide variety of more specific concepts. The formulation of a fundamental definition of anxiety would encompass a number of existing specific definitions, with each specific definition being weighted according to how much of the fundamental definition it accounts for (which in turn is determined by the amount of extraneous variation which it is free from, and by its independence from other components or indicators.) For instance:

Fundamental concept of anxiety $= f(\text{anxiety}_1, \text{anxiety}_2, \ldots \text{anxiety}_N)$

In this case each component of the fundamental concept may be considered an indicator of the generalized concept of anxiety. In any given experiment it would be advisable for the researcher to measure as many of the indicators as possible and, by studying their interrelationships, to further contribute to the development of the fundamental concept.

Growth Measures

More often than not in psychology experimenters deal with variables that change with time. This is universally true with learning studies. For example, we may be interested in how a skill grows with repeated practice under two different methods. Frequently a statistical test is run on terminal data, i.e., data obtained on only the last trial. However, the learning curves of the two groups could provide considerable information about how the two methods led to their terminal points; participants using one method might have been "slower starters" but gained more rapidly at the end. And, in addition to providing such valuable information, you may want to run statistical tests that compare the two curves at specific points or even as whole units.[8]

[8] This is known as trend analysis; see, for example, Edwards (1968) or Lindquist (1953).

Another important question concerns the possible retention of experimental effects. For example, we might find that one method leads to better learning than a second method, but we might also be interested in whether that advantage is maintained over a period of time (with various kinds of intervening activity). Suppose that your task is to train mechanics in the performance of a highly technical job. The training you give them is to be followed by training on something else, so that it will be quite a while before they will actually use the training they received from you. In this case, you would not only be interested in which of several methods is more efficient for learning but also which method leads to the best retention. If you conducted an experiment, you might have the men return to you for another test just before they started their on-job duty. On the basis of this delayed test, you could decide which of the several methods would be best to use. Delayed measures are especially important in the evaluation of educational curricula, especially in colleges and universities. Evaluations of instructors too often change over the years. Although student opinionnaires are relatively poor evaluations of teachers (see McGuigan, 1974) they conceivably could be improved if taken some years after the students receive their college degree. It has been said that a student's opinion of a teacher at the end of a course, perhaps prior to the stressful consequences of taking a final examination, correlate negatively with that student's opinion of teacher effectiveness some years after graduation. Unfortunately, psychologists and other researchers seldom take delayed measures of their experimental effects, even when such a practice would be quite easy for them.

Review of Chapter 7

1. If you were to make an independent start in the study of the behavior of organisms, what kind of approach would you develop? As you ponder this question, you might begin to integrate the approach developed by psychologists, as detailed in this chapter and elsewhere in the book. You can note that psychologists have applied the concepts of variables to their subject matter such that they characterize relevant features of the organism (response and organismic or mediating variables) as they relate to the external environment (stimulus variables). In review you might detail for yourself relationships among these variables, and then return to the initial question above — what might be a better approach?

2. Contrast independent variable control by selection of values with independent variable control through purposive manipulation. In the latter specify the two ways in which randomization is used. Discuss

in this context the difference between an experiment and the method of systematic observation.

3. The quasi-experimental design known as the method of systematic observation always entails confounding! True or false? Be sure to defend your answer.

4. What is the importance in distinguishing between a control versus a comparison group?

5. Be sure that you automatically, without hesitation, can define "independent" and "dependent variables." Relate these concepts to that of a hypothesis.

6. Discuss quantification of the dependent variable. What are some of the standard measures?

7. Define "validity" and "reliability" and relate these concepts to that of the "dependent variable."

8. Discuss the advantages of multiple and delayed dependent variable measures. Are there any disadvantages?

9. What is a "fundamental concept"?

8

Experimental Design
the case of two matched groups

The type of design that we have considered up to this point is the two-groups design, which requires that participants be assigned to each group in a random fashion. The two-randomized-groups design is based on the assumption that the chance assignment will result in two essentially equal groups (which we usually determine by comparing their means). The extent to which this assumption is justified, we have also said, increases with the number of participants used.

The basic logic of all experimental design is the same: Start with groups that are essentially equal, administer the experimental treatment to one and not the other, and note the changes on the dependent variable. If the two groups had equivalent means on the dependent variable before the administration of the experimental treatment, and if a reliable difference between the means of the groups on the dependent variable results after the administration of the experimental treatment, and if extraneous variables have been adequately controlled, then that difference on the dependent variable may be attributed to the experimental treatment. The matched-groups design is simply one way of helping to satisfy the assumption that the groups have essentially equal dependent variable values prior to the administration of the experimental treatment.

A Simplified Example
of a Two-Matched-Groups Design

Let us say that we are interested in testing the hypothesis that both reading and reciting material lead to better retention than reading alone. We might form two groups of participants, one to learn the material by reading and reciting, the second to spend all their time reading. If we were using a randomized-groups design, we would assign participants to the two groups at random, regardless of what we might know about them. With the matched-groups design, however, we use scores on an initial measure called the matching variable to help assure equivalence of groups. Let us say that we use intelligence test scores as our matching variable. We might have scores for 10 participants such as those presented in Table 8-1.

Table 8-1   Scores of a Sample of Participants on a Matching Variable

Participant number	Intelligence Test Score
1	120
2	120
3	110
4	110
5	100
6	100
7	100
8	100
9	90
10	90

Our strategy is to construct the groups so that they are equal in intelligence. To accomplish this we need to pair the participants who have equal scores, assigning one member of each pair to each group. It is apparent that the following participants can be paired: 1 and 2, 3 and 4, 5 and 6, 7 and 8, and 9 and 10. The method we shall use for dividing these pairmates into two groups is randomization. This assignment by randomization is necessary in order to prevent possible experimenter biases for interfering with the matching. For example, the experimenter may, even though unaware of such actions, assign more highly motivated participants to one group in spite of each pair having the same intelligence score. By a flip of a coin we might determine that participant 1 goes in the reading and reciting group; number 2 goes in the reading group. The next coin flip might determine that participant 3 goes into the reading group and number 4 into the reading and reciting group. And so on for the remaining pairs (see Table 8-2).

Table 8-2 The Construction of Two Matched Groups on the Basis of Intelligence Scores

READING GROUP		READING AND RECITING GROUP	
Participant Number	*Intelligence Score*	*Participant Number*	*Intelligence Score*
2	120	1	120
3	110	4	110
6	100	5	100
7	100	8	100
10	90	9	90
	520		520

We may note that the sums (and therefore the means) of the intelligence scores of the two groups in Table 8-2 are equal. Let us now assume that the two groups are subjected to their respective experimental treatments and that we obtain the retention scores for them indicated in Table 8-3 (the higher the score, the better they retain the learning material). Note that we have placed the pairs in rank order according to their initial level of ability on the matching variable, i.e., the most intelligent pair is placed first, and the least intelligent pair is placed last.

Table 8-3 Dependent Variable Scores for the Pairs of Participants Ranked on the Basis of Matching Variable Scores

INITIAL LEVEL OF ABILITY	READING GROUP		READING AND RECITING GROUP	
	Participant Number	*Retention Score*	*Participant Number*	*Retention Score*
1	2	8	1	10
2	3	6	4	9
3	6	5	5	6
4	7	2	8	6
5	10	2	9	5

Statistical Analysis of a Two-Matched-Groups Design

The two groups of scores in Table 8-3 suggest that the group that both read and recited their material is superior to the reading-only group, but as before, we must ask, are they reliably superior? To answer this question we may apply the t-test, although the application will be a bit different for a matched-groups design. The equation is:

$$(8\text{-}1) \qquad t = \frac{\bar{X}_1 - \bar{X}_2}{\sqrt{\dfrac{\Sigma D^2 - \dfrac{(\Sigma D)^2}{n}}{n(n-1)}}}$$

The symbols are the same as those previously used, with the exception of D, which is the difference between the dependent variable scores for each pair of participants. To find D we subtract the retention score for the first member of a pair from the second. For example, the first pair consists of participants 2 and 1. Their scores were 8 and 10, respectively, and $D = 8 - 10 = -2$. Since we will later square the D scores (to obtain ΣD^2), it makes no difference which group's score is subtracted from which. We could just as easily have said: $D = 10 - 8 = 2$. The only caution to observe is that we need to be consistent, i.e., we must always subtract the reading group's score from the reading-reciting group's score, or vice versa. Completion of the D calculations is shown in Table 8-4.

Table 8-4 Computation of the Value of D for Equation 8-1

Initial Level of Ability	Reading Group	Reading and Reciting Group	D
1	8	10	-2
2	6	9	-3
3	5	6	-1
4	2	6	-4
5	2	5	-3

Equation (8-1) instructs us to perform three operations with respect to D: First, to obtain ΣD, the sum of the D scores, i.e.,

$$\Sigma D = (-2) + (-3) + (-1) + (-4) + (-3) = -13$$

Second, to obtain ΣD^2, the sum of the squares of D, i.e., to square each value of D and to sum these squares as follows:

$$\Sigma D^2 = (-2)^2 + (-3)^2 + (-1)^2 + (-4)^2 + (-3)^2$$

$$= 4 + 9 + 1 + 16 + 9 = 39$$

Third, to compute $(\Sigma D)^2$, which is the square of the sum of the D scores, i.e.,

$$(\Sigma D)^2 = (\Sigma D)\,(\Sigma D)$$

Recall that n is the number of participants in a group (not the *total* number of participants in the experiment). In a design where we match (pair) participants we may safely assume that the number of participants in each group is the same. In our example $n = 5$. The numerator is the difference between the (dependent variable) means of the two groups, as with the previous application of the *t*-test. The means of the two groups are 4.6 and 7.2. Substitution of all these values in Equation (8-1) results in the following:

$$t = \frac{7.2 - 4.6}{\sqrt{\dfrac{39 - \dfrac{(-13)^2}{5}}{5(5-1)}}} = 5.10$$

The equation for computing the degrees of freedom for the matched *t*-test is: $df = n - 1$. (Note that this is a different equation for df from that for the two-randomized-groups design). Hence, for our example, $df = 5 - 1 = 4$. Consulting our Table of t (p. 437), as before, with a t of 5.10 and 4 degrees of freedom we find that our t is reliable at the 1 percent level ($P < 0.01$). We may thus reject our null hypothesis (that there is no difference between the population means of the two groups) and conclude that the groups differ significantly. If these were real data we would note that the mean for the reading-reciting group is higher than that for the reading group and we would conclude that the former is significantly superior; the hypothesis would be confirmed.

Correlation and the Two-Matched-Groups Design

The Meaning of Correlation

To adequately understand this design we shall have to consider the topic of correlation. A correlation is a measure of the extent to which two variables are related. The measure of correlation that we shall be concerned with is symbolized by r. This symbol stands for the Pearson Product Moment Coefficient of Correlation. Since the value of r tells us the extent to which two variables are (linearly) related, it is a very valuable statistic that is used in a variety of ways. The value that r may assume varies between $+1.0$ and -1.0. A value of $+1.0$ indicates a perfect positive correlation and -1.0 indicates a perfect negative correlation. To illustrate this let us say that a group of people have been administered two different intelligence tests. Since both tests presumably measure the same thing, we may assume that the scores are highly correlated. They might be as indicated in Table 8-5.

Table 8-5 Fictitious Scores on Two Intelligence Tests Received by Each Person

Person Number	Score on Intelligence Test A	Score on Intelligence Test B
1	120	130
2	115	125
3	110	120
4	105	115
5	100	110
6	95	105

We may note that the individual who received the highest score on Test A also received the highest score on Test B. And so on down the list, person 6 receiving the lowest score on both tests. A computation of r for this very small sample would yield a value of +1.0. Hence, the scores on the two tests are perfectly correlated; notice that whoever is highest on one test is also highest on the other test, whoever is lowest on one is lowest on the other, and so on with *no exception being present.*[1] Now let us say that there are one or two exceptions in the ranking of test scores. Suppose that person 1 had the highest score on Test A but the third highest score on Test B; that number 3 had the third highest score on Test A but the highest score on Test B; and that all other relative positions remained the same. In this case the correlation would not be perfect (1.0) but would still be rather high (it would actually be .77).

Moving to the other extreme let us see what a perfect negative correlation would be, i.e., one where $r = -1.0$. We might administer two tests, one of democratic characteristics and one that measures amount of prejudice (see Table 8-6). The person who scores highest on the first test receives the lowest score on the second. This inverse relationship may be observed to hold for all participants without exception, resulting in a computed r of -1.0. Again if we had one or two exceptions in the inverse relationship the r would be something like $-.70$ indicating a negative relationship between the two variables, but one short of being perfect.

To summarize, given measures on two variables for each individ-

[1] Actually, another characteristic of the scores must also be present for this type of correlation to be perfect. That is that the interval between successive pairs of scores on one variable must be proportional to that for corresponding pairs on the other variable. In our example five IQ points separate each person on each test. However, this requirement is not crucial to the present discussion.

Table 8-6   Fictitious Scores on Two Personality Measures for Each Participant

Participant Number	Score on Test of Democratic Characteristics	Score on Test of Prejudice
1	50	10
2	45	15
3	40	20
4	35	25
5	30	30
6	25	35

ual a positive correlation exists if, as the value of one variable increases, the value of the other one also increases. If there is no exception the correlation will be high and possibly even perfect; if there are relatively few exceptions it will be positive but not perfect. Thus, as test scores on intelligence Test A increase, the scores on Test B also increase. On the other hand, if the value of one variable decreases while that of the other variable increases, a negative correlation exists. No exception indicates that the negative relation is high and possibly perfect. Hence as the extent to which people exhibit democratic characteristics increases the amount of their prejudice decreases; and this, of course, is what we could expect.

One final point concerning the value that r may assume. If $r = 0$ one may conclude that there is a total lack of (linear) relationship between the two measures. In other words as the value of one variable increases the value of the other variable varies in a random fashion. Examples of situations where we would expect r to be zero would be where we would correlate height of forehead with intelligence, or number of books that a person reads in a year with the length of toenails.[2] Additional examples of positive correlations would be the height and weight of a person, one's IQ and ability to learn, and one's grades in college and high school grades. We would expect to find negative correlations between the amount of heating fuel a family uses and the outside temperature, or the weight of a person and success as a jockey.

In science we seek to find relationships between variables. And a negative relationship (correlation) is just as important as a positive relationship. Do not think that a negative correlation is undesirable or that it indicates a lack of relationship. To illustrate, for a fixed sample, a correlation of $-.50$ indicates just as strong a relationship as a corre-

---

[2] However, it has been argued that this would actually be a positive correlation on the grounds that excessive book reading cuts into a person's toenail cutting time. Resolution of the argument must await relevant data.

lation of +.50, and a correlation of −.90 indicates a stronger relationship than does one of +.80.

The Computation of a Correlation Coefficient

The Pearson Product Moment Coefficient of Correlation, symbolized by r as we noted, is the most frequently used correlation coefficient.[3] We shall illustrate the computation of a different, but related, correlation coefficient—it is called The Spearman Rank Correlation Coefficient, which is symbolized by r_S; r_S has the advantage that it is quicker and easier to compute, and it can be conveniently computed without a calculator. By and large, what we have said for r is true for r_S—the only difference of note is that Spearman Rank Correlation Coefficient is slightly less powerful than the Pearson r. The equation for computing a Spearman-Correlation Coefficient is:

$$(8\text{-}2) \qquad r_S = 1 - \frac{6\Sigma d^2}{n^3 - n}$$

We shall illustrate the computation of r_S by using the values of Table 8-6. Equation (8-2) tells us that there are two basic values that we need: (1) d, which is the difference between the ranks of the two measures that we are correlating; and (2) n, which is the number of participants in the sample. The value of n is, obviously, six, so let us turn to d. To compute d, we need to rank order the scores for each variable separately, i.e., we assign a rank of one to the highest score for the first variable, a rank of two to the second highest score for that variable, and so on. Then we similarly rank the scores for the second variable. The ranks for the two variables of Table 8-6 are presented in Table 8-7.

We can note, for example, that participant 1 scored the highest on the test of democratic characteristics and the lowest on the test of prejudice. Participant 1 thus received ranks of 1 and 6 on these two tests. To compute d we subtract the second rank from the first, i.e., 1 − 6 = −5; "−5" is thus entered under the column labeled "d." And so on for the other differences in ranks for the remaining participants. The value of d is then squared in the final column, and the sum of

---

[3] An equation for computing r directly from raw data is:

$$r_{XY} = \frac{n\Sigma XY - (\Sigma X)(\Sigma Y)}{\sqrt{[n\Sigma X^2 - (\Sigma X)^2][n\Sigma Y^2 - (\Sigma Y)^2]}}$$

Where r_{XY} is the correlation between two variables X and Y and ΣXY is the sum of the cross products of the values of X and Y for each subject.

Table 8-7 Ranks of the Scores in Table 8-6 and the Computation of d^2

Participant Number	Rank on Test of Democratic Characteristics	Rank on Test of Prejudice	d	d^2
1	1	6	-5	25
2	2	5	-3	9
3	3	4	-1	1
4	4	3	1	1
5	5	2	3	9
6	6	1	5	25
				$\Sigma d^2 = 70$

the squares of d is entered at the bottom of the column, viz., $\Sigma d^2 = 70$. We are now ready to substitute these values into Equation (8-2) and to compute r_s:

$$r_S = 1 - \frac{6(70)}{6^3 - 6}$$

$$= 1 - \frac{420}{216 - 6}$$

$$= 1 - 2 = -1.00$$

And, as we already knew, these two arrays of scores are perfectly, though negatively, correlated. You are now in a position to quickly compute a Spearman Rank Correlation Coefficient between any two sets of scores that are of interest to you.

Selecting the Matching Variable

Recall that in matching participants we have attempted to equate our two groups with respect to their mean values on the dependent variable. In other words, we have selected some initial measure of ability by which to match the participants and have assigned them to two groups on the basis of these values so that the two groups are essentially equal on this measure. If the matching variable is highly correlated with the dependent variable scores, our matching has been successful.[4] For in this event we largely equate the groups on their dependent variable values by using the indirect measure of the matching variable. If the scores on the matching variable and the dependent

---

[4] Let us emphasize that r (and r_S) is a measure of the extent to which two variables (in our case the matching and the dependent variables) are linearly related. We are, thus, simplifying our discussion by considering only linear relationships between our two variables. Curvilinear relationships are excluded from the above discussion because our knowledge about the several possible correlations involved in Equation (8-4) (p. 221) is considerably limited.

variable do not correlate to a noticeable extent, however, then our matching is not successful. In short, the degree to which the matching variable values and the dependent variable values correlate is an indication of our success in matching.

How can we find a matching variable that correlates highly with our dependent variable? It might be possible to use the dependent variable itself. For example, if we are studying the process of learning to throw darts at a target, there might be two methods of throwing that we wish to evaluate. The design would call for one group of participants to use Method A, the other Method B. To assign participants to the groups by matching, we might first have all participants throw darts for five trials. We could use their scores on these five trials as the basis for pairing them off into two groups. They would then be trained by the two methods, respectively, and a later proficiency score computed. The *t*-test for matched groups would then be conducted on that later proficiency score. Our matching would be judged successful to the extent that the first set of scores correlated with the later set of scores. Since both sets of scores are obtained on the same task, we would expect the correlation to be rather high. Thus, it is clear *that the matching variable that is most likely to show a correlation with the dependent variable is that dependent variable itself.*

However, it should be apparent that this technique is not always feasible. Suppose that the dependent variable is a measure of rapidity in solving a problem. If practice on the problem is first given to obtain matching scores, then everyone would know the answer when it is administered later as a dependent variable. Or take another example where we create an artificial situation to see how people react under stress. Using the same situation to take initial measures for the purpose of matching participants would destroy the novelty of the situation after the independent variable is administered. In such situations we must find other measures that are highly correlated with dependent variable performance.

In the problem-solving example we might give the subjects a different, but similar, problem to solve and match on that. Or, if our dependent variable is a list of problems to solve, we might select half of that list to use as a matching variable and use the other half as a dependent variable. In the stress example it might be reasonable to assume that a psychophysiological measure of stress would be related to performance during stress. For example we might take a measure of how much the participants sweat under normal conditions and assume that those who normally sweat a lot are highly anxious individuals. Matching on such a test might be feasible.

A widely used matching variable in human learning studies is a measure of intelligence. The assumption is that the higher the intelligence the better the learning capacity. Intelligence test scores are quite easy to obtain or may perhaps already be on file in the case of college students.

Another general possibility is to match participants on more than one variable. In a learning experiment, we might match participants on initial learning scores and intelligence. Further consideration might suggest additional measures that could be combined with these two.[5]

Now we have said that if a matching variable does not correlate rather highly with the dependent variable, a matched-groups design should probably not be used. For this reason you should be rather certain that a high correlation exists between both variables. You might consult previous studies, for they may provide information on correlations between your dependent variable and various other variables. You could then make a selection from among those that correlate most highly. Of course, you should be as sure as possible that the same correlation holds for your participants with the specific techniques that you use.

You might also conduct a pilot study where you would make a number of measures on some participants, including your dependent variable measure. Selection of the most highly correlated measure with the dependent variable would afford a fairly good criterion, if it is sufficiently high. It if is too low, you should pursue other possibilities or consider abandoning the matched-groups design.

One procedural disadvantage of matching occurs in many cases. When using initial trials on a learning task as our matching variable, we need to bring the participants into the laboratory to obtain the data on which to match them. Then, after computations have been made and the matched groups formed, the participants must be brought back for the administration of the independent variable. The requirement that participants be present twice in the laboratory is sometimes troublesome. It is more convenient to use measures that are already available, such as intelligence tests scores or college board scores. It is also easier to administer group tests, such as intelligence or personality tests, which can be accomplished in the classroom. On the basis of such tests appropriate participants can be selected and assigned to groups before they enter the laboratory.

---

[5] However, it is frequently advisable to use a special technique for combining the various measures, discussion of which is probably not too fruitful here. For further information on one way to use more than one matching variable you are referred to Peters and Van Voorhis (1940).

The example of a matched-groups design and its statistical analysis that we previously used was constructed so that we could "breeze through" it in order to observe the general principles involved. There are, however, a number of details that prove somewhat troublesome when using this design, so let us illustrate it with a more realistic problem.

The data that we shall use were taken from a study in which a principle from S-R theory was subjected to test (McGuigan, Calvin, & Richardson, 1959). The aspect of the research with which we shall be concerned dealt with performance of participants at difficult and easy choice points in a stylus maze. The participants were blindfolded and required to learn a maze (for the hand) that contained 10 choice points. Their stylus, held in the hand, was placed at the starting point, and the participants progressed through the maze until they arrived at the end. Each time they took the wrong path at a choice point, an error was scored. The participants continued to practice the maze until they learned it perfectly. The five easiest and the five most difficult choice points had previously been determined, and the number of errors made by each participant for each of these two categories was counted. The principle from S-R theory that was tested made a prediction about the performance of the participants as a function of the difficulty of the choice points. More particularly, it said that the participants with high drive should not perform as well (should make more errors) at the difficult choice points as participants with low drive. The drive level of the participants was measured by administering the Taylor Manifest Anxiety Scale (MAS) (Taylor, 1953). The participants with the highest anxiety levels were the high-drive participants and those with the lowest anxiety levels were the low-drive participants. In short, two groups of participants were formed—high- and low-drive participants. And it was predicted that the high-drive participants would make more errors at the difficult choice points than would the low-drive participants. There was, however, one final qualification for selecting participants for the particular aspect of the research with which we shall be concerned—that the two groups of participants did not differ as far as learning ability was concerned. That is, it was predicted that differences in drive (anxiety) would lead to differences in performance, so to ascertain that any obtained differences in performance were due to variation of drive, learning ability should be held constant. Equalization of learning ability was accomplished by selecting pairs of participants in the high- and low-drive groups who made the same total number of errors in learning the maze.

With this general understanding of the rationale of the research, which was ingeniously developed from general S-R theory by Farber and Spence (1953), let us progress in a more specific manner through each step. First, it was necessary to measure drive level, so 56 participants were administered the Taylor Manifest Anxiety Scale. These participants then practiced the maze until they learned to progress through it with no errors, during which time the number of errors made at each choice point was tallied. To select the specific high- and low-drive participants used, we shall consider the 10 participants who had the highest MAS scores and the 10 participants who had the lowest. Table 8-8 presents the MAS scores and the total number of errors for these two classes of participants.

Table 8-8 Anxiety Scores and Total Numbers of Errors to Learn the Maze for High- and Low-Drive Participants

HIGH-DRIVE PARTICIPANTS			LOW-DRIVE PARTICIPANTS		
Participants Number	*MAS Score*	*Number of Errors*	*Participant Numbers*	*MAS Score*	*Number of Errors*
1	36	11	11	1	17
2	35	18	12	4	67
3	35	44	13	6	10
4	33	26	14	7	18
5	30	6	15	7	20
6	29	13	16	8	28
7	29	12	17	8	14
8	28	11	18	10	12
9	28	21	19	10	63
10	28	5	20	10	28

Now, having ascertained high- and low-drive groups, we next need to pair the participants from each group according to the total number of errors that they made. This task well illustrates why the present is a "more realistic" example than the previous one. To proceed, let us consider participant 1 who made 11 errors. Who in the low-drive group should this participant be paired with? None of the low-drive subjects made precisely this number of errors, but we can note that participant 13 made 10 errors and that participant 18 made 12 errors; either of these two participants would be satisfactory (though not perfect) as a pairmate. Participant 2 can be perfectly matched with participant 14, for they both made 18 errors. When we look at participant 3, who made 44 errors, we can find no reasonable pairmate and thus exclude that participant from further consideration. By further examining the participants in this manner, the original researchers finally arrived at five pairs of participants who were satisfactorily matched; there was no "mismatch" of more than one error. The remaining 10 participants could not be matched in a reasonable manner, and thus were not studied further. The resulting matched groups are presented in Table 8-9.

Table 8-9 High- and Low-Drive Groups Matched on Total Number of Errors*

HIGH-DRIVE PARTICIPANTS			LOW-DRIVE PARTICIPANTS		
No.	*MAS Score*	*Number of Errors*	*No.*	*MAS Score*	*Number of Errors*
9	28	21	15	7	20
2	35	18	14	7	18
6	29	13	17	8	14
7	29	12	18	10	12
1	36	11	13	6	10
	$X = 15.00$			$\bar{X} = 14.80$	
	$s = 4.30**$			$s = 4.15$	

*Pairs of participants are ranked according to number of errors.
**s stands for standard deviation and is discussed on page 218.

By excluding 10 participants we have been able to achieve a good matching between the two groups, as seen by comparing their means.[6] Incidentally, we may note one difference between this and the previous example as far as matching is concerned. In the previous example we paired participants and randomly determined which of each pair went in which group. In the present example, however, groups were formed on the basis of a personality characteristic; the MAS score determined to which group they were assigned. Either procedure is legitimate, and we simply have one more example of an experiment versus a systematic observation study.

Now, to turn to our empirical question: Did the high-drive group make more errors at the difficult choice points than did the

[6] But not without some cost. For by discarding participants we are possibly destroying the representatives of our sample. Hence the confidence that we can place in our generalization to our population is reduced (chapter 15). We might also add that we would be interested in comparing the groups on the basis of a measure of variability of the scores. In this case the groups would be rather well matched with regard to their variability as evidenced by the standard deviations of 4.30 and 4.15 for the high- and low-drive groups respectively. These considerations also point out a limitation of the matched-group design in that it is difficult or impossible to adequately match participants where there are a large number of experimental conditions.

low-drive group? To answer this question let us consider the number of errors made by each group at the easy and at the difficult choice points (Table 8-10). There we can see, for instance, that the high-drive participants who ranked highest in total number of errors made 10 errors at the easy choice points and 11 at the difficult choice points. The difference between the latter and the former is entered in the "Difference" column of Table 8-10. We can also see that the pair-mate for this participant made 6 errors at the easy choice points and 14 at the difficult choice points, the difference being 8 errors.

Table 8-10 Number of Errors Made at the Easy and Difficult Choice Points as a Function of Drive Level

	HIGH-DRIVE PARTICIPANTS			LOW-DRIVE PARTICIPANTS		
Level on	*Choice Point*			*Choice Point*		
Initial Measure	*Easy*	*Difficult*	*Difference*	*Easy*	*Difficult*	*Difference*
1	10	11	1	6	14	8
2	8	10	2	4	14	10
3	4	9	5	2	12	10
4	4	8	4	3	9	6
5	4	7	3	4	6	2

Think about these data for a minute. If the high-drive group made more errors at the difficult choice points than did the low-drive group, then the difference scores in Table 8-10 should be greater for the high-drive group. And, to test the prediction, they should be *significantly* greater. Consequently, we need to obtain the difference between these difference scores and to compute a matched *t*-test on them. We have entered the difference scores of Table 8-10 in Table 8-11 and computed the difference between these difference scores under the column labeled "*D*."

Table 8-11 Difference Between Number of Errors on the Easy and Difficult Choice Points as a Function of Drive Level

Level on Initial Measure	*Difference for High-Drive Participants*	*Difference for Low-Drive Participants*	*D*
1	1	8	-7
2	2	10	-8
3	5	10	-5
4	4	6	-2
5	3	2	1
	$\bar{X}_{HD} = 3.00$	$\bar{X}_{LD} = 7.20$	$\Sigma D = -21$
			$\Sigma D^2 = 143$

That is, the difference between the number of errors at the easy and difficult choice points for the top-ranked participant of the high-drive group was one, and for the pairmate it was eight. The difference between these two values is -7. And so on for the remaining pairs of participants. We now seek to test the scores under *"D"* to see if their mean is reliably different from zero. Equation (8-1) requires the following values, computed from Table 8-11:

$$\bar{X}_{HD} = 3.00$$

$$\bar{X}_{LD} = 7.20$$

$$\Sigma D = -21$$

$$\Sigma D^2 = 143$$

$$n = 5$$

Substituting these values into Equation (8-1):[7]

$$t = \frac{7.20 - 3.00}{\sqrt{\dfrac{143 - \dfrac{(-21)^2}{5}}{5\,(5 - 1)}}} = 2.53$$

Entering our table of t (p. 131 or p. 437 in the Appendix) with a value of 2.53 and 4 df, we can see that a t of 2.776 is required for significance at the .05 level. Hence, we cannot reject the null hypothesis and thus cannot assert that variation in drive level resulted in different performance at the difficult choice points. In fact, we can even observe that the direction of the means is counter to that of the prediction, i.e., the low-drive group actually made more errors than did the high-drive group. Speculations about why the prediction was not confirmed were offered by the original authors. The interested student is encouraged to visit the library to consult the articles dealt with in that publication, but our primary purpose here is accomplished by this realistic consideration of a matched groups design.

---

[7] Remember that we compute the absolute difference between the means in the numerator, so that it is easiest for us to place the largest mean first. We will then interpret the results according to which group has the highest mean. Incidentally, we might make use of a general principle of statistics in computing the numerator of the t-test for the matched-groups design: that the difference between the means is equal to the mean of the differences of the pair observations. Therefore, as a shortcut, instead of computing the means of the two groups and subtracting them, as we have done, we could divide the sum of the differences (ΣD) by n and obtain the same answer:

$$\frac{\Sigma D}{n} = \frac{21}{5} = 4.20.$$

Statistical Analysis with the A-Test

The statistical test that we have presented for analyzing the two-matched-groups design has been the t-test. However, several different statistical tests may be appropriate for analyzing the results obtained from any particular experimental design. The t-test has been used for the matched-groups design for several reasons: (1) because we used it for the randomized groups design, it was convenient to expand it for the present design; (2) because it is the test that is generally used; and (3) because we later want to consider several important characteristics of experimentation by referring to the generalized formula for t as it applies to the matched-groups design. There is, however, a computationally simpler test that can be used for this design—the A-test (Sandler, 1955).[8]

This statistic Equation (8-3) has been rigorously derived from the t ratio; hence the two tests always yield the same conclusions as far as computed probability for rejecting the null hypothesis is concerned.[9] The equation for computing A is:

$$8\text{-}3 \qquad\qquad A = \frac{\Sigma D^2}{(\Sigma D)^2}$$

To illustrate the computation of A, we need merely refer to the data presented in Table 8-11. There we found that $\Sigma D = -21$, and $\Sigma D^2 = 143$. Hence

$$A = \frac{143}{(-21)^2} = 0.32$$

We now need to determine the probability level for A. To do this we compute the degrees of freedom just as we did for Equation (8-1), i.e., $df = n - 1$. Since $n = 5$ in this example, $df = 4$. And entering Table 8-12 with four degrees of freedom we read across the rows. We note that as we move to the right, the value of A, unlike the value of t in the t table, decreases. That is, the *smaller* the value of A, the smaller the probability associated with it. In the t table the *larger* the value of t, the smaller the associated value of P. Hence, we find that with $df = 4$, a value of $A = 0.32$ has a P of less than 0.10 but greater than 0.05. And this is precisely what the results of the t-test told us. For further

---

[8] Appropriate for a two-tailed test and for the usual null hypothesis that $\mu_1 - \mu_2 = 0$. Do not, therefore, use this application of the A-test for a one-tailed test.

[9] The relationship between t and A is:

$$t^2 = \frac{n - 1}{An - 1}$$

practice with the A-test, you might check the results of the first experiment (Table 8-4) that we used as an example. (You should find that $A = 0.23$ and thus $P < 0.05$). The computational advantages of the A-test over, the t-test for the two-matched-groups design should be apparent. Not only is A quicker and easier to compute than t, and thus less conducive to error, but it is also more accurate since fewer rounding errors are involved. Certainly you should prefer the A-test in a two-matched-groups design.

Table 8-12 Table of A

For any given value of $n - 1$, the table shows the values of A corresponding to various levels of probability. A is significant at a given level if it is equal to or *less than* the value shown in the table.

			PROBABILITY			
$n - 1$	*0.10*	*0.05*	*0.02*	*0.01*	*0.001*	$n - 1$
1	0.5125	0.5031	0.50049	0.50012	0.5000012	1
2	0.412	0.369	0.347	0.340	0.334	2
3	0.385	0.324	0.286	0.272	0.254	3
4	0.376	0.304	0.257	0.238	0.211	4
5	0.372	0.293	0.240	0.218	0.184	5
6	0.370	0.286	0.230	0.205	0.167	6
7	0.369	0.281	0.222	0.196	0.155	7
8	0.368	0.278	0.217	0.190	0.146	8
9	0.368	0.276	0.213	0.185	0.139	9
10	0.368	0.274	0.210	0.181	0.134	10
11	0.368	0.273	0.207	0.178	0.130	11
12	0.368	0.271	0.205	0.176	0.126	12
13	0.368	0.270	0.204	0.174	0.124	13
14	0.368	0.270	0.202	0.172	0.121	14
15	0.368	0.269	0.201	0.170	0.119	15
16	0.368	0.268	0.200	0.169	0.117	16
17	0.368	0.268	0.199	0.168	0.116	17
18	0.368	0.267	0.198	0.167	0.114	18
19	0.368	0.267	0.197	0.166	0.113	19
20	0.368	0.266	0.197	0.165	0.112	20
21	0.368	0.266	0.196	0.165	0.111	21
22	0.368	0.266	0.196	0.164	0.110	22
23	0.368	0.266	0.195	0.163	0.109	23
24	0.368	0.265	0.195	0.163	0.108	24
25	0.368	0.265	0.194	0.162	0.108	25
26	0.368	0.265	0.194	0.162	0.107	26
27	0.368	0.265	0.193	0.161	0.107	27
28	0.368	0.265	0.193	0.161	0.106	28
29	0.368	0.264	0.193	0.161	0.106	29
30	0.368	0.264	0.193	0.160	0.105	30
40	0.368	0.263	0.191	0.158	0.102	40
60	0.369	0.262	0.189	0.155	0.099	60
120	0.369	0.261	0.187	0.153	0.095	120
∞	0.370	0.260	0.185	0.151	0.092	∞

Which Design to Use: Randomized Groups or Matched Groups?

Sometimes the results from a randomized-groups design seem unreasonable, and the experimenter wonders whether random assignment actually resulted in equivalent groups. The pretests provided with matching assures the advantage of the matched-groups design over the randomized-groups design that there was approximate equality of the two groups prior to the start of the experiment. That equality is not helpful to us, however, unless it is equality as far as measures of the dependent variable are concerned. Hence, if the matching variable is highly correlated with the dependent variable, then the equality of groups is beneficial. If not, then it is not beneficial — in fact, it can be detrimental. If a high correlation obtains, we should prefer a matched-groups design. But how high is high? We shall now offer some guiding considerations to help answer this question. To do this let us point out a general disadvantage of the matching design. We have said that the formula for computing degrees of freedom is $n - 1$. The formula for degrees of freedom with the randomized-groups design is $N - 2$. In other words when using the matched-groups design you have fewer degrees of freedom available than with the randomized-groups design, assuming equal numbers of participants in both designs. For instance, if there are seven participants in each group, $n = 7$, or $N = 14$. With the matched-groups design we would have $7 - 1 = 6$ degrees of freedom whereas for the randomized-groups design, we would have $14 - 2 = 12$. And we may recall that the greater the number of degrees of freedom available, the smaller the value of t required for statistical reliability, other things being equal. For this reason the matched-groups design suffers a disadvantage compared to the randomized-groups design.

It may happen that a given t would have been reliable with the randomized-groups design but not with the matched-groups design. Suppose we obtained the same value of t regardless of the design used. The t, we might say, is 2.05 obtained with 16 participants per group. With a matched-groups design we would have 15 df and find that a t of 2.131 is required for reliability at the 5 percent level — hence the t is not reliable; but with the 30 df available we would need a value of only 2.042 for reliability at the five percent level.

To summarize this point concerning the choice of a matched-groups or a randomized-groups design — the advantage of the former is that the value of t may be increased if there is a positive correlation between the matching variable and the dependent variable. On the other hand, one loses degrees of freedom when using the matched-groups design; half as many degrees of freedom are available with it as with the randomized-groups design. Therefore, if the correlation is going to be large enough to more than offset the loss of degrees of freedom, then one should use the matched-groups design.[10] If it is not, then the randomized-groups design should be used.[11] In short, if you are to use the matched-groups design you should be rather sure that the correlation between your matching and your dependent variable is rather high and positive.

At this point a bright student might say: "Look here, you have made so much about this correlation between the matching and the dependent variable, and I understand the problem. You say to try to find some previous evidence that a high correlation exists. But maybe this correlation doesn't hold up in your own experiment. I think I've got this thing licked. Let's match our participants on what we think is a good variable and then actually compute the correlation. If we find that the correlation is not sufficiently high, then let's forget that we matched participants and simply run a t-test for a randomized-groups design. If we do this, we can't lose; either the correlation is pretty high and we offset our loss of degrees of freedom using the matched-groups design or it is too low so we use a randomized-groups design and don't lose our degrees of freedom."

"This student," we might say, "is thinking, and that's good. But what he's thinking is wrong." An extended discussion of what is wrong with the thinking must be left to a course in statistics, but we can say that the error is similar to that previously referred to in setting the probability level for t as a criterion for rejecting the null hypothesis. There we said that the experimenter may set whatever level is desired, providing it is set before the conduct of the experiment. Analogously, the experimenter may select whatever design is desired, providing it is selected before the experiment is conducted. And in either case the decision must be adhered to. For where one chooses a matched-groups design there is also a mortgage to a certain type of statistical test (e.g., the matched t-test, which has a certain probability attached to its results). And if one changes the design the probability that he can assign to his t through the use of the t table is disturbed. Hence, if an experimenter decides to use a matched-groups design, that de-

---

[10] It might be observed that if the number of participants in a group is large (e.g., if $n = 30$), then one can afford to lose degrees of freedom by matching. That is, there is such a small difference between the value of t required for reliability at any given level with a large df that one would not lose much by matching participants even if the correlation between the independent and dependent variables is zero. Hence the loss of df consideration is only an argument against the matched groups design when n is relatively small.

[11] An elaboration of these statements is offered in the appendix.

cision must be adhered to. Perhaps the following experience might be consoling to you in case you ever find yourself in the unlikely situation described. The author once helped conduct an experiment in which a matched-groups design was used (McGuigan & MacCaslin, 1955b). Previous research had shown that the correlation between the variable that was used to match participants and the dependent variable was 0.72. This was an excellent opportunity to use a matched-groups design. However it turned out that the correlation was −0.24 for the data collected. And we shall see in the appendix what a negative correlation does to the value of t. Hence in the author's experiment not only were degrees of freedom lost, but (the value of the statistical test that would correspond to) the value of t was actually decreased. Another disadvantage of matching is that there is a statistical regression effect if the matching involves two different populations. The regression effect is a statistical artifact that occurs in repeated testings such that the value of the second test score regresses toward the mean of the population. This effect may suggest a change in dependent variable scores when in fact there is none.

In conclusion, the matched-groups design can be quite useful in selected situations, but its disadvantages can be sizable. Traditionally the matched-groups design was used relatively frequently, perhaps because of the intuitive security it gave the experimenter in knowing that one had equivalent groups. However, over the years it has become a less popularly used design and is quite possibly heading for a very remote position in the researchers' arsenal of experimental designs.

Appendix to Chapter 8

It is particularly advantageous to consider several additional matters that should facilitate an understanding of experimentation in general, and of the two-matched-groups design in particular. To do this we shall consider the *generalized* equation for the t-test, *generalized* in that it is applicable to either the randomized-groups or the matched-groups design. It may be written:[12]

[12] It is important not to forget that sample characteristics (statistics) are used as estimates of corresponding population characteristics (parameters). The (population) parameters are fixed, but unknown, while the (sample) statistics can be expected to vary from sample to sample. Since our emphasis is on how to compute statistics, we are using notations for sample statistics in our discussions. Your further work will lead to a greater appreciation of the distinction, which can be more clearly made by contrasting the notation for statistics and parameters, such as X and μ for the mean, s and σ for the standard deviation, and r and ρ for correlation.

(8-4)
$$t = \frac{\bar{X}_1 - \bar{X}_2}{\sqrt{\frac{s_1^2}{n_1} + \frac{s_2^2}{n_2} - 2(r_{12})\left(\frac{s_1}{\sqrt{n_1}}\right)\left(\frac{s_2}{\sqrt{n_2}}\right)}}$$

The two previous equations for t [Equations (5-2) and (8-1)] are derivatives of Equation (8-4). To understand Equation (8-4) we need to understand the following new symbols: s (the standard deviation) and s^2 (the variance), both of which are measures of variability, and r_{12}.

The Standard Deviation and Variance

Suppose someone asks us about the intelligence of the students at a given college. With 1,000 students in the college, 1,000 scores is a very cumbersome number. If we start reading the scores, before we could reach the thousandth score, however, our inquirer undoubtedly would have withdrawn his question. A much more reasonable procedure for telling one about the intelligence scores of our college students would be to resort to certain summary statements about them. We could, for instance, compute a mean and tell our inquirer that the mean intelligence of the student body is 125, or whatever. Although this would be accurate, it would not be adequate, for there is more to the story than that. Whenever we seek to describe a group of data we need to offer two kinds of statistics — *a measure of central tendency* and *a measure of variability*. Measures of central tendency tell us something about the central point value of a group of data. They are kinds of averages that tell us what the typical score in a distribution of data is. The most common measure of central tendency is the mean. Measures of variability tell us how the scores are spread out; they indicate something about the nature of the distribution of scores. In addition to telling us this, they also tell us about the range of scores in the group. The most frequently used measure of variability, probably because it is usually the most reliable of these measures (in the sense that it varies least from sample to sample), is the standard deviation. The standard deviation is symbolized by s.

To illustrate the importance of measures of variability we might imagine that our inquirer says to us: "Fine. You have told me the mean intelligence of your student body, but how homogeneous are your students? Do their scores tend to concentrate around the mean, or are there many that are considerably below the mean?" To answer this, we might resort to the computation of the standard deviation. The larger the standard deviation, the more variable are our scores. To illustrate, let us assume that we have collected the intelligence scores of students at two different colleges. Plotting the number of people who obtained each score at each college we might obtain the distributions shown in Figure 8-1.

By computing the standard deviation[13] for the two groups, we might find their values to be 20 for College A and 5 for College B. Comparing the distributions for the two colleges, we note that there is considerably more variability in College A than in College B. That is, the scores for College A are more spread out or scattered than for College B. And this is precisely what our standard deviation tells us; the larger the value for the standard deviation, the greater the variability of the distribution of scores. The standard deviation (for a normal distribution) also gives us the more precise bit of information that about two-thirds of the scores fall within the interval that is one standard deviation above and one standard deviation below the mean. To illustrate, let us first note that the mean intelligence of the students of the two colleges is the same, 125. If we subtract one standard deviation (i.e., 20) from the mean for College A and add one standard deviation to that mean, we obtain two values: 105 ($125 - 20 = 105$) and 145 ($125 + 20 = 145$). Therefore about two-thirds of the students in College A have an intelligence score between 105 and 145. Similarly, about two-thirds of the students at College B have scores between 120 ($125 - 5$) and 130 ($125 + 5$). Hence, we have a further illustration that the scores at College A are more spread out than at College B. We might for a moment speculate about these student bodies. College

---

[13] Note again that we are primarily concerned with values for samples. From the sample values the population values may be inferred. This is another case where we must limit our consideration of statistical matters to those that are immediately relevant to the conduct of experiments. But you are again advised to pursue these important topics by further work in statistics. An equation for computing the standard deviation for a sample is $s = \sqrt{\dfrac{n\Sigma X^2 - (\Sigma X)^2}{n(n-1)}}$. Or, if

you have already computed the SS, the equation is $s = \sqrt{\dfrac{SS}{(n-1)}}$. As Guilford

(1965) points out, the estimate of the population value for the standard deviation would be slightly different put when N is large (30 or greater) the two computed values are practically identical. You should be able to compute s on the basis of the knowledge that you now possess. To check yourself you might compute s for the dependent variable scores of the two groups in Table 8-3. You should find that for the reading group $s = 2.61$ and for the reading and reciting group $s = 2.17$.

Incidentally, with regard to the second equation above it is worth your while to examine Equation (5-2) on p. 124. There you can see that the denominator of the t ratio contains the components for computing s. In computing t you can, therefore, quickly compute s for your two groups. They are valuable to have in studying and reporting your data and, furthermore, the two values of s^2 can be easily compared to see if you have satisfied the assumption of homogeneity of variance (cf. p. 418).

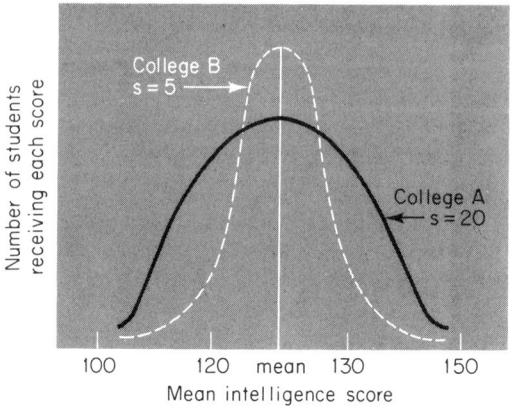

Figure 8-1. Distribution of intelligence scores at two colleges.

A, we might guess, is rather lenient in its selection of students, as might be the case in some state universities. College B is much more selective, having a rather homogeneous student body. Such a sample might occur for a private institution with high tuition costs. In any event we wish to make only one point here, that the larger the value of the standard deviation, the more variable (spread out) the scores.

The symbol s^2 is known as the *variance* of a set of values. It has essentially the same characteristics as the standard deviation and is merely the square of the standard deviation. Hence, if $s = 5$, then $s^2 = 25$. To illustrate these statistics further, let us assume that we have obtained the dependent variable scores for the two groups in an experiment shown in Table 8-13.

Table 8-13   Some Dependent Variable Scores for Two Groups of Participants

EXPERIMENTAL GROUP		CONTROL GROUP	
Participant	*Score*	*Participant*	*Score*
1	10	1	7
2	1	2	6
3	0	3	7
4	5	4	5
5	3	5	6
6	7	6	7
7	9	7	6
8	6	8	5
9	8	9	7
10	2	10	7

The scores for the experimental group vary from 0 to 10; the standard deviation here is 3.48. The scores for the control group, on the other hand, are much less variable. In fact, all the scores are either 5, 6, or 7. We should expect that the standard deviation for the control group is considerably smaller than for the experimental group. We shall not be disappointed, for its computation yields a value of .82. In turn, the variance is 12.10 for the experimental group and .68 for the control group. Using the variances as indices of variability, we can see that they also show that the variability of the experimental group is greater than that of the control group. Incidentally, we might note that if all the scores for one group were the same, say seven, both the standard deviation and the variance would be zero, for there would be zero variability among the scores.

The Nature of r_{12}

With this understanding of the standard deviation and variance, let us now turn to the symbol r_{12}, the last unfamiliar symbol in Equation (8-4). We have already discussed the general nature of a correlation, but it remains for us to specify this particular correlation. It stands for the (linear) correlation between the dependent variable scores of the two groups of participants in a matched-groups design. (Let us observe that r_{12} is read "the correlation between the dependent variable scores of Group 1 and Group 2" and not "r-twelve.") That is, in this type of design, we have paired participants in one group with participants in the other group. And these pairs are ranked on the basis of their matching variable values. Thus, the highest pair of values on the matching variable is ranked first, the second highest pair ranked second, and so on. If the dependent variable values of the first pair of participants are the highest scores in each group, if the second pair of subjects provided the second highest dependent variable values in their respective groups, and so on down the rank of pairs without exception, then the correlation between these two sets of dependent variable values would be perfect. That is, r_{12} would equal 1.0.[14] Similarly, if there are only a few exceptions in this order, r_{12} would be high but less than perfect. And so on for the other possibilities, as discussed in the previous section on correlation.

To cement our understanding of the nature of r_{12} refer to Table 8-3 and note the sets of dependent variable values for each of our two

---

[14] Again on the assumption that the increase in scores for each group is proportional (see footnote 1, p. 202). If they are not, then the correlation will be somewhat less than 1.0.

groups. If we correlated these two sets of scores, we would find that $r_{12} = 0.90$.[15] That is, the highest pair of participants on the matching variable has the highest set of dependent variable values in its group, the lowest pair of participants has the lowest dependent variable values, and so on.

The nature of r_{12} should now be clear. And you could compute it for any set of data that you want. But what is its significance? To answer this question, let us restate a principle that was stressed earlier. For the matched-groups design to be successful, you should have a reasonably high correlation between the matching variable and the dependent variable scores. Strictly speaking, this latter correlation does not enter into your statistical analysis, but is taken account of only indirectly. That is, rather than using an actual value for the matching variable-dependent variable correlation in our t-test, we use the value of r_{12}. And this is possible because the value of r_{12} is an indication of the value of the matching variable-dependent variable correlation.[16] That is, if the matching variable-dependent variable correlation is high, r_{12} will be high; and if the matching variable-dependent variable correlation is low, r_{12} will also be low. The reasonableness of these statements should be apparent after a little reflection. The only reason that we would expect a high value of the correlation between the paired dependent variable values of our two groups is that they were matched together on the basis of a variable that correlates with the dependent variable. Thus, the reason that the pair of participants who were ranked first on the matching variable should both exhibit the highest scores on the dependent variable is that there is a correlation between the matching variable and the dependent variable. On the other hand, if the correlation between the matching and the dependent variable scores is zero, we would expect the value of r_{12} to be zero, for there would be no reason to expect that the top-ranked pair of participants on the matching variable should both exhibit the highest dependent variable scores.

The situation at this point is that r_{12} is the correlation between the dependent variable scores of pairs of participants in a matched-groups design. Since r_{12} is an indication of the value of the correlation between the matching and the dependent variable scores, we use it in conducting our statistical analysis.

Let us now take a broader look at Equation (8-4). If we want to

[15] Here is a good example of what we see in footnote 1, i.e., although there is no exception in the rankings of the two sets of values, the intervals between the values in each set are not proportional, and hence the value of r only approaches 1.0.

[16] Again emphasizing that we are considering only linear relationships, ignoring curvilinearity.

increase our chances of obtaining a value of t that would allow us to reject the null hypothesis, there are two courses of action that can be followed. First, we can attempt to increase the value of the numerator (the difference between the means of the two groups) or to decrease the value of the denominator. As the value of the numerator increases, or as the value of the denominator decreases, the value of t increases. And we know that the larger t is, the more likely it is to indicate a statistically reliable mean difference between groups. To illustrate, let us say that the numerator is 5 and the denominator is 10. In this case t is:

$$t = \frac{5}{10} = 0.50$$

But if we are able to decrease the value of the denominator to 2, with no change in the numerator, we would have:

$$t = \frac{5}{2} = 2.50$$

And a t of 2.50 is likely to be reliable, whereas one of 0.50 is not.

Our question should now be how we can decrease the value of the denominator of Equation (8-4). This matter is discussed more thoroughly in chapter 16, but let us consider one possible way here. We may note that the larger the variance of the two groups, the larger is the denominator. If s_1^2 and s_2^2 are each 10, the denominator will be larger than if they are both 5. But we may note that from the variances we subtract r_{12} (and also s_1 and s_2, but these need not concern us here). And any subtraction from the variances of the two groups will result in a smaller denominator with, as we said, an attendant increase in t. Furthermore, we said, the size of r_{12} depends on the correlation between the matching variable and the dependent variable. Hence, if that correlation is large and positive, we may note that the denominator is decreased.

By way of illustration, assume that the difference between the means of the two groups is 5 and that there are 9 participants in each group (n_1 and n_2 both equal 9). Further, assume that s_1 and s_2 are both 3 (hence s_1^2 and s_2^2 are both 9) and that r_{12} is 0.70. Substituting these values in Equation (8-4) we obtain:

$$t = \frac{5}{\sqrt{\frac{9}{9} + \frac{9}{9} - 2(0.70)\left(\frac{3}{\sqrt{9}}\right)\left(\frac{3}{\sqrt{9}}\right)}} = 6.49$$

It should now be apparent that the larger the positive value of r_{12}, the larger is the term that is subtracted from the variances of the two groups. In an extreme case of the above illustration, where $r_{12} = 1.0$, we may note that we would subtract 2.00 from the sum of the variances (2.00); this leaves a denominator of zero, in which case t might be considered to be infinitely large. On the other hand, suppose that r_{12} is rather small—say it is 0.10. In this case we would merely subtract 0.20 from 2.00, and the denominator would be only slightly reduced. Or if $r_{12} = 0$, then it can be seen that zero would be subtracted from the variances, not reducing them at all. The lesson should now be clear: *the larger the value of r_{12} (and hence the larger the value of the correlation between the matching variable and the dependent variable), the larger the value of t.*

Now let us consider the heart of the matter: How large should be the correlation between the matching and the dependent variable in order to prefer a matched-groups design to a randomized-groups design? Well, we can't answer this question, at least not in this form. But we can give a very good answer by changing it somewhat. This is, in fact, the principal reason why we have stressed the nature of r_{12} and its relationship with the correlation between the matching and the dependent variable. Thus, we shall answer the question: How large should r_{12} be before a matched-groups design should be preferred? And the answer is given in Figure 8-2.[17] To use Figure 8-2 you merely select the number of subjects in each group *(n)* and enter the figure at that value on the horizontal axis. Then read up until you reach the curve. The value of the curve at that point on the vertical axis indicates the necessary minimum value of r_{12} in order for a matched-groups design to be preferred. For instance, if you want to know the minimal value of r_{12} that you will need in the event that you have 12 participants in each group, you would find "12" on the horizontal axis. Then reading up to the curve and over to the vertical axis you find that r_{12} is about 0.17. *In order to prefer a matched-groups design,* then, you should have a minimal value of 0.17 for r_{12}.

One final consideration of the value of r_{12} is what the effect of a negative correlation would be on the value of t. A little reflection should reveal that a negative correlation *increases* the denominator, thus decreasing t. In this case, instead of subtracting from the variances, we would have to add to them ("a minus times a minus gives us a plus"). Furthermore, the larger the negative correlation, the larger our denominator becomes. For example, suppose that in the previous example instead of having a value of $r_{12} = 0.70$, we had $r_{12} = -0.70$. In this case we can see that our computed value of t would decrease from 6.49 to 2.72. That is,

---

[17] We are simplifying the situation by presenting a rather conservative case. It can rather safely be said that if the value of your r_{12} is at least that indicated for a certain number of subjects in Figure 8-2, then you should prefer a matched-groups design.

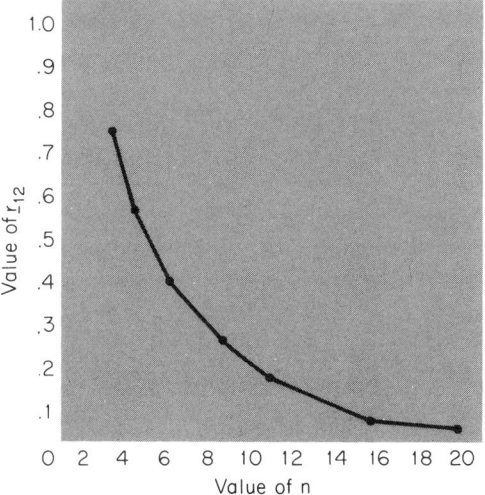

Figure 8-2. A relationship between n and r_{12}. Enter with a value of n and read the expected value of r_{12} that intersects the curve at that point. If your expected value of r_{12} exceeds the value obtained from the curve, a matched-groups design is to be preferred.

$$t = \frac{5}{\sqrt{\dfrac{9}{9} + \dfrac{9}{9} - 2(-0.70)\left(\dfrac{3}{\sqrt{9}}\right)\left(\dfrac{3}{\sqrt{9}}\right)}} = 2.72$$

We previously said that Equation (8-4) is a generalized formula, applicable to either of the two designs that we have discussed. One might ask, however, in what way it is applicable to the randomized-groups design, for it contains a correlation term and we have not referred to any correlation when using it; it is absurd, for instance, to talk about the correlation between pairs of participants on the dependent variable when using the randomized-groups design, for by its very nature participants are not paired. The answer to this is that since participants have not been paired, the correlation between any random pairing of participants in the long run is zero. That is, if we randomly selected any participant in an experimental group, paired that value with a randomly selected participant in the control group, and continued this procedure for all participants, we would expect the correlation between the dependent variable values to be zero (or more precisely, the correlation would not be reliably different from zero). There simply would be no reason to expect other than a zero correlation since the participants were not paired together on more than a chance basis. When using the randomized-groups design, we assume that r_{12} of Equation (8-4) is zero. And being zero, the term that includes r_{12} "drops out." Thus, Equation (8-4) assumes the following form for the randomized-groups design:

$$(8\text{-}5) \qquad t = \frac{\bar{X}_1 - \bar{X}_2}{\sqrt{\dfrac{s_1^2}{n_1} + \dfrac{s_2^2}{n_2}}}$$

One final note: We have labeled the type of design discussed in this chapter as the *matched-groups design* where we have limited our discussion to the case of two groups. The two groups may be said to be matched because we paired participants with similar scores. Since all participants were paired together, the groups had to be approximately equivalent. This fact may be determined by comparing the distribution of matching scores for the two groups. The best such comparison would probably be to compare the means and standard deviations of the two groups. We would expect to find that the two groups would be quite similar on these two measures. However, the same result could also be achieved in other ways. That is, two groups could be formed in other ways so that they would have similar means and standard deviations. For example, we could simply assign participants to two groups so that their total scores would be similar; no participants would be paired together, but the means and standard deviations of the two groups would be approximately the same. Therefore, it may be considered that the technique of pairing participants together is a specific type of design that results in matched groups. For this reason it could as well be called the *paired-groups design* to distinguish it from alternative procedures that result in matched groups. Alternative procedures, however, require different statistical analyses from those presented here (McNemar, 1962). Since they are not so generally used, nor judged to be as effective, they will not be considered further.

The two-matched-groups design (or, if you prefer, the two-paired-groups design) implies that the design could be extended to more than two groups. For a discussion of a matched-groups design for more than two groups you are referred to Edwards (1968).

Summary of the Computation of t for a Two-Matched-Groups Design

Assume that two groups of participants have been matched on an initial measure as indicated, and that the following dependent variable scores have been obtained for them.

Initial Measure	Group 1	Group 2
1	10	11
2	10	8
3	8	6
4	7	7
5	7	6
6	6	5
7	4	3

1. The equation for computing t, Equation (8-1), is:

$$t = \frac{\bar{X}_1 - \bar{X}_2}{\sqrt{\dfrac{\Sigma D^2 - \dfrac{(\Sigma D)^2}{n}}{n(n-1)}}}$$

2. Compute the value of D for each pair of participants, and then the sum of D (ΣD), the sum of the squares of D (ΣD^2), the sum of D squared [$(\Sigma D)^2$], and n.

Initial Measure	Group 1	Group 2	D
1	10	11	−1
2	10	8	2
3	8	6	2
4	7	7	0
5	7	6	1
6	6	5	1
7	4	3	1

$$\Sigma D = 6$$
$$\Sigma D^2 = 12$$
$$n = 7$$

3. Determine the difference between the means. This may be done by computing the mean of the differences. Since the latter is easier, we shall do this.

$$\text{Mean of the differences} = \frac{\Sigma D}{n} = \frac{6}{7} = 0.86$$

4. Substitute the above values in Equation (8-1):

$$t = \frac{0.86}{\sqrt{\dfrac{12 - \dfrac{(6)^2}{7}}{7(7-1)}}}$$

5. Perform the operations as indicated and determine the value of t.

$$t = \frac{0.86}{\sqrt{0.16}} = 2.15$$

6. Determine the number of degrees of freedom associated with the computed value of t.

$$df = n - 1 = 7 - 1 = 6$$

7. Enter the table of t with the computed values of t and df. Determine the probability associated with this value of t. In this example, $0.1 > P > 0.05$. Therefore, assuming a criterion of 0.05, the null hypothesis is not rejected.

Problems

1. A psychologist seeks to test the hypothesis that the western grip for holding a tennis racket is superior to the eastern grip. Participants are then matched on the basis of a physical fitness test; they are then trained in the use of these two grips, respectively, and the following scores on their tennis playing proficiency are obtained. Assuming adequate controls, that a 0.05 level for rejecting the null hypothesis is set, and that the higher the score the better the performance, what can be concluded with respect to the empirical hypothesis?

Rank on Matching Variables	Score on Dependent Variable	
	Eastern Grip Group	Western Grip Group
1	10	2
2	5	8
3	9	3
4	5	1
5	0	3
6	8	1
7	7	0
8	9	1

2. To test the hypothesis that the higher the induced anxiety, the better the learning, an experimenter formed two groups of participants by matching them on an initial measure of anxiety. Next, considerable anxiety was induced into the experimental group, but not into the control group. The following scores on a learning task were then obtained, where the higher the score, the better the learning. Assum-

ing adequate controls were exercised and that a criterion of 0.05 was set, was the hypothesis confirmed?

Rank on Matching Variable	Dependent Variable Scores Experimental Group	Control Group
1	8	6
2	8	7
3	7	4
4	6	5
5	5	3
6	3	1
7	1	2

3. A military psychologist wishes to evaluate a training aid that was designed to facilitate the teaching of soldiers to read a map. Two groups of participants were formed, matching them on the basis of a visual perception test (an ability that is important in the reading of maps). A criterion of 0.02 for rejecting the null hypothesis was set and proper controls were exercised. Assuming that the higher the score, the better the performance, did the training aid facilitate map reading proficiency?

Rank on Matching Variable	Scores of Group That Used the Training Aid	Scores of Group That Did Not Use the Training Aid
1	30	24
2	30	28
3	28	26
4	29	30
5	26	20
6	22	19
7	25	22
8	20	19
9	18	14
10	16	12
11	15	13
12	14	10
13	14	11
14	13	13
15	10	6
16	10	7
17	9	5
18	9	9
19	10	6
20	8	3

9

Experimental Design
the case of more than two randomized groups

Concerning the Value
of Using More Than Two Groups

Among the more frequently used designs in psychological research are those that employ more than two groups. Suppose a psychologist had two methods of remedial reading available. The methods are both presumably helpful to students who have not adequately learned to read by the usual method, but which method is superior? Furthermore, is either of these methods actually superior to the normal method for such problem cases? To answer these questions, one might design an experiment that involves three groups of participants.

If 60 students who show marked reading deficiencies are available, the first step would be randomly to assign them to three groups. Assume that an equal number of participants is assigned to each group, although this need not be the case. The first group would be taught to read by using Method A and the second group by Method B. A comparison of the results from these two groups would tell which, if either, is the superior method. One also would want to know if either method is superior to the normal method of teaching, which has heretofore been ineffective with this group. So, the third group would continue training under the normal method, as a control group. After a certain period of time, perhaps nine months, a standard reading test might be administered to the three groups. A comparison of the reading proficiency of the three groups on this test should answer the questions.

It is also possible to answer these questions by conducting a series of separate two-groups experiments. It would be possible, for instance, to conduct one experiment in which Method A is compared to Method B, a second in which Method A is compared to the control condition, and a third experiment in which Method B is compared to the control condition. Such a procedure is obviously less desirable, for not only would more work be required but the problem of controlling extraneous variables would be sizable. For example, we would wish to hold the experimenter variable constant, so the same experimenter should conduct all three experiments. Even so, it is likely that the experimenter would behave differently in the first and last experiments, perhaps due to improvement in teaching proficiency, or even because of boredom or fatigue. Therefore, the design in which three groups are used simultaneously is superior in that less work is required, fewer participants are used, and experimental control is better.

The randomized-groups design for the case of more than two groups may be applied to a wide variety of problems. Some problems that are amenable to this type of design would be the influence of different amounts of drive upon learning; the influence of number of reinforcements upon conditioning; or the influence of various kinds of interpolated activities upon learning.

The procedure for applying a multigroup design (i.e., a design with more than two groups) to any of the above problems would be to select several values of the independent variable and randomly assign a group of participants to each value. For example, to study the influence of different amounts of drive (perhaps defined as length of food deprivation) upon performance, we might choose the following values of the independent variable: 0 hours, 1 hour, 12 hours, 24 hours, 36 hours, and 48 hours of deprivation. Having selected six values of the independent variable, we would have six different groups of participants, probably animals. To study the influence of different periods of practice upon learning we might select four values of the independent variable: 0, 5, 10, and 15 trials. We would then randomly assign our participants to four groups and train one group under each condition.

These considerations now make apparent yet another advantage of a multigroups over a two-groups design. That is that if you attempted to attack any of the above problems by means of a two-groups design, you would have to decide which two of many values of the independent variable to employ. A sizable advantage of a multigroups design becomes apparent when one tries to decide which two values of the independent variables to use. Consider the above example concerning the influence of different periods of practice upon performance. We previously selected four values of the independent variable to study. Which two would we use for a two-groups design? It would be advisable to have a control condition, so we would probably choose a zero value for one group. The second group might be trained under a five-trial condition.

Now, let us imagine that the four-groups design yields the following results: no difference in performance among the 0-, 5-, and 10-trial conditions, but the 15-trial condition is superior to the first three. The conclusion from this four-group experiment would be that variation of the length of practice from 0 to 10 trials does not affect performance; however, greater periods of practice increase proficiency. But if the two-groups design (using only 0 and 5 trials for practice) were applied to this problem, the results would *suggest* that variation of the length of practice does not effect performance, a conclusion that would be in error. Thus, in general, it should be apparent that the more values of the independent variable sampled, the better our evaluation of its influence on a given dependent variable.

Research in any given area usually progresses through two stages: first, we seek to determine which of many possible independent variables influences a given dependent variable; and second, when a certain independent variable has been identified as influential on a dependent variable, we attempt to establish the precise relationship between them. Even though it is possible that a two-groups design would accomplish the first purpose, it could not accomplish the second. For an adequate relationship cannot usually be established with only two values of the independent variable (and therefore also only two values of the dependent variable). To illustrate this point refer to Figure 9-1, where the values of an independent variable are indicated on the horizontal axis, and the corresponding values of the dependent variable are read on the vertical axis.[1] The two plotted points (obtained from a two-groups design) indicate that as the value of the independent variable increases, the mean value of the depen-

---

[1] The range of the independent variable in the following discussions should be clear from the context, e.g., from zero to infinity. We shall also assume that the data points are highly reliable and thus not the product of random variation.

dent variable also increases. But this is a crude picture, for it tells us nothing about what happens between (or beyond) the two plotted points. See Figure 9-2 to illustrate a few of the infinite number of possibilities.

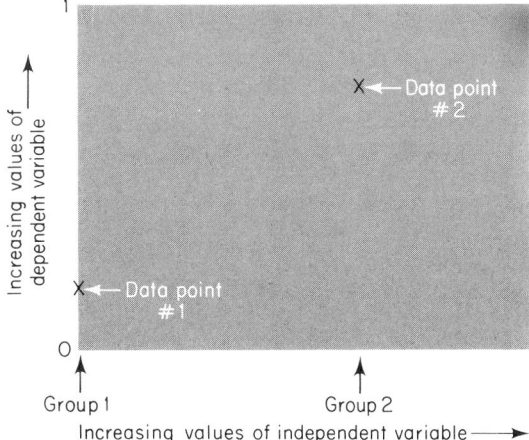

Figure 9-1. Two data points obtained from a two-groups design. Group 1 was given a zero value of the independent variable while Group 2 was given a positive value. The value of the dependent variable is less for Group 1 (data point #1) than for Group 2 (data point #2).

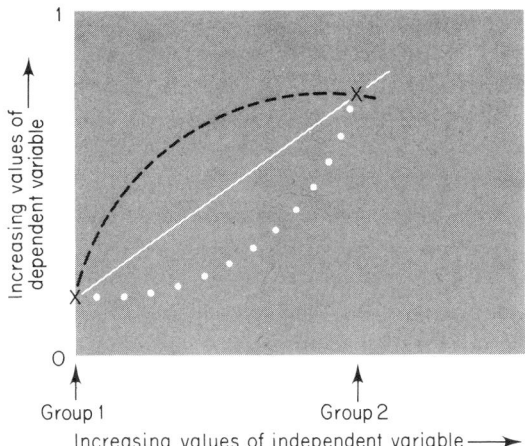

Figure 9-2. The actual relationship between the independent and the dependent variable is partially established by the two data points. However, the curves that may pass through the two points are infinite in number. Three possible relationships are shown.

By using a three-groups design the relationship may be established more precisely. Let us say that we have the same two groups as in Figure 9-1, but in addition we have a third group that received a value of the independent variable halfway between those of the other two groups. Assuming that the mean dependent variable value for Group 3 is that depicted in Figure 9-3, we would conclude that the relationship is probably a linear (straight-line) function. Of course, we might be wrong. That is, the relationship is not necessarily the

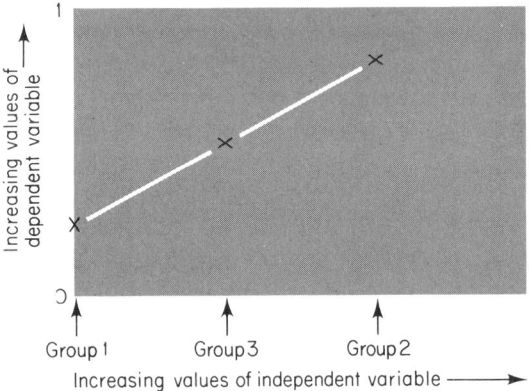

Figure 9-3. The addition of a third data point (Group 3) suggests that the relationship is a linear function.

straight line indicated in Figure 9-3, for it is possible that some other relationship is actually the "true" one, such as one of those shown in Figure 9-4. Nevertheless, with only three data points we prefer to bet that the straight line is the "true" relationship because it is the simplest of the several possible relationships. And experience suggests that it is reasonable to assume that the simplest curve yields the best predictions. That is, we would predict that if we obtain a data point for a new value of the independent variable (in addition to the three already indicated in Figure 9-3) the new data point would fall on the straight line. Different predictions would be made from the other curves of Figure 9-4.

To illustrate, suppose that we add a fourth group whose independent variable value is halfway between those of Groups 1 and 3. On the basis of the four relationships depicted in Figure 9-4 we could make four different predictions about the dependent variable value of this fourth group. First, using the straight-line function, we would predict that the data point for the fourth group would be that indicated by X_1 in Figure 9-5, i.e., if the straight line is the "true" relationship, the data point for the fourth group should fall on that line.

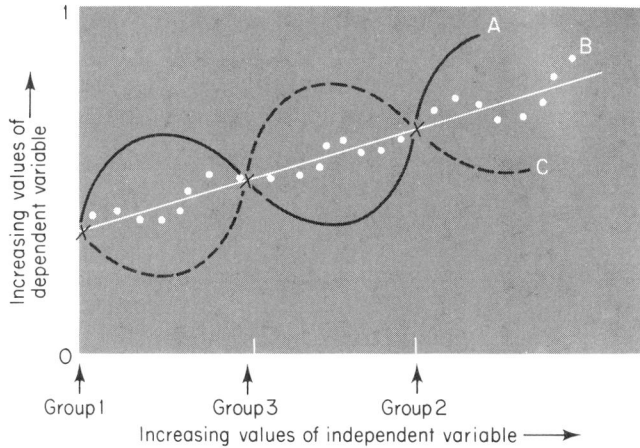

Figure 9-4. Other curves may possibly pass through the three data points.

The three curves of Figure 9-4, however, lead to three additional (and different) predictions.

Assume that when the data for the fourth group are analyzed, they indicate that the mean value is actually that indicated by the X_1 of Figure 9-5. This increases our confidence in the straight-line function; it, rather than the other possible functions, is probably the "true" one. If these were actually the results, our procedure of preferring

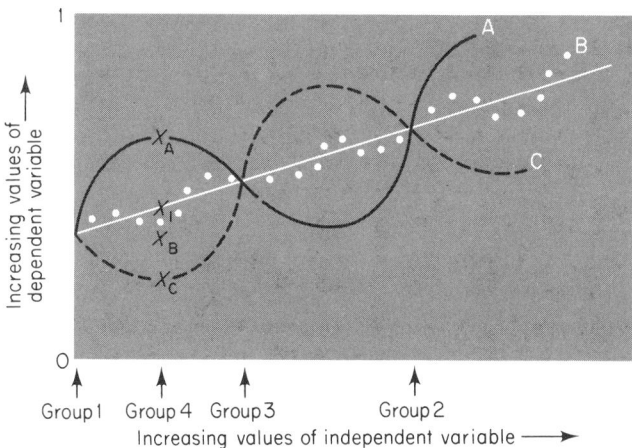

Figure 9-5. Four predictions of a data point for Group 4. From the straight line of Figure 9-3, we would predict that the dependent variable value would be indicated by X_1. From the three curves of Figure 9-4 (curves A, B, and C), we would predict that the data point would be that indicated by the X_A, the X_B, and the X_C respectively.

the simplest curve as the "true" one (at least until contrary results are obtained) is justified. This procedure is what Reichenbach (1938) called *inductive simplicity*, the selection of the simplest curve that fits the data points. The safest induction is that the simplest curve provides the best prediction of additional data points. With the randomized design for more than two groups you can establish as many points as you like, consistent with the effort you wish to expend.

One general principle of experimentation when using a two-groups design is that it is advisable to choose rather extreme values of the independent variable.[2] If we had followed this principle, we would not have erred in the example concerning the influence of the period of practice upon performance. For instead of choosing 0- and 5-trial conditions, as we did, we perhaps should have selected 0 and 15 trials. In this event the two-groups design would have led to a conclusion more in line with that of the four-groups design. However, it should still be apparent that the four-groups design yielded considerably more information, allowing us to establish the relationship between the two variables with a high degree of confidence. Even so, the selection of extreme values for two groups can lead to difficulties in addition to those already considered. To illustrate, assume that two data points are obtained, such as those indicated by the X's in Figure 9-6.

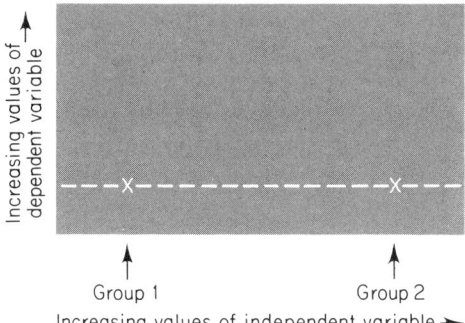

Figure 9-6. Two data points for extreme values of the independent variable using a two-groups design. These points suggest that the independent variable does not affect the dependent variable.

[2] Let us emphasize the word "rather," for seldom would we want to select independent variable values for two groups that are really extreme. This is so because it is likely that all generalizations in psychology break down when the independent variable values are unrealistically extreme. Weber's law, which you probably studied in introductory psychology, is a good example. For although Weber's law holds rather well for weights that you can conveniently hold in your hand it would obviously be absurd to state that it is true for extreme values of weights such as those of atomic size or those of several tons.

Our conclusion would probably be that manipulation of the independent variable does not influence the dependent variable, for the dependent variable values for the two groups are the same. The best guess is that there is a lack of relationship as indicated by the horizontal straight line fitted to the two points.[3] Yet the actual relationship may be that indicated in Figure 9-7, a relationship that probably would have been uncovered by the use of a three-groups design. The corresponding principle with a three-groups design would be to select two rather extreme values of the independent variable and also one value midway between them. Of course if the data point for Group 3 had been the same value as for Groups 1 and 2, then we would be more confident that the independent variable did not affect the dependent variable.

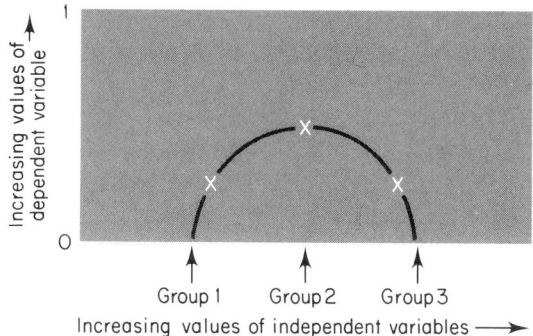

Figure 9-7. Postulated actual relationship for the data point of Figure 9-6. This relationship would be uncovered by a suitable three-groups design.

To summarize what we have said, we may note that psychologists seek to determine which of a number of independent variables influence a given dependent variable and also attempt to establish the relationship between them. With a two-groups design one is never sufficiently sure that the appropriate values of the independent variable were selected in the attempt to determine whether or not that variable is effective. By using more than two groups, however, we increase our chances of: (1) accurately determining whether a given independent variable is effective; and (2) specifying the relationship between the independent and the dependent variable. For these reasons there has been a noticeable decrease in the relative frequency of the

---

[3] One might quibble and say that the horizontal line is still a relationship. However, from our point of view this "relationship," if true, would indicate that variation of the independent variable does not affect the dependent variable.

use of two-groups design, with the greater effectiveness of multi-groups designs of various kinds being increasingly recognized.

With these principles behind us, let us now be more concrete with regard to the potential pitfalls in using a two-groups design. To illustrate, consider an experiment by Bersh (1951) on the development of secondary reinforcement. The general principle that accounts for the development of this important phenomenon states that any neutral stimulus (such as a light) that is associated with a primary reinforcer (such as food) will itself acquire the power of acting as a reinforcer. To determine that a stimulus has become a secondary reinforcer, one can present it after an organism makes a response to see if that response increases in strength. For instance, after a light has been associated with food (and presumably thereby acquired secondary reinforcing properties) an experimenter could place a rat in a Skinner Box that contains a bar that can be pressed. Then, whenever the rat presses the bar the light is presented. If the bar-pressing response increases in frequency it may be concluded that the light has acquired the status of a secondary reinforcer.

Bersh's problem was set by the failure of Schoenfeld, Antonitis, and Bersh (1950) to establish a light as a secondary reinforcer, even though they *had* associated the light with food. The reason for the failure, they suggested, was because of an ineffective time interval between the onset of the light and the presentation of the food. Bersh's problem, therefore, was to ascertain the effect of varying this temporal interval; his independent variable was the length of time that the light was on prior to the delivery of the pellet. His procedure was to place a rat in a Skinner Box with the bar temporarily removed. He then presented a light to the animal and, after the lapse of some specific amount of time, he delivered a pellet of food. Once the light and food had been associated a number of times, the bar was replaced in the Skinner Box and the animal was allowed to press it. Each depression of the bar resulted in the onset of the light. The dependent variable was the number of bar pressing responses that occurred within a ten minute period, so that the greater the number of responses, the stronger the secondary reinforcing properties of the light.

Now, place yourself in the position of the experimenter as he designed this experiment. In the training phase you present a light to the rat, after which you deliver a pellet of food. If you use a two-groups design, what two time values would you select to separate these two presentations? As more or less of a control condition you would probably want to use a zero value, i.e., you would probably present the light and food simultaneously with no time intervening. But what would be the value for your second condition? Suppose that, because you had to do something,[4] you decided to turn on the light one second before the delivery of the food.

If you actually conducted this experiment your results should resemble those reported by Bersh, i.e., the animals who had a 0.0-second delay between light onset and delivery of food would make approximately 19 bar presses within the 10-minute test period. But approximately 25 responses would be made by the animals for whom light preceded food by 1.0 second during training. Hence, the light acquires stronger secondary reinforcing properties when it precedes food by one second than when it occurs simultaneously with food. May it now be concluded that the longer the time interval between presentation of light and food, the stronger the reinforcing properties of the light? To study this question we have plotted the number of responses made for these two conditions in Figure 9-8, and have fitted a straight line to them. Before we can have confidence in this conclusion, we must face gnawing questions such as what would have happened had there been a 0.5-second delay or a 2.0-second delay.

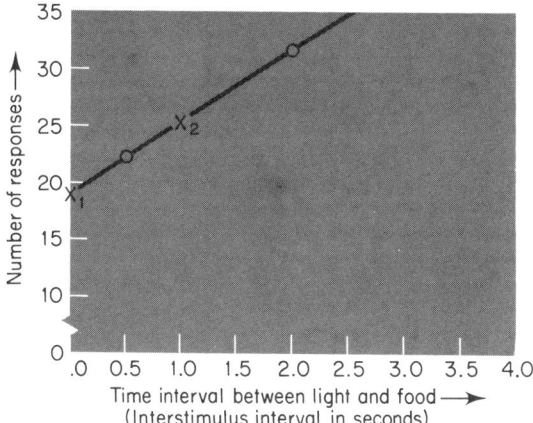

Figure 9-8. Two data points for a two-groups design. Data point #1 (indicated by X_1) resulted from a zero-second time interval during secondary reinforcement training, and data point #2 (X_2) resulted from a one-second delay. The suggestion is that the longer the time interval, the larger the number of resulting responses. Hence the prediction for other time interval values, such as for 0.5 and 2.0 seconds, are indicated by the circles.

---

[4] In research, as in many phases of life, one frequently faces problems for which no appropriate response is available. A principle that the author has found useful was given by a college mathematics teacher (Dr. Bell) to be applied when confronted with an apparently unsolvable math problem: "If you can't do anything, do something." You will be delightfully amazed at the frequency with which this principle leads, if not directly, at least indirectly, to success.

Would dependent variable values for these conditions have fallen on the straight line, as suggested by the two circles in Figure 9-8? The answer, of course, is that we would never know unless there were an experiment involving such conditions. Fortunately, in this instance, relevant data are available. In addition to the 0.0 second and the 1.0 second delay conditions, Bersh also ran groups of rats with delays of 0.5 seconds, 2.0 seconds, 4.0 seconds and 10.0 seconds, and the complete curve is presented in Figure 9-9. By studying Figure 9-9 we can see how erroneous would be the conclusion based on the two-groups experiment. Instead of a 0.5-second delay resulting in about 22 responses, as was predicted by Figure 9-8, this value led to results about the same as a 1.0-second delay, after which the curve, instead of continuing to rise, falls rather dramatically. In short, then, the conduct of a two-groups experiment on this problem would have resulted in an erroneous conclusion—strength of secondary reinforcement increases from a 0.0- to a 0.5-second delay, is approximately the same for a 0.5- to a 1.0-second delay, after which it decreases. And this complex relationship could not possibly have been determined by means of a single two-groups design. The more values of the independent variable sampled, the better our estimation of its influence on a given dependent variable.

Statistical Analysis of a Randomized-Groups Design with More than Two Groups

As in previous designs, we need to determine whether our groups reliably differ. However, we now have several groups to compare. As before we shall use mean values to compare groups. But what statistical procedure is most appropriate for this type of problem? Unfortunately for our present purposes, there is much disagreement among statisticians and among psychologists as to the correct answer to this question. In part, though only in part, the disagreements stem from different types of hypotheses that are being tested, and different aspects of the question that are emphasized. We here wish to minimize the extent to which we enter the various controversies, and to this end shall say that we are interested in doing the following: (1) in making comparisons only between pairs of individual groups; that is, we are not interested in combining two or more groups to test these combined groups against some other group or combination of groups; and (2) in making comparisons between all

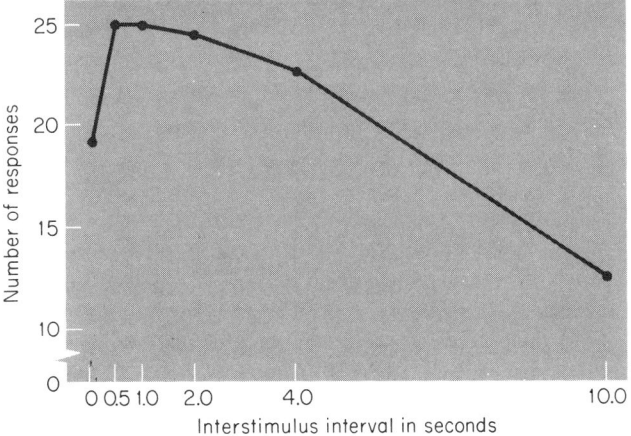

Figure 9-9. Strength of secondary reinforcement as a function of the inter-stimulus interval during training. The measures are the number of responses during the first 10 minutes of the test session (Bersh, 1951).

possible combinations of the separate groups taken two at a time. To emphasize these points, let us say that you conduct an experiment in which you compute the means of three groups. In this event, we are saying that you would be primarily interested in determinging whether Group 1 differs (mean-wise) from Group 2, whether Group 1 differs from Group 3, and whether Group 2 differs from Group 3. You would not, for example, be interested in determining whether Groups 1 and 2, considered together as one group, differ from Group 3. With this limitation of our interest (although we shall comment briefly on procedures where our interest might be otherwise), we shall consider two types of statistical analysis for the more-than-two-randomized-groups design. After prolonged investigation of the numerous statistical procedures available, it was the author's opinion that Duncan's Range Test is the most appropriate.[5] However an alternative procedure that is often used by psychologists starts with the analysis of variance. By presenting both of these procedures you can make your own decision as to which you prefer to apply.

[5] Perhaps the most popular alternative test for multiple comparisons is that by Scheffé (1953). Other procedures are those by Tukey, and the Newman-Keuls method (cf. following Ryan, 1959). The statisticians who addressed themselves to this problem have apparently met an impasse, for little work by them has been conducted since these multiple comparison methods were first developed some years ago. Psychologists more recently have studied the differences, such as Perlmutter and Myers (1973) who advocate yet another procedure, the Bonferroni t statistic, and Keselman, Toothaker, and Shooter (1975). However, little evidence of a consensus as to which method to be used has developed.

To apply Duncan's Range Test let us consider an experiment by Sulzbacher (1967) on programmed learning. The standard method of studying a programmed textbook is for the student to read each frame and to write in the missing words. For example, a frame may be:

> "Underwood is an experimental psychologist, Lewis is an experimental psychologist, and   Experimental
> Duncan is an ——————— ———————.   Psychologist

Immediately on writing in the appropriate words the student receives confirmation of the response by uncovering the correct answer, here placed at the right.

Sulzbacher's question concerned the value of writing the answer in each frame as against just thinking the answer without actually writing it in. Accordingly he formed three groups of elementary school pupils and had each study a mathematics program. The first group took the program in the normal manner and thus was called the Overt Responding (OR) Group. The second group just "thought" the answers without writing them [The Covert Responding (CR) Group] and then received confirmation. A control group merely read through the program with the answers already filled in for them, and the response confirmation portion was blacked out [Non-Responders (NR) group]. Amount of learning for each student was computed with the G ratio, and the resulting scores are presented in Table 9-1.[6]

We seek to test for significant differences among the three groups or, more precisely, among the means of the three groups. The first step is to compute the sum of squares *(SS)* of the dependent vari-

---

[6] The advantage of *G* is that it takes into account the amount of knowledge that the students had prior to their study of the experimental material. A straight gain score does not do this. For instance, if the students score 90 percent on the pretest, the actual gain score is artificially restricted to a maximum of 10 percent. *G* has been more widely applied to measure learning that just in the evaluation of programmed texts (cf. McGuigan & Peters, 1965). For instance, *G* has been used to evaluate effectiveness of teachers (McGuigan, 1967, 1974). *G* is a ratio of amount actually learned to amount that could possibly have been learned. The higher the *G* value, the greater the amount learned from the program. *G* is computed as follows: The mean score on the pretest is calculated. After the students study, the test is again administered and the mean posttest score is computed. The difference between the pre and the post test scores indicates the *actual gain*. The difference between the pretest score and the possible score is a measure of *possible gain*. By dividing the possible gain into the actual gain one can tell how much the students learned relative to how much they could have learned, as measured by the test.

Table 9.1 *G* Ratios for Three Methods of Studying a Programmed Text

Overt Responders (OR)		*Covert Responders (CR)*		*Non-Responders (NR)*	
.46	.75	.26	.42	.51	.83
.36	.87	.57	.81	.78	.45
.56	.24	.82	.40	.65	.53
.28	.29	.94	.78	.64	.55
.72	.50	.85	.89	.33	.05
.80	.41	.57	.43	.65	.62
.59	.87	.00	.32	.58	.57
.60	.48	.90	.56	.50	.58
.91	.78	.62	.58	.56	.48
.93	.86	.70	.71	.38	.68

$n = 20$	$n = 20$	$n = 20$
$\Sigma X = 12.26$	$\Sigma X = 12.13$	$\Sigma X = 10.92$
$\Sigma X^2 = 8.5072$	$\Sigma X^2 = 8.5147$	$\Sigma X^2 = 6.4922$
$\bar{X} = .613$	$\bar{X} = .606$	$\bar{X} = .546$

able scores for each group. Recall that the equation for the sum of squares of any group is:

(9-1)
$$SS = \Sigma X^2 - \frac{(\Sigma X)^2}{n}$$

Hence it can be seen that we need ΣX^2, ΣX, and n for each group. These values are computed as before and are presented in Table 9-1. We merely need to substitute them separately for each group in Equation (9-1). For Group OR the SS is:

$$SS_{OR} = 8.5072 - \frac{(12.26)^2}{20} = .9918$$

The SS of Groups CR and NR can be seen to be:

$$SS_{CR} = 8.5147 - \frac{(12.13)^2}{20} = 1.1579$$

$$SS_{NR} = 6.4922 - \frac{(10.92)^2}{20} = .5299$$

Next we compute the square root of the error variance (s_e) (see p. 421), which for three groups is given by equation:

(9-2)
$$s_e = \sqrt{\frac{SS_A + SS_B + SS_C}{3(n - 1)}}$$

where n is still the number of participants in each group. Substituting the sum of squares for the three groups and n in Equation (9-2) we find that s_e is:

$$s_e = \sqrt{\frac{.9918 + 1.1579 + .5299}{3(20 - 1)}} = .2168$$

We now determine the degrees of freedom for s_e appropriate for Duncan's Range Test, which is given by the equation:

$$(9\text{-}3) \qquad\qquad df = N - r$$

N is the total number of participants in the experiment, and r is the number of groups. This is found to be:

$$df = 60 - 3 = 57$$

Assuming that we have set a criterion of $P < .05$ for testing the difference between two means, we now need to refer to Table 9-2.[7] There we see that the columns are labeled "Number of Groups" and the rows "df." The values in the table are the "least significant standardized ranges," which are symbolized r_p.[8] We shall require various values of r_p for each test between two means that we shall make. We have three means in the present example: .613, .606, 546. We shall make tests between: (1) the extreme means of the three groups (which is a test between .613 versus .546); (2) between the highest and middle means (.613 versus .606); and (3) between the middle and lower means (.606 versus .546). The symbol for the least significant standardized ranges of Table 9-2 we recall is r_p. The subscript p to the r indicates the three tests that we seek to make. For the first test, $p = 3$ and hence $r_p = r_3$. For the other two tests $p = 2$ and hence for both $r_p = r_2$. We need now to enter Table 9-2 to obtain the value of r_p for the first test above, viz., when the extreme value of a group of three means is being tested; the appropriate column for a test between the extreme means of the three groups (r_3) is labeled "3." Finding column 3 we read down rows until we come to the appropriate degrees of freedom. Recall in this example that our degrees of freedom is 57 df. However, 57 df is not in the table. The best that we can do is to find a row for 40 df and for 60 df. The values of r_p for three groups (i.e., r_3) for 40 and 60 df are 3.01 and 2.98, respectively. By interpolating (linearly) we find that the desired value of r_3 is 2.98. This value shall be used for our first test, that of comparing the groups with the high-

---

[7] See Duncan (1955) for an elaboration of the precise nature of this probability value.

[8] Do not confuse r_p with r, which is the number of groups. It is unfortunate that we cannot take the time to lay a broad base for our techniques of statistical analysis and for discussing these concepts in greater detail. The student will correct this deficiency in statistics courses and by referring to other sources. For an elaboration of the concepts in this chapter you might refer to Duncan (1955), Li (1957), or to the more psychologically oriented works of Edwards (1968), Hays (1963), Ray (1960), and Winer (1971).

Table 9-2 Values of r_p for Duncan's Range Test (significance level = 5 percent).

df	2	3	4	5	6	7	8	9	10	12	14	16	18	20	50	100
1	18.0	18.0	18.0	18.0	18.0	18.0	18.0	18.0	18.0	18.0	18.0	18.0	18.0	18.0	18.0	18.0
2	6.09	6.09	6.09	6.09	6.09	6.09	6.09	6.09	6.09	6.09	6.09	6.09	6.09	6.09	6.09	6.09
3	4.50	4.50	4.50	4.50	4.50	4.50	4.50	4.50	4.50	4.50	4.50	4.50	4.50	4.50	4.50	4.50
4	3.93	4.01	4.02	4.02	4.02	4.02	4.02	4.02	4.02	4.02	4.02	4.02	4.02	4.02	4.02	4.02
5	3.64	3.74	3.79	3.83	3.83	3.83	3.83	3.83	3.83	3.83	3.83	3.83	3.83	3.83	3.83	3.83
6	3.46	3.58	3.64	3.68	3.68	3.68	3.68	3.68	3.68	3.68	3.68	3.68	3.68	3.68	3.68	3.68
7	3.35	3.47	3.54	3.58	3.60	3.61	3.61	3.61	3.61	3.61	3.61	3.61	3.61	3.61	3.61	3.61
8	3.26	3.39	3.47	3.52	3.55	3.56	3.56	3.56	3.56	3.56	3.56	3.56	3.56	3.56	3.56	3.56
9	3.20	3.34	3.41	3.47	3.50	3.52	3.52	3.52	3.52	3.52	3.52	3.52	3.52	3.52	3.52	3.52
10	3.15	3.30	3.37	3.43	3.46	3.47	3.47	3.47	3.47	3.47	3.47	3.47	3.47	3.48	3.48	3.48
11	3.11	3.27	3.35	3.39	3.43	3.44	3.45	3.46	3.46	3.46	3.46	3.46	3.47	3.48	3.48	3.48
12	3.08	3.23	3.33	3.36	3.40	3.42	3.44	3.44	3.46	3.46	3.46	3.46	3.47	3.48	3.48	3.48
13	3.06	3.21	3.30	3.35	3.38	3.41	3.42	3.44	3.45	3.45	3.46	3.46	3.47	3.47	3.47	3.47
14	3.03	3.18	3.27	3.33	3.37	3.39	3.41	3.42	3.44	3.45	3.46	3.46	3.47	3.47	3.47	3.47
15	3.01	3.16	3.25	3.31	3.36	3.38	3.40	3.42	3.43	3.44	3.45	3.46	3.47	3.47	3.47	3.47
16	3.00	3.15	3.23	3.30	3.34	3.37	3.39	3.41	3.43	3.44	3.45	3.46	3.47	3.47	3.47	3.47
17	2.98	3.13	3.22	3.28	3.33	3.36	3.38	3.40	3.42	3.44	3.45	3.46	3.47	3.47	3.47	3.47
18	2.97	3.12	3.21	3.27	3.32	3.35	3.37	3.39	3.41	3.43	3.45	3.46	3.47	3.47	3.47	3.47
19	2.96	3.11	3.19	3.26	3.31	3.35	3.37	3.39	3.41	3.43	3.44	3.46	3.47	3.47	3.47	3.47
20	2.95	3.10	3.18	3.25	3.30	3.34	3.36	3.38	3.40	3.43	3.44	3.46	3.46	3.47	3.47	3.47
22	2.93	3.08	3.17	3.24	3.29	3.32	3.35	3.37	3.39	3.42	3.44	3.45	3.46	3.47	3.47	3.47
24	2.92	3.07	3.15	3.22	3.28	3.31	3.34	3.37	3.38	3.41	3.44	3.45	3.46	3.47	3.47	3.47
26	2.91	3.06	3.14	3.21	3.27	3.30	3.34	3.36	3.38	3.41	3.43	3.45	3.46	3.47	3.47	3.47
28	2.90	3.04	3.13	3.20	3.26	3.30	3.33	3.35	3.37	3.40	3.43	3.45	3.46	3.47	3.47	3.47
30	2.89	3.04	3.12	3.20	3.25	3.29	3.32	3.35	3.37	3.40	3.43	3.44	3.46	3.47	3.47	3.47
40	2.86	3.01	3.10	3.17	3.22	3.27	3.30	3.33	3.35	3.39	3.42	3.44	3.46	3.47	3.47	3.47
60	2.83	2.98	3.08	3.14	3.20	3.24	3.28	3.31	3.33	3.37	3.40	3.43	3.45	3.47	3.48	3.48
100	2.80	2.95	3.05	3.12	3.18	3.22	3.26	3.29	3.32	3.36	3.40	3.42	3.45	3.47	3.53	3.53
∞	2.77	2.92	3.02	3.09	3.15	3.19	3.23	3.26	3.29	3.34	3.38	3.41	3.44	3.47	3.61	3.67

est and the lowest means. More specifically, it is used for comparing the extreme means of a group of three means. But for the second and third tests we will need to compare adjacent groups. Since adjacent groups involve the comparison of only two means, we need to enter column 2 of Table 9-2 in order to make tests between two adjacent groups (hence $r_p = r_2$). Reading down column 2 until we come to 40 and 60 df we find, by interpolating, that r_2 is 2.83 for 57 df. These two values of r_p are written in the top row of Table 9-3. If the procedure for selecting and using the values for r_p is not clear, you should proceed through this section, temporarily accepting our statements on faith, for we discuss the matter more thoroughly later.

Table 9-3 Values of r_p and R_p for 2 and 3 Groups with 57 df

	Number of Groups	
	2.	3
r_p	2.83	2.98
R_p	.137	.145

Our next step is to compute the "least significant ranges" for our values, which is symbolized by R_p, where:

(9-4)
$$R_p = s_e\, r_p \sqrt{\frac{1}{n}}$$

Therefore, to find R_p for each number of groups we need to multiply our previously computed value of s_e by the appropriate value of r_p and $\sqrt{1/n}$. Since our computed value of $s_e = .217$ (rounded off), r_p for two groups is 2.83, and $n = 20$, R_2 is:

$$R_3 = (.217)\,(2.98)\,\sqrt{\frac{1}{20}} = .145$$

Similarly R_3 is:

$$R_2 = (.217)\,(2.83)\,\sqrt{\frac{1}{20}} = .137$$

These values of R_p have been entered in the bottom row of Table 9-3. The rank order of the means for our three groups from the lowest to the highest was for Groups OR, CR, and NR respectively. Hence, our ordering is:

	Group		
	NR	CR	OR
Mean	.546	.606	.613

Now, the final step is to compare the differences between our ordered means and the values of R_p. Starting with highest and the lowest means it can be seen that the difference between Groups OR and NR is .067. By comparing Group OR with NR we have compared the extreme means of three groups. Therefore, we read the value of $R_3 =$ .145 from Table 9-3. This means that the difference between the group with the highest mean (Group OR) and the group with the lowest mean (Group NR) must exceed .145 for that difference to be reliable at the five percent level. We just found that the difference between Group OR and Group NR was .067. Since the obtained value of .067 does not exceed the required value of .145 we may conclude that the sample mean of Group OR is not reliably different from the sample mean of Group NR; had, on the other hand, the mean difference between these two groups been .150 or .146 say, they would have been reliably different.

Should we now determine whether the mean of Group OR is reliably different from that of Group CR? Obviously not in this example, but to illustrate for other cases, the procedure would be as follows. Since these means are adjacent in our ranked order, we are comparing two means. Hence, we read from Table 9-3 that the appropriate value of $R_2 = .137$. The obtained difference between the means of these two groups is .007. If the obtained difference exceeds the value of .137, these two groups differ beyond the .05 level of probability. Since .007 does not exceed .137, our conclusion is in the negative; the sample mean of Group OR is not reliably different from the sample mean of Group CR.

At this point we may note a rule of general procedure. Since the difference between the extreme means (Groups OR and NR) was not reliable there was no necessity for testing the lesser difference between Groups OR and CR. For if there is no reliable difference between the most extreme means (here between Groups OR and CR), any lesser difference can not possibly be statistically reliable.

As another example, the remaining possible comparison would be between Groups CR and NR. The difference between their means is .060. Since we would now be comparing two groups, we would read the R_p value of .137 from Table 9-3. As before, the necessary difference between a group of two means must exceed .137. Since .060 does not exceed .137, we conclude that the means of Groups CR and NR do not differ reliably, a fact that we already knew. On the other hand, if the extreme means do differ reliably, the only way to find out whether the lesser differences are reliable is to proceed with the test. This is the reason that we illustrated the procedure even though it was inappropriate for this case.

The major finding of this study, thus, is that the three groups did not differ reliably, when considered pairwise, on the criterion measure. Consequently variation of modes of responding failed to influence amount learned.

The general rule for selecting and using the various values of r_p is: after the group means have been ordered, count the number of groups *between* the two that you are testing and add that number to those two. Then, enter the table of r_p (Table 9-2) for that number of groups. In the three-group design just discussed we have the following situation:

Group NR	Group CR	Group OR
$\bar{X}_1$	$\bar{X}_2$	$\bar{X}_3$

If we are comparing Groups OR and NR, we count one group between these two, giving us a total of three. Hence, we enter column 3 of Table 9-2. A test between Groups OR and CR, however, has no groups intervening between them. Therefore, we add zero to our count of two, indicating that we enter column 2 of Table 9-2. And likewise, when we are comparing Groups CR and NR, the number is two. Let's say that we have four groups and are testing the highest mean against the lowest. In this case, two groups intervene. Therefore, we add two to the groups that we are testing, indicating that we should enter Table 9-2 at the column labeled 4. However, we will also need to compare the highest group with the second lowest. In this case, one group will intervene, making our count three. The next comparison will be between the two highest groups, a situation in which no groups intervene. Hence, the count would be two, and we should record that value of r_p for further use. Similarly, if we have a five groups design and are testing the highest mean against the lowest, the count would be five. For testing the highest against the middle group, the count would be three. And so forth.

A Five-Groups Design

The statistical analysis of a multigroups design using Duncan's Range Test should now be clear. The same procedure is followed for experiments involving more than three groups with several minor computational differences. Its application to a design involving five groups is illustrated by the following problem. A psychologist is hired for the purpose of constructing a training program for assemblyline workers. Their job is to assemble parts of a rather complicated electronic device. Management asks the psychologist how many times the workers should practice assembling the device before they are placed on the actual production line. To be safe in answering the question the psychologist decides to conduct an experiment. The independent variable, the number of times for practicing assembly of the device, is varied in five ways: 0 trials, 10 trials, 30 trials, 70 trials, and 100 trials.[9] From the 50 trainees available for the experiment, the psychologist decides to assign randomly ten to each condition. The dependent variable is the number of parts that the workers can correctly assemble in one hour, a test for which is administered after the training. Assume that the number of parts that the five groups correctly assembled are as shown in Table 9-4.

The values of ΣX, ΣX^2, n, and $\bar{X}$ have been computed for each group. To compute the sum of squares for each group we merely substitute the values required by Equation (9-1):

$$SS_1 = 25 - (11)^2/10 = 12.90$$
$$SS_2 = 39 - (15)^2/10 = 16.50$$
$$SS_3 = 546 - (72)^2/10 = 27.60$$
$$SS_4 = 5533 - (235)^2/10 = 10.50$$
$$SS_5 = 5727 - (239)^2/10 = 14.90$$

Table 9-4 Number of Parts Correctly Assembled During Test

	GROUP			
1	2	3	4	5
		Number of Training Trials		
0	*10*	*30*	*70*	*100*
0	2	4	24	24
1	2	5	25	24
3	1	7	23	25
0	4	9	23	25
0	0	8	25	22
2	0	7	22	24
1	0	6	24	22
0	3	8	23	26
3	2	9	22	24
1	1	9	24	23
ΣX: 11	15	72	235	239
ΣX^2: 25	39	546	5533	5727
n: 10	10	10	10	10
$\bar{X}$: 1.10	1.50	7.20	23.50	23.90

[9] Actually in an experiment such as this one where the independent variable (number of trials) is quantified along some dimension, a regression analysis (such as the method of orthogonal polynomials, cf. Cochran & Cox, 1957) is more powerful. This does not mean, however, that Duncan's Range Test is inappropriate for this type of situation. Rather it means that there is a slight possibility that we may fail to find differences among groups with Duncan's Range Test, when in fact they exist.

The previous formula for s_e was specialized for the case of three groups. An equation applicable to any number of groups is:

$$(9\text{-}5) \qquad s_e = \sqrt{\frac{SS_1 + SS_2 + SS_3 + \cdots SS_r}{(n_1 - 1) + (n_2 - 1) + \cdots (n_r - 1)}}$$

The numerator of this general equation for computing s_e merely indicates that you should add the sums of squares for all of the groups together. That is, you add the SS for the first group to the SS for the second group, add these to the SS for the third group, and continue in this manner until you have added the SS for all r groups, where of course r indicates the number of groups. The denominator merely indicates that you add the number of participants in each group minus one together, continuing for all r groups. Since we have five groups in the present example, we have five sums of squares to add, and since $n_1 = n_2 = n_3 = n_4 = n_5$ we may merely multiply $n - 1$ by 5. Hence Equation (9-5) may be written for this case as:

$$(9\text{-}6) \qquad s_e = \sqrt{\frac{SS_1 + SS_2 + SS_3 + SS_4 + SS_5}{5(n - 1)}}$$

Substituting the sum of squares and n for the 5 groups in Equation (9-6) we obtain:

$$s_e = \sqrt{\frac{12.90 + 16.50 + 27.60 + 10.50 + 14.90}{5(10 - 1)}} = 1.35$$

The appropriate degrees of freedom is:

$$df = 50 - 5 = 45$$

For a 5 percent level test we enter Table 9-2 to obtain the necessary values of r_p. We have five groups so we enter the column labeled 5. Reading down to 40 and 60 df we find (by interpolating) that the value of r_5 for 45 df is 3.16. Similarly the value of $r_4 = 3.10$, or $r_3 = 3.00$, and of $r_2 = 2.85$. These values are entered in the top row of Table 9-5.

Table 9-5   Values of r_p and R_p for Five Groups with 45 df

	2	3	4	5
r_p	2.85	3.00	3.10	3.16
R_p	1.23	1.30	1.34	1.37

To obtain the R_p values we multiply the appropriate value of r_p by s_e and $\sqrt{1/n}$ (Equation 9-4). These computations are:

For 2 groups: $R_p = (2.85)\ (1.35)\ \sqrt{1/10} = 10.5$

For 3 groups: $R_p = (3.00)\ (1.35)\ \sqrt{1/10} = 1.11$

For 4 groups: $R_p = (3.10)\ (1.35)\ \sqrt{1/10} = 11.5$

For 5 groups: $R_p = (3.16)\ (1.35)\ \sqrt{1/10} = 11.7$

These values have been entered in the bottom row of Table 9-5. We next order our means from lowest to highest:

	Group				
	1	2	3	4	5
Mean	1.10	1.50	7.20	23.50	23.90

Following our previous procedure we first must test the difference between the extreme means of five groups. The extreme means are for Groups 5 and 1, their difference being 22.80. Since we are considering five groups, we read the value of $R_5 = 1.37$ from Table 9-5 (for five groups: three intervening groups, plus the two we are testing). The difference of 22.80 is larger than 1.37 and, therefore, we may conclude that Group 5 is reliably different from Group 1 at the 5 percent level. The next comparison is between Groups 5 and 2. The difference between their means is 22.40. Now we are comparing the difference between extreme values of 4 groups. Hence, we read R_p from Table 9-5 as 1.34. Since 22.40 exceeds 1.34, we may conclude that Group 5 is reliably different from Group 2. The next comparison is between Group 5 and Group 3. Three means enter our consideration. Hence, the appropriate R_p from Table 9-5 is 1.30. The difference in means between Groups 5 and 3 is 16.70. Since this difference exceeds 1.30, these two groups are reliably different.

The difference in means between Groups 5 and 4 is 0.40. This comparison involves consideration of only two means. Therefore, the appropriate R_p is 1.23. But 0.40 does not exceed 1.23, from which we conclude that there is no reliable difference between Groups 5 and 4. In short, we have found that Group 5 is reliably different from Groups 3, 2, and 1, but is *not* reliably different from Group 4. Since Group 5 has a higher mean than Groups 3, 2, and 1, we may further conclude that it is reliably superior to them. The fact may be indicated by the following scheme. Any two means that *are* underscored by the same line are *not* reliably different. Any two means that are *not* underscored by the same line *are* reliably different. You might wish to make a special effort to memorize these. Since Group 5 is not reliably different from Group 4, we draw a line under those two means. But since Group 5 is reliably different from Groups 3, 2 and 1, we do not extend the line under them. That is:

	Group				
	1	2	3	4	5
Mean	1.10	1.50	7.20	23.50	23.90

We have now tested for reliable differences between Group 5 and the other groups. Our next step is to compare Group 4 with Groups 3, 2, and 1. The extreme values among these four groups occur for Groups 4 and 1, their mean difference being 22.40. From Table 9-5 we find that the R_p for comparing 4 groups is 1.34. Since 22.40 is larger than the necessary value of 1.34, we may say that Groups 4 and 1 differ reliably. Now to move on to Group 2. The mean difference between Groups 4 and 2 is 22.00. From Table 9-5 we find that the appropriate value of R_p is 1.30. Since 22.00 exceeds 1.30, Groups 4 and 2 differ reliably. Do Groups 4 and 3 differ reliably? The R_p for a two-groups comparison is 1.23. The difference in means of these two groups is 16.30. Therefore, the answer is in the affirmative and we may conclude that Group 4 is reliably superior to Groups 3, 2, and 1. This finding is indicated by *not* drawing a common line under Groups 4, 3, 2, and 1. We now proceed to test for reliable differences between Group 3 and Groups 2 and 1. The difference between the means of Groups 3 and 1 is 6.10. From Table 9-5 we find that the value of R_p for comparing three groups is 1.30. Our obtained difference exceeds this value. We, therefore, conclude that Group 3 is reliably different from Group 1. To compare Groups 3 and 2 we find that their mean difference is 5.70. R_p for comparing two groups is 1.23. Therefore, these two groups differ reliably. These findings are indicated by not drawing a common line under Groups 3, 2, and 1:

	Group				
	1	2	3	4	5
Mean	1.10	1.50	7.20	23.30	23.40

Our final comparison is between Groups 2 and 1. The difference between their means may be seen to be 0.40. From Table 9-5 we find a value of 1.23 for R_p when two groups are being tested. Since 0.40 does not exceed 1.23, we conclude that there is no reliable difference between Groups 2 and 1. This finding is indicated by drawing a line under Groups 2 and 1:

Group

	1	2	3	4	5
Mean	1.10	1.50	7.20	23.20	23.40

The above lines under the means of the five groups constitute a summary of our findings. The line under Groups 4 and 5 indicates that these groups are not reliably different. But since that line does not extend to the other groups, we know that both Groups 4 and 5 are reliably superior to the other groups. The common line under Groups 1 and 2 indicates that there is no reliable difference between them. The lack of a common line under Groups 3, 2, and 1 indicates that Group 3 is reliably superior to Groups 1 and 2. In short, we reach the following conclusions:

1. Groups 4 and 5 do not differ reliably.
2. Both Groups 4 and 5 are reliably superior to Groups 1, 2, and 3.
3. Group 3 is reliably superior to Groups 1 and 2.
4. Groups 1 and 2 do not differ reliably.

In Figure 9-10 we have plotted the relationship between the independent and the dependent variables. The number of practice trials used by the five groups is indicated on the horizontal axis and the mean number of parts assembled by each group in the hour test is indicated on the vertical axis. From our statistical analysis and Figure 9-10, the psychologist's answer to the problem should be obvious.

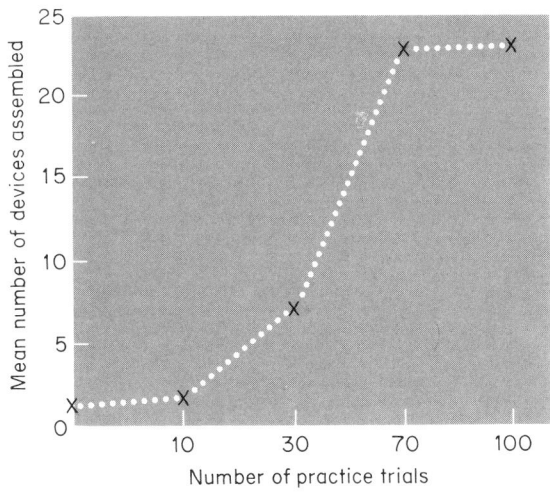

Figure 9-10. Fictitious relationship between number of practice trials and mean number of devices assembled.

Practice beyond 70 trials does not seem to increase the participants' scores, but up to that point practice is beneficial. Hence, he would recommend 70 practice trials. It may be, of course, that practice beyond 100 trials would lead to a further increase in test scores, but such a possibility could only be verified by further experimentation.

In the preceding examples we have used a 5 percent level test. Table 9-6 presents the necessary values for testing the differences between groups at the one percent level. The procedure for using Table 9-6 is precisely the same as that for Table 9-2.

In order to ensure that the scheme for indicating significant differences between group means is clear, let us consider several additional examples. For instance, suppose we have three groups and find that Group 3 is not significantly different from Group 2, but has a significantly higher mean than does Group 1. However, we find that there is no significant difference between Groups 1 and 2. These findings would be indicated as follows:

$$\text{Group}$$
$$(\textit{Increasing Means} \longrightarrow)$$

	1	2	3
Mean	$\overline{X}_1$	$\overline{X}_2$	$\overline{X}_3$

Taking another example with three groups, assume that Group 3 is significantly higher than Groups 2 and 1, but there is no significant difference between Groups 2 and 1:

$$\textit{Group}$$

1	2	3
$\overline{X}_1$	$\overline{X}_2$	$\overline{X}_3$

Consider a four-groups design. Assume that Group 4 is significantly superior to the other groups; that Group 3 is not significantly superior to Group 2, but is significantly superior to Group 1; and that there is no difference between Groups 1 and 2:

$$\textit{Group}$$

	1	2	3	4
Mean	$\overline{X}_1$	$\overline{X}_2$	$\overline{X}_3$	$\overline{X}_4$

With four groups again, assume that Group 4 does not differ significantly from Groups 3 and 2, but is significantly different from Group 1. Furthermore, Groups 3 and 2 are not significantly different,

Table 9-6 Values of r_p for Duncan's Range Test (significance level = 1 percent).

df	NUMBER OF GROUPS																
	2	3	4	5	6	7	8	9	10	12	14	16	18	20	50	100
1	90.0	90.0	90.0	90.0	90.0	90.0	90.0	90.0	90.0	90.0	90.0	90.0	90.0	90.0	90.0	90.0
2	14.0	14.0	14.0	14.0	14.0	14.0	14.0	14.0	14.0	14.0	14.0	14.0	14.0	14.0	14.0	14.0
3	8.26	8.5	8.6	8.7	8.8	8.9	8.9	9.0	9.0	9.0	9.1	9.2	9.3	9.3	9.3	9.3
4	6.51	6.8	6.9	7.0	7.1	7.1	7.2	7.2	7.3	7.3	7.4	7.4	7.5	7.5	7.5	7.5
5	5.70	5.96	6.11	6.18	6.26	6.33	6.40	6.41	6.5	6.6	6.6	6.7	6.7	6.8	6.8	6.8
6	5.24	5.51	5.65	5.73	5.81	5.88	5.95	6.00	6.0	6.1	6.2	6.2	6.3	6.3	6.3	6.3
7	4.95	5.22	5.37	5.45	5.53	5.61	5.69	5.73	5.8	5.8	5.9	5.9	6.0	6.0	6.0	6.0
8	4.74	5.00	5.14	5.23	5.32	5.40	5.47	5.51	5.5	5.6	5.7	5.7	5.8	5.8	5.8	5.8
9	4.60	4.86	4.99	5.08	5.17	5.25	5.32	5.36	5.4	5.5	5.5	5.6	5.7	5.7	5.7	5.7
10	4.48	4.73	4.88	4.96	5.06	5.13	5.20	5.24	5.28	5.36	5.42	5.48	5.54	5.55	5.55	5.55
11	4.39	4.63	4.77	4.86	4.94	5.01	5.06	5.12	5.15	5.24	5.28	5.34	5.38	5.39	5.39	5.39
12	4.32	4.55	4.68	4.76	4.84	4.92	4.96	5.02	5.07	5.13	5.17	5.22	5.24	5.26	5.26	5.26
13	4.26	4.48	4.62	4.69	4.74	4.84	4.88	4.94	4.98	5.04	5.08	5.13	5.14	5.15	5.15	5.15
14	4.21	4.42	4.55	4.63	4.70	4.78	4.83	4.87	4.91	4.96	5.00	5.04	5.06	5.07	5.07	5.07
15	4.17	4.37	4.50	4.58	4.64	4.72	4.77	4.81	4.84	4.90	4.94	4.97	4.99	5.00	5.00	5.00
16	4.13	4.34	4.45	4.54	4.60	4.67	4.72	4.76	4.79	4.84	4.88	4.91	4.93	4.94	4.94	4.94
17	4.10	4.30	4.41	4.50	4.56	4.63	4.68	4.72	4.75	4.80	4.83	4.86	4.88	4.89	4.89	4.89
18	4.07	4.27	4.38	4.46	4.53	4.59	4.64	4.68	4.71	4.76	4.79	4.82	4.84	4.85	4.85	4.85
19	4.05	4.24	4.35	4.43	4.50	4.56	4.61	4.64	4.67	4.72	4.76	4.79	4.81	4.82	4.82	4.82
20	4.02	4.22	4.33	4.40	4.47	4.53	4.58	4.61	4.65	4.69	4.73	4.76	4.78	4.79	4.79	4.79
22	3.99	4.17	4.28	4.36	4.42	4.48	4.53	4.57	4.60	4.65	4.68	4.71	4.74	4.75	4.75	4.75
24	3.96	4.14	4.24	4.33	4.39	4.44	4.49	4.53	4.57	4.62	4.64	4.67	4.70	4.72	4.74	4.74
26	3.93	4.11	4.21	4.30	4.36	4.41	4.46	4.50	4.53	4.58	4.62	4.65	4.67	4.69	4.73	4.73
28	3.91	4.08	4.18	4.28	4.34	4.39	4.43	4.47	4.51	4.56	4.60	4.62	4.65	4.67	4.72	4.72
30	3.89	4.06	4.16	4.22	4.32	4.36	4.41	4.45	4.48	4.54	4.58	4.61	4.63	4.65	4.71	4.71
40	3.82	3.99	4.10	4.17	4.24	4.30	4.34	4.37	4.41	4.46	4.51	4.54	4.57	4.59	4.69	4.69
60	3.76	3.92	4.03	4.12	4.17	4.23	4.27	4.31	4.34	4.39	4.44	4.47	4.50	4.53	4.66	4.66
100	3.71	3.86	3.98	4.06	4.11	4.17	4.21	4.25	4.29	4.35	4.38	4.42	4.45	4.48	4.64	4.65
∞	3.64	3.80	3.90	3.98	4.04	4.09	4.14	4.17	4.20	4.26	4.31	4.34	4.38	4.41	4.60	4.63

but Group 3 is significantly superior to Group 1. And Group 2 is significantly different from Group 1:

Group

	1	2	3	4
Mean	$\overline{X}_1$	$\overline{X}_2$	$\overline{X}_3$	$\overline{X}_4$

Statistical Analysis for Unequal n's

The preceding discussion has assumed that the number of participants in each of the several groups is equal. However, it is frequently the case that different numbers of participants are assigned to the groups. Kramer (1956) has extended Duncan's Range Test so that it is applicable to this situation (see also Duncan, 1957, and Kramer, 1957, for this and additional extensions). The same general procedure as that for equal n's is followed with only minor exceptions. To illustrate, consider some pilot data from college students who were asked to engage in one of five kinds of activities: Group 1 listened to a story presented by means of a tape recorder; Group 2 similarly listened to classical music; Group 3 memorized a portion of the same story that was visually presented to them; Group 4 read a selection from the story, and Group 5 listened to a blank tape on the tape recorder with instructions to pay attention in case they heard anything (which, of course, they didn't). Prior to these activities, all participants rested for one minute, following which they engaged in their activity for five minutes. Throughout the session several measures of covert behavior were recorded. The values that we shall here study are integrated electromyograms from the chin. The empirical question was whether or not covert oral language behavior (as measured by the slight electrical activity of chin muscles in the form of electromyograms) was different under these various stimulating conditions. In particular, it was hypothesized that there would be greater covert chin activity during the silent presentation of language stimuli (Groups 1, 3, and 4) than under the nonlanguage stimulus (control) conditions (Groups 2 and 5). To approach the answer, the amount of chin activity during the rest condition was subtracted from the amount during each of the five minutes of the activity periods. The resulting curves for each of the groups is presented in Figure 9-11. There we can see that Group 1, who listened to the story, gave the largest chin response throughout the five-minute activity period. The next highest amounts of covert activity were given by Group 4 (reading), followed by Group 3, who memorized the story. The lowest levels of responding were made by Groups 5 and 2, the control groups. It is especially interesting to note

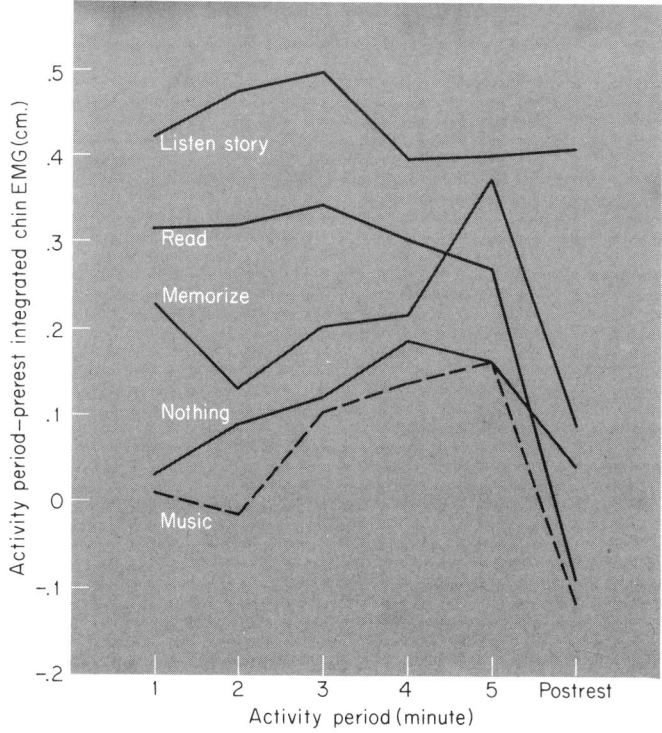

Figure 9-11. Amplitude of covert oral behavior during five minutes of several kinds of activity. Mean chin EMG during rest was subtracted from values during each of the five minutes and plotted on the vertical axis.

that the students who received language stimuli (regardless of whether they listened, read, or memorized) yielded larger amounts of covert oral activity than did those who did not receive language stimuli (the "nothing" and the "music" groups). Let us now use these data to illustrate the application of Duncan's Range Test to the situation where we have unequal numbers of participants in the experimental groups. The mean amount of covert oral language behavior for each participant is presented in Table 9-7. Let us also use this example to illustrate a major point made in chapter 5, viz., when you have repeated measures on each participant, for this type of statistical analysis we need to obtain one dependent variable measure for each participant. That is, as can be noted in Figure 9-11, we have measures for each individual over a five-minute period, and we need to obtain a single dependent variable value representative for the entire interval. The solution here was to obtain the mean chin electromyographic amplitude for each participant during the five minutes of activity; then, a

Table 9-7 Measures of Covert Oral Behavior During Various Activities

GROUP				
1 (*Listen Story*)	*2* (*Listen Music*)	*3* (*Memorize*)	*4* (*Read*)	*5* (*Nothing*)
1.7	−1.2	12.4	7.1	−2.2
−.6	2.0	2.4	−2.3	.6
28.9	.1	−.8	4.3	−.1
1.1	.5	0.0	0.0	.6
0.0	−.3	0.0	.6	3.1
−.5	3.7	2.2	2.4	2.1
2.1		.2	6.1	4.9
			8.1	.2
$\Sigma X = 32.70$	$\Sigma X = 4.80$	$\Sigma X = 16.40$	$\Sigma X = 26.30$	$\Sigma X = 9.20$
$\Sigma X^2 = 844.33$	$\Sigma X^2 = 19.48$	$\Sigma X^2 = 165.04$	$\Sigma X^2 = 183.13$	$\Sigma X^2 = 43.64$
$n = 7$	$n = 6$	$n = 7$	$n = 8$	$n = 8$
$\bar{X} = 4.67$	$\bar{X} = .80$	$\bar{X} = 2.34$	$\bar{X} = 3.29$	$\bar{X} = 1.15$

comparable value of the mean chin amplitude during the resting (pre-activity) was subtracted from that value during activity. These mean differences are the values presented in Table 9-7. For example, the first individual in Group 1 had a covert chin response value of 1.7 cm greater while listening to the story than while resting prior to the experiment. You can note that the numbers of participants in the five groups are different.

The first step is to compute ΣX, ΣX^2, n, and $\bar{X}$ for each group (see Table 9-7). With these values we shall compute the sum of squares of the five groups:

$$SS_1 = 844.33 - \frac{(32.70)^2}{7} = 691.58$$

$$SS_2 = 19.48 - \frac{(4.80)^2}{6} = 15.64$$

$$SS_3 = 165.04 - \frac{(16.40)^2}{7} = 126.62$$

$$SS_4 = 183.13 - \frac{(26.30)^2}{8} = 96.67$$

$$SS_5 = 43.64 - \frac{(9.20)^2}{8} = 33.06$$

We need next to compute s_e. However since the n's of the groups are not equal, we shall have to use Equation (9-5) instead of Equation (9-6) which, for five groups, becomes:

$$(9\text{-}7) \quad s_e = \sqrt{\frac{SS_1 + SS_2 + SS_3 + SS_4 + SS_5}{(n_1 - 1) + (n_2 - 1) + (n_3 - 1) + (n_4 - 1) + (n_5 - 1)}}$$

Substituting the appropriate values of SS and n in Equation (9-7):

$$s_e = \sqrt{\frac{691.58 + 15.64 + 126.62 + 96.67 + 33.06}{(7-1) + (6-1) + (7-1) + (8-1) + (8-1)}} = 5.57$$

The df are computed as before:

$$df = 36 - 5 = 31$$

We now write down the values of r_p for five, four, three, and two groups. Assuming a 5 percent level test, use Table 9-2. We enter the column marked 5. Reading down that column we find the r_p for five groups that corresponds to 31 df. It is 3.20. For four groups $r_p = 3.12$, and so forth (Table 9-8).

Table 9-8 Values of r_p for a Five-Groups Design with 31 df

	Groups			
	2	3	4	5
r_p	2.98	3.04	3.12	3.20

Up to this point the procedure for unequal n's has been essentially the same as for equal n's. We shall now depart to some extent. Instead of using Equation (9-4) we now use Equation (9-8) for computing R_p when we have unequal n's:

(9-8)
$$R_p = (s_e)(r_p)\sqrt{\frac{1}{2}\left(\frac{1}{n_a} + \frac{1}{n_b}\right)}$$

Where n_a and n_b are the n's for whatever two groups are being compared, as we shall shortly see. Let us start by ordering the means and testing the difference between the extreme groups:

	Group				
	2	5	3	4	1
Mean	.80	1.15	2.34	3.29	4.67

The difference in means between Group 1 and Group 2 is 3.87. We next determine the value of R_p for this test; s_e will be the same for all tests, but r_p will depend on the number of groups being compared. In this comparison we are considering five means, so r_p will be that for comparing the extreme means of five groups. From Table 9-8 we read: $r_5 = 3.20$. The n's are those for groups 1 and 2, i.e., 7 and 6 respectively. Hence, R_p is:

$$R_p = (5.57)(3.20)\sqrt{\frac{1}{2}\left(\frac{1}{7} + \frac{1}{6}\right)} = 17.82\sqrt{.1548} = 7.01$$

Since the mean difference between Groups 1 and 2 (3.87) does not exceed the R_p for this comparison (7.03), we may conclude that Group 1 is not reliably different from Group 2. We immediately know that because the groups with the largest mean difference do not differ reliably none of the other group means differ reliably; we need not conduct further tests. The statistical findings may be summarized as follows: The common line under all five groups indicates that there is no reliable difference between any pair of them.

	Group				
	2	5	3	4	5
Mean	.80	1.15	2.34	3.29	4.67

In this preliminary investigation, the mean amounts of covert oral language behavior were larger for the language conditions than they were for the nonlanguage conditions. Typically in pilot investigations, we use relatively small numbers of subjects, as in this case, and thus should not expect reliable differences, as were found later in the more complete investigation (McGuigan & Bailey, 1969). The immediate point to be illustrated is that this procedure can be extended to any number of groups. Besides substituting the appropriate n's in Equation (9-8), one merely needs to be cautious about selecting the appropriate value of r_p.

Analysis of Variance

We shall now consider the classical statistical procedure applied to the multigroups design prior to the development of specialized tests like Duncan's new Multiple Range Test. However, we shall explain only enough of this procedure to allow you to statistically analyze the type of design covered in this and the next chapter.[10]

As with the preceding designs, we set up a null hypothesis. Our statistical test will allow us to either reject the null hypothesis or fail to reject it. For the present design, consider the null hypothesis to be that there is no difference among the means of the several groups. Keeping this null hypothesis in mind, let us return to it after our consideration of analysis of variance.

---

[10] Simplified but more thorough treatments of analysis of variance by psychologists are readily available in a number of good textbooks. For more detailed treatments see the references cited on page 244. In recent years the development of the small computer has increased the availability of computer programs for statistical analysis. Students are encouraged to study these procedures, such as an analysis of variance of one-, two- and three-treatment designs for a PDP-8 (Breaux, 1972).

You already have some acquaintance with the term variance (p. 218), which will help in the ensuing discussion. It would be helpful to review it now.

The simplest application of analysis of variance would be in testing the mean difference between two randomized groups. We have already discussed the *t*-test for this purpose. However, equivalent results would be obtained by conducting an analysis of variance on a two-groups design. That is, we could analyze a two-groups design by using either the *t*-test or analysis of variance (with the *F*-test, to be explained shortly) and obtain precisely the same conclusions. The same statement cannot be made if more than two groups are used, for the obvious reason that the *t*-test cannot be used for testing more than two means simultaneously. Let us say that the dependent variable values that result from a two-groups design are those plotted in Figure 9-12. That is, the curve to the left represents values for the participants in Group 1, and the frequency distribution to the right is for Group 2.

Now, are the means of these groups reliably different? To answer this question by using analysis of variance we first need to note that the *total sum of squares* may be determined. The total sum of squares is a value that results when we take all participants in the experiment into account as a whole. The total sum of squares is computed from the dependent variable values of all the participants ignoring the fact that some were under one experimental condition while others were under another experimental condition. The important point for us to observe here is that the total sum of squares may be partitioned (analyzed) into parts. In particular, the total sum of squares may be partitioned into two major components: the sum of

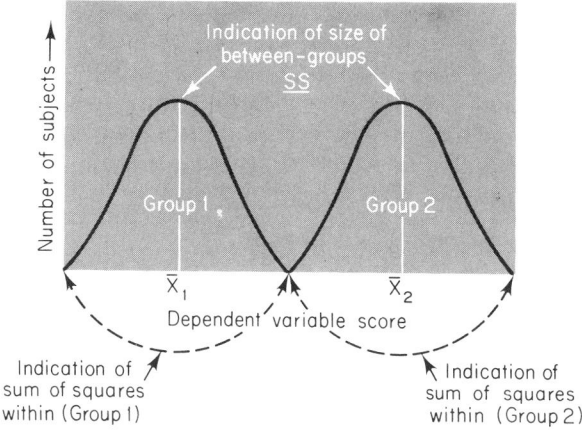

Figure 9-12. A crude indication of the nature of within- and between-groups sum of squares using only two groups.

squares *between groups* and the sum of squares *within groups*. *Roughly*, the sum of squares between groups may be thought of as determined by the extent to which the means of the two groups differ.

In Figure 9-12 the size of the between-groups sum of squares is crudely indicated by the distance between the two means. More accurately we may say that the larger the difference between the means, the larger is the between-groups sum of squares. The within-groups sum of squares, on the other hand, is determined by the extent to which the participants in each group differ among themselves. If the participants in Group 1 differ sizably among themselves, and/or if the same is true for participants in Group 2, the within-groups sum of squares is going to be large. And the larger the within-groups sum of squares, the larger the "error" in the experiment. By way of illustration, assume that all of the participants in Group 1 have been treated precisely alike. Hence, if they were precisely alike when they went into the experiment, they should all receive the same value on the dependent variable. If this happened, the within-groups sum of squares (as far as Group 1 is concerned) would be zero, for there would be no variation among their values. Of course, the within-groups sum of squares is almost never zero, since all the participants are not the same before the experiment and the experimenter is never able to treat all of them precisely alike.

Let us now reason by analogy to the *t*-test. You will recall that the numerator of Equation (5-2) (p. 124) is a measure of the difference between the means of two groups. It is thus analogous to our between-groups sum of squares. The denominator of Equation (5-2) is a measure of the "error" in the experiment and is thus analogous to our within-groups sum of squares. This should be apparent when one notes that the denominator of Equation (5-2) is large if the variances of the groups are large, and small if the variances of the groups are small (see p. 223). Recall that the larger the numerator and the smaller the denominator of the *t* ratio, the greater the likelihood that the two groups are reliably different. The same is true in our analogy: The larger the between-groups sum of squares and the smaller the within-groups sum of squares, the more likely our groups are to be reliably different. Looking at Figure 9-12 we may say that the larger the distance between the two means and the smaller the within (internal) variances of the two groups, the more likely they are to be reliably different. For example, the difference between the means of the two groups of Figure 9-13 is more likely to be statistically reliable than the difference between the means of the two groups of Figure 9-12. This is so because the difference between the means in Figure 9-13 is represented as greater than that for Figure 9-12 and also because the

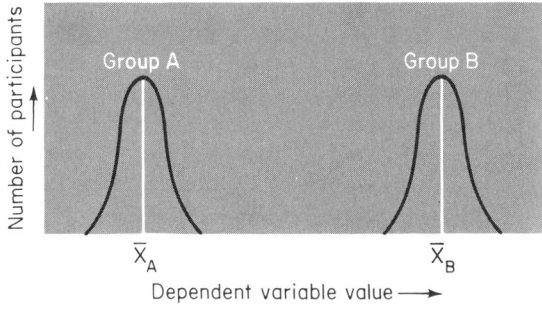

Figure 9-13. A more extreme difference between two groups than that shown in Figure 9-12. Here the between-groups sum of squares is greater but the within-groups sum of squares is less.

sum of squares within the groups of Figure 9-13 is represented as less than for Figure 9-12.

We have discussed the case of two groups. Precisely the same general reasoning applies when there are more than two groups: the total sum of squares in the experiment is analyzed into two parts, the within- and the among-groups sum of squares. ("Between" is used for two groups; "among" is the same concept applied to more than two). If the difference among the several means is large, the among-groups sum of squares will be large. If the difference among the several means is small, the among-groups sum of squares will be small. If the participants who are treated alike differ sizably, then the within (internal) sum of squares of each group will be large. And if the individual group variances are large, the within-group sum of squares will be large. The larger the among-groups sums of squares and the smaller the within-group sum of squares the more likely it is that the mean of the groups differ reliably.

We have attempted to present, in a surface fashion, the major rationale underlying analysis of variance. As we now turn to the computation of the several sums of squares we shall be more precise. The equations to be given are based on the following reasoning and their computation automatically accomplishes what we are going to say. First, a mean is computed that is based on all of the dependent variable values in the experiment taken together (ignoring the fact that some participants were under one condition and others under another condition). Then, the total sums of squares measures the deviation of all of the values from this overall mean. The among-groups sum of squares is a measure of the deviation of the means of the several groups from the overall mean. And the within-groups sum of squares is a pooled sum of squares based on the deviation of the scores in each group from the mean of that group. As we proceed we shall continue to enlarge on these introductory statements.

Our purpose will be to compute the total SS and then analyze it into its parts. A generalized equation for computing the total SS is:

(9-9)
$$\text{Total } SS = (\Sigma X_1^2 + \Sigma X_2^2 + \cdots + \Sigma X_r^2) - \frac{(\Sigma X_1 + \Sigma X_2 + \Sigma X_3 + \cdots + \Sigma X_r)^2}{N}$$

As before, the subscript r simply indicates that we continue adding the values indicated (the sum of X-squares, and the sum of X respectively) for as many groups as we have in the experiment.

Our next step is to analyze the total SS into components. There are two major components, that among groups and that within groups. A generalized equation for computing the among-groups SS is:

(9-10)
$$\text{Among } SS = \frac{(\Sigma X_1)^2}{n_1} + \frac{(\Sigma X_2)^2}{n_2} + \frac{(\Sigma X_3)^2}{n_3} + \cdots + \frac{(\Sigma X_r)^2}{n_r}$$
$$- \frac{(\Sigma X_1 + \Sigma X_2 + \Sigma X_3 + \cdots + \Sigma X_r)^2}{N}$$

The within-groups component of the total SS may be computed by subtraction. That is:

(9-11)
$$\text{Within } SS = \text{Total } SS - \text{Among } SS$$

In a more-than-two-randomized-groups design, of course, there may be any number of groups. To compute the several SS we must compute the ΣX and ΣX^2 separately for each group. The subscripts, as before, indicate the different groups. Hence ΣX_1 is the sum of the dependent variable values for Group 1, ΣX_3^2 is the sum of the squares of the dependent variable values for Group 3, and so forth. N remains the total number of participants in the experiment. To illustrate the analysis of variance procedure, consider an experiment related to one previously analyzed (pp. 124–134). In this study Jacobson, Fried, and Horowitz (1966) classically conditioned one group of planarians to a light; more specifically this group (Group CC for "classically conditioned") received paired presentations of a light and a shock. The planarians normally contract when shocked, but after conditioning they also contracted to the conditional stimulus, the light. Group PC (for "pseudoconditioning") was treated in the same way as Group CC, but the light and shock were not paired, i.e., these planarians were shocked and received light on their trials, but the light and shock were not associated so that one would not expect conditioning to occur. The third group (NC for "nonconditioned") simply remained in their home containers and were not exposed to the experimental situation.

After the above procedure was followed, untrained planarians were injected with ribonucleic acid (RNA, cf. p. 124) from the above groups. More specifically, a new group of planarians received RNA that was extracted from Group CC, a second naive group received injections of RNA from Group PC, and a third with injections from Group NC. These new groups were then tested to see how often they would give the conditional response (contraction) to the conditional stimulus (light). The number of conditional responses made by each animal during 25 test trials is presented in Table 9-9.

Table 9-9 Number of Responses on the 25 Test Trials for Each Injected Planarian

PLANARIANS INJECTED WITH RNA FROM:		
Group 1 (NC)	*Group 2 (PC)*	*Group 3 (CC)*
0	0	6
0	0	6
0	0	6
0	0	7
1	0	7
1	0	7
1	0	7
1	0	7
1	0	8
1	1	8
1	1	8
1	2	8
1	2	8
1	2	9
1	2	9
2	3	9
2	3	9
2	3	9
2	3	9
3	3	9
3	3	9
3	4	10
3	4	10
4	5	10
5	5	10
ΣX: 40	46	205
ΣX^2: 104	154	1721
n: 25	25	25
$\bar{X}$: 1.60	1.84	8.20

To compute the total SS, we may write the specialized form of Equation (9-9) for three groups as follows:

(9-12)   Total $SS = (\Sigma X_1^2 + \Sigma X_2^2 + \Sigma X_3^2) - \dfrac{(\Sigma X_1 + \Sigma X_2 + \Sigma X_3)^2}{N}$

We can see that the sum of X for Group NC is 40, for Group PC it is 46, and for Group CC it is 205. Or written in terms of Equation (9-12) we may say that $\Sigma X_1 = 40$, $\Sigma X_2 = 46$, and $\Sigma X_3 = 205$. Similarly, $\Sigma X_1^2 = 104$, $\Sigma X_2^2 = 154$, $\Sigma X_3^2 = 1721$, and $N = 75$. Substituting these values in Equation (9-12), we find the total SS to be:

$$\text{Total } SS = (104 + 154 + 1721) - \frac{(40 + 46 + 205)^2}{75} = 849.92$$

To compute the among-groups SS for three groups, we substitute the appropriate values in Equation (9-13), the specialized form of Equation (9-10) for three groups. This requires that we merely substitute the value of ΣX for each group, square it, and divide by the number of participants in each group. The last term, we may note, is the same as the last term in Equation (9-12). For this reason it is not necessary to compute it again, providing there was no error in its computation the first time. Making the appropriate substitutions from Table 9-9 and performing the indicated computations we find that:

(9-13)
$$\text{Among-groups } SS = \frac{(\Sigma X_1)^2}{n_1} + \frac{(\Sigma X_2)^2}{n_2} + \frac{(\Sigma X_3)^2}{n_3} - \frac{(\Sigma X_1 + \Sigma X_2 + \Sigma X_3)^2}{N}$$

$$= \frac{(40)^2}{25} + \frac{(46)^2}{25} + \frac{(205)^2}{25} - 1129.08 = 700.56$$

Substituting in Equation (9-11), we find that:

$$\text{Within-groups } SS = 849.92 - 700.56 = 149.36$$

We have said that we are conducting an analysis of variance and you may have wondered where the variances are that we are analyzing. We shall consider them now, but under a different name, for they are referred to in our sample values not as variances, but as *mean squares*. That is, we are computing sample values as estimates of population values. The mean squares (sample values) are estimates of the variances (population values). For example, the mean square within groups is an estimate of the within groups variance. The rule of computing mean squares is simple: Divide a given sum of squares by the appropriate degrees of freedom.

In introducing the equations that we use to determine the three degrees of freedom that we need, let us emphasize what we have done with regard to sums of squares. We have computed a total SS and partitioned it into two parts, the among SS and the within SS. The same procedure is followed for df. First we determine that:

(9-14) $$\text{Total } df = N - 1$$

then that:

(9-15) $$\text{Among } df = r - 1$$

and that:

(9-16) $$\text{Within } df = N - r$$

For our example we then find that, with $N = 75$ and $r = 3$,

$$\text{Total } df = 75 - 1 = 74$$
$$\text{Among } df = \ 3 - 1 = 2$$
$$\text{Within } df = 75 - 3 = 72$$

And we may note that the among df plus the within df equals the total df ($72 + 2 = 74$).

There are two mean squares that we need to compute, a mean square for the among-groups source of variation, and that for the within-groups. To compute the former we divide the among-groups SS by the among-groups df, and similarly for the latter. Hence, the within-groups mean square is 149.36 divided by 72. We shall enter these values in a summary table (Table 9-10 on the following page).

Now, as we have previously indicated, if the among-groups mean square is sizable, relative to the within-groups mean square, then we may conclude that the dependent variable values for the groups are different.

However, we must again face the problem: how sizable is "sizable," i.e., how large must the among component be in order for us to conclude that a given independent variable is effective? To answer this we apply a suitable statistical test. The test that is considered most appropriate is the F-test, which was developed by Professor Sir Ronald Aymer Fisher, one of the outstanding statisticians of all time. It was so named in his honor by another outstanding statistician, Professor George W. Snedecor. The F statistic for this design may be defined as follows:[11]

(9-17) $$F = \frac{\text{Mean square among groups}}{\text{Mean square within groups}}$$

This statistic is obviously easy to compute, and we may note the

---

[11] This is only one of a number of applications of the F-test. F is always a ratio of two variances and the fact that variances other than those in Equation (9-17) are sometimes used should not be a source of confusion. Simply realize that F may also be used in ways different than here.

Table 9-10 Summary Table for an Analysis of Variance

Source of Variation	Sum of Squares	df	Mean Square	F
Among Groups	700.56	2	350.28	169.22
Within Groups	149.36	72	2.07	
Total	849.92	74		

similarity between it and the t-test. For in both cases the numerator is an indication of the differences between or among groups (plus experimental error) and the denominator is an indication of the experimental error, or, as it is also called, the error variance of the experiment. More particularly, in this simple (of many) application of the F-test, the numerator contains an estimate of the error variance plus an estimate of the "real" effect (if any) of the independent variable. The denominator is only an estimate of the error variance. Now you can see what happens when you divide the numerator by the denominator: The computed value of F reflects the effect that the independent variable had in producing a difference between means. For example, suppose the independent variable is totally ineffective in influencing the dependent variable. In this case we would expect (at least in the long run) that the numerator would not contain any contribution from the independent variable (there would be no "real" among-groups mean square). Hence, the value for the numerator would only be an estimate of the error variance; a similar estimate of the error variance is in the denominator. Therefore, in the long run (with a large number of df), if you divide a value for the error variance by a value for the error variance, you obtain an F in the neighborhood of 1.0.

Thus, any time we obtain an F of approximately one, we can be rather sure that variation of the independent variable did not produce a difference in the dependent variable means of our groups. However, the numerator may be somewhat larger than the denominator — the among-groups mean square may be somewhat larger than the within-groups mean square. Now the question is, how large must the numerator of the F ratio be before we conclude that the means of the groups are really different. For, if the numerator is large (relative to the denominator), the value of F will be large. You will note that this is the same question we asked concerning the t-test. That is, how large must t be before we can reject our null hypothesis? And we shall answer this question for the F-test in a manner similar to that for the t-test. First, however, we should actually compute our values of F. Following our example, we have divided the mean square within groups

(2.07) into the mean square between groups (350.28), and inserted the resulting value (169) of F in Table 9-10.

Just as with the t-test, we next must determine the value of P that is associated with our computed value of F. Assume that we have set a criterion of $P = 0.05$. If the value of F has a probability of less than 0.05, then we may reject the null hypothesis; we may assert that there is a statistically reliable difference among the means of the groups. If, however, the P associated with our F is greater than 0.05, then we fail to reject the null hypothesis. We conclude that there is no reliable difference among the group means (or, more precisely, that not a single pair of means reliably differ).

To ascertain the value of P associated with our F, refer to Table 9-11, which fulfills the same function as the Table of t, although it is a bit different to use. Let us initially note that: (1) across the top we find "df associated with the numerator" and (2) down the left side we find "df associated with the denominator." Therefore, we know that we need two df values to enter the table of F. In this example we have 2 df for among groups (the numerator of the F-test) and 72 df for within groups (the denominator of the F-test). Hence we find the column labeled "2" and read down to find a row labeled "72." There is none, but there are rows for 60 and 120 df; 72 falls between these two values. We find a row for a P of 0.01, a row for a P of 0.05, and rows for P's of 0.10 and 0.20. We are making a 0.05 level test, so we shall ignore the other values of P. With 2 and 72 df, we interpolate between 3.15 and 3.07 and find that we must have an F 3.13 for significance at the 5 percent level. Since the computed F (169.22) exceeds this value,[12] the null hypothesis is rejected—we conclude that the groups do reliably differ. And, on the assumption that proper experimental techniques have obtained, it is concluded that variation of the independent variable reliably influenced the dependent variable. More specifically, with regard to this experiment, injection of RNA from groups with various training resulted in the groups differing reliably on the dependent variable measure.

Just as with the table of t, you should study the table of F sufficiently to make sure that you have an adequate understanding of it. To provide a little practice, say that you have six groups in your experiment with 10 participants per group. In this event you have five degrees of freedom for your among source of variation and 54 df for the within. Assume a 1 percent level test. What is the value of F that you must obtain in order to reject the null hypothesis? To answer this question, enter the column labeled "5" and read down until you

---

[12] A bit of an understatement.

Table 9-11 Table of F for Degrees of Freedom from 29– ∞*

df Associated with Denominator	P	df ASSOCIATED WITH NUMERATOR 1	2	3	4	5	6	8	12	24	∞
29	0 01	7.60	5.42	4.54	4.04	3.73	3.50	3.20	2.87	2.49	2.03
	0.05	4.18	3.33	2.93	2.70	2.54	2.43	2.28	2.10	1.90	1.64
	0.10	2.89	2.50	2.28	2.15	2.06	1.99	1.89	1.78	1.65	1.47
	0.20	1.72	1.70	1.65	1.60	1.57	1.54	1.50	1.45	1.39	1.29
30	0.01	7.56	5.39	4.51	4.02	3.70	3.47	3.17	2.84	2.47	2.01
	0.05	4.17	3.32	2.92	2.69	2.53	2.42	2.27	2.09	1.89	1.62
	0.10	2.88	2.49	2.28	2.14	2.05	1.98	1.88	1.77	1.64	1.46
	0.20	1.72	1.70	1.64	1.60	1.57	1.54	1.50	1.45	1.38	1.28
40	0.01	7.31	5.18	4.31	3.83	3.51	3.29	2.99	2.66	2.29	1.80
	0.05	4.08	3.23	2.84	2.61	2.45	2.34	2.18	2.00	1.79	1.51
	0.10	2.84	2.44	2.23	2.09	2.00	1.93	1.83	1.71	1.57	1.38
	0.20	1.70	1.68	1.62	1.57	1.54	1.51	1.47	1.41	1.34	1.24
60	0.01	7.08	4.98	4.13	3.65	3.34	3.12	2.82	2.50	2.12	1.60
	0.05	4.00	3.15	2.76	2.52	2.37	2.25	2.10	1.92	1.70	1.39
	0.10	2.79	2.39	2.18	2.04	1.95	1.87	1.77	1.66	1.51	1.29
	0.20	1.68	1.65	1.59	1.55	1.51	1.48	1.44	1.38	1.31	1.18
120	0.01	6.85	4.79	3.95	3.48	3.17	2.96	2.66	2.34	1.95	1.38
	0.05	3.92	3.07	2.68	2.45	2.29	2.17	2.02	1.83	1.61	1.25
	0.10	2.75	2.35	2.13	1.99	1.90	1.82	1.72	1.60	1.45	1.19
	0.20	1.66	1.63	1.57	1.52	1.48	1.45	1.41	1.35	1.27	1.12
∞	0.01	6.64	4.60	3.78	3.32	3.02	2.80	2.51	2.18	1.79	1.00
	0.05	3.84	2.99	2.60	2.37	2.21	2.09	1.94	1.75	1.52	1.00
	0.10	2.71	2.30	2.08	1.94	1.85	1.77	1.67	1.55	1.38	1.00
	0.20	1.64	1.61	1.55	1.50	1.46	1.43	1.38	1.32	1.23	1.00

*Table 9-11 is abridged from Table V of Fisher and Yates: *Statistical Tables of Biological, Agricultural, and Medical Research,* published by Oliver and Boyd Ltd, Edinburgh, by permission of the author and publishers. Complete table is in the Appendix.

find the rows for 54 df. There is no row for 54 df so you must interpolate. The 54 df falls between the tabled values of 40 df and 60 df. If you had had 40 df for your within groups, then you would have needed a computed F of 3.51 in order to reject the null hypothesis; similarly if you had had 60 df, you would have required an F of 3.34. By linearly interpolating we find that an F of 3.39 is required for significance at the 1 percent level. Try some additional problems for yourself.

Now, if you had conducted the above experiment you might feel quite happy with yourself; you would have succeeded in rejecting the null hypothesis. But wait a moment. What null hypothesis did you reject? Your conclusion is that there is a (at least one) mean difference between your groups. But where does the difference lie? Is it between the means of Groups 1 and 2, between Groups 1 and 3, between Groups 2 and 3, or are two, or all, of these differences reliable? The traditional (but, as we shall see, inappropriate) answer to this question is to run t-tests between the groups as indicated above; this involves running three t-tests. Obviously this involves testing three additional null hypotheses, set up for our three t-tests, e.g., there is no difference between the means of Groups 1 and 2. Recall that for the F-test there was only one null hypothesis, viz., there is no difference between any pair of means when all means are simultaneously compared. Using the data in Table 9-9, for your own practice compute the values of t; you will find them to be:

Between Groups 1 and 2: $t_{12} = \quad .58$
Between Groups 1 and 3: $t_{13} = 18.85$
Between Groups 2 and 3: $t_{23} = 18.17$

We find that the t between Groups 1 and 3 and between Groups 2 and 3 is reliable with the usual criterion of $P < 0.05$; our question as to where among the three groups the reliable difference lies is answered. That is, Group 3 (the Group that received RNA from the planarians that were classically conditioned) yielded reliably more contractions to the light during the test trials than did the other two (control) groups. The "... data would seem to suggest that a specific learned response was transferred by way of the injection of the RNA preparation" (Jacobson et al., 1966, p. 5).

To briefly summarize the general approach of this section, we may say that on the assumption that only comparisons between pairs are to be made, and on the assumption that all possible pairs are to be compared, we first run an F-test. If the value of this "overall" F is reliable, we then conduct all possible t-tests in order to ascertain which specific groups are significantly different. If, however, the overall F is not reliable, then we conclude that there are no reliable differences between the various pairs of groups, thereby not running additional t-tests.

Let us briefly comment on a different approach that you might wish to take in an experiment. Suppose that you are not really interested in making all possible comparisons between your groups. For instance, suppose that you conduct a three-group experiment, and that your empirical hypothesis suggests that Groups 1 and 2 should differ and that Groups 1 and 3 should also differ; but in this event the comparison between Groups 2 and 3 is rather uninteresting to you; your hypothesis says nothing about this comparison. In this event you need not conduct your overall F-test; you go directly to your t-test analysis, computing the two indicated values of t.

Appendix to Chapter 9

The traditional approach of conducting an analysis of variance and an F-test, possibly followed by t-tests, was prominent in the 1940s and certainly was a relatively advanced procedure for the time. However, statisticians later became aware of the serious shortcomings of this procedure. In a classic article on this topic, Ryan (1959) commented on this problem as follows:

> In the older psychological literature, this problem has been dealt with in a haphazard manner, without recognizing the issues involved. More recently, statistical procedures specifically designed for multiple comparisons have become available. . . . It has not been clear to many psychologists, however, that there are several different methods with different basic assumptions or approaches . . . these issues have not been clearly faced in the psychological literature (p. 26).

As discussed above, the traditional reasoning was as follows: conduct the F-test and if it is not reliable, then conclude that there are no reliable differences between any pairs of means. However, if the F allows you to reject the null hypothesis, then conduct all possible t-tests to ascertain where the reliable difference(s) lie. We have already noted that it is possible to legitimately obtain a reliable value of t, when the overall F-test is not reliable. Let us now illustrate the inappropriateness of applying the t-test to the multi-randomized groups design. Suppose that we conduct a two-groups experiment on rats. We set our reliability level at 0.05. Recall that, assuming that the null hypothesis is true, this reliability level means that if we obtain a t that has a P of 0.05, the odds are five in 100 that a t of this size or larger could have occurred by chance. Since this would happen only rarely (5 percent of the time) we reason that the *t* was not the result of random fluctuations. Rather, we prefer to conclude that the two groups are "really" different as measured by the dependent variable. We thus reject our null hypothesis and conclude that variation of the independent variable was effective in producing the difference between our two groups. Now, after completing the above work, say that we conduct a new two-groups experiment, for example one on schizophrenics. Note that the two experiments are independent of each other. In the experiment on schizophrenics we also set our reliability level at 0.05, and follow the same procedure as before. Again our reliability level means that the odds are 5 in 100 that a *t* of the corresponding size could have occurred by chance.

But let us ask a question. Given a reliability level of 0.05 in each of the two experiments, what are the odds that by chance the *t* in one, the other, or both experiments will be reliable? Before you reach a hasty conclusion, let us caution you that the probability is *not* 0.05. Rather, the joint probability could be shown to be 0.0975.[13] That is, the odds of obtaining a *t* reliable at the 0.05 level in either or both experiments are 975 out of 10,000. And this is certainly different from 0.05.

To illustrate, we might develop an analogy: What is the probability of obtaining a head in two tosses of a coin? On the first toss it is one in two, and on the second toss it is one in two. But the probability of obtaining two heads on two successive tosses (before your first toss) is $1/2 \times 1/2 = 1/4$. To develop the analogy further, the probability of obtaining a head on the first toss, or on the second toss, or on both tosses (again, computed before *any* tosses) is $P = 0.75$.

In a different situation, suppose that we conduct a three-groups experiment. In this case there are three *t*-tests in which we would probably be interested: a *t* between Groups 1 and 2, between Groups 1 and 3, and between Groups 2 and 3. Assume that we set a reliability level of 0.05 *for each t*. If the first *t*-test yields a value reliable beyond the 0.05 level, we reject the null hypothesis. And likewise for the other two *t*-tests. But what are the odds of obtaining a reliable *t* when we consider all three *t*-tests? That is, what are the odds of obtaining a reliable *t* in at least one of the following situations:

First:	Between Groups 1 and 2
or Second:	Between Groups 1 and 3
or Third:	Between Groups 2 and 3

[13] By the following formula: $P_j = 1 - (1 - a)^k$ where P_j is the joint probability, *a* is the reliability level, and *k* is the number of independent experiments. For instance in this case $a = .05$, $k = 2$. Therefore $P_j = 1 - (1 - 0.05)^2 = 0.0975$.

or Fourth:	Between Groups 1 and 2 and also between Groups 1 and 3
or Fifth:	Between Groups 1 and 2 and also between Groups 2 and 3
or Sixth:	Between Groups 1 and 3 and also between Groups 2 and 3
or Seventh:	Between Groups 1 and 2 and also between Groups 2 and 3 and also between Groups 1 and 3.

The answer to this question is more complex than before. The reason for this is that these t-tests are not independent. For example, the computed value of t between Groups 1 and 2 would be related to the computed value of t between Groups 1 and 3 because Group 1 occurs in both t-tests. Similarly, t-tests between Groups 2 and 3 and between Groups 1 and 3 would not be independent. And lacking independence in the t-tests, it would be difficult to say just what the joint or overall reliability level is (as we were able to say in the previous case that it was 0.9075). About the best we can say for the general case is that the reliability level for all possible t-tests is less than that which would obtain if the t-tests were independent. This is not much help. But the moral is: the running of all possible t-tests (between pairs of means) is not a satisfactory technique of statistical analysis. It simply does not provide a reasonable probability level when all possible t's are considered. However, this is not the worst of it. If, for instance, we had seven groups in our experiment, we would have to run 21 t-tests in order to consider all possible combinations between pairs of means. Although we would not shirk the work involved in running all these tests if it were necessary, we still have a limited amount of energy and would prefer to expend it in ways other than running a great many t-tests, if possible.

One appropriate solution might be to run fewer t-tests. Assuming that we are only interested in t-tests between pairs of means, we could select those t-tests that are independent and run them. Following the seven-groups example, however, it could be shown that only three such t-tests *between pairs* would be independent. And we are usually interested in comparing more than three pairs of groups in such an experiment.

The importance of this discussion is that we have demonstrated our objections to the older procedure of analyzing the multi-randomized-groups design, that of an analysis of variance followed by all possible t-tests. These criticisms are not directed toward the analysis of variance phase, for that by itself is perfectly legitimate. Thus you may conduct your analysis of variance and run your F-test. If you are able to reject the null hypothesis, then you know that some reliable difference exists between your groups, but that is all that the F-test tells you for you do not know where the difference lies.

Duncan's Range Test seems considerably more appropriate, for it: (1) allows us to make all possible comparisons between pairs of our groups, just as the 21 t-tests in the previous example would; (2) is less work than running an F-test and a large number of t-tests; and (3) provides a more reasonable probability level than for all possible t-tests, considered jointly.

You may well ask why we would present a procedure for statistical analysis that we consider inappropriate. The reasons are two: (1) because some psychologists still prefer the traditional approach, which is certainly their privilege; and (2) because analysis of variance is so important in itself and forms a basis for our procedures in the next chapter.

Summary of the Computation of Duncan's Range Test for a Randomized-Groups Design with More than Two Groups

Assume that the following dependent variable values have been obtained for four groups of participants.

Group 1	Group 2	Group 3	Group 4
1	2	8	7
1	3	8	8
3	4	9	9
5	5	10	9
5	6	11	10
6	6	12	11
7	6	12	11

1. First we wish to compute ΣX, ΣX^2, n, and X values for each group.

ΣX	28	32	70	65
ΣX^2	146	162	718	617
n	7	7	7	7
$\overline{X}$	4.00	4.57	10.0	9.29

2. Compute the sum of squares (SS) for each group. These values are determined by substituting in Equation (9-1) and performing the indicated operations.

$$SS = \Sigma X^2 - \frac{(\Sigma X)^2}{n}$$

$$SS_1 = 146 - \frac{(28)^2}{7} = 34.00$$

$$SS_2 = 162 - \frac{(32)^2}{7} = 15.71$$

$$SS_3 = 718 - \frac{(70)^2}{7} = 18.00$$

$$SS_4 = 617 - \frac{(65)^2}{7} = 13.43$$

3. Using Equation (9-5), compute the square root of the error variance. Substituting the above values and performing the appropriate operations, we find that:

$$s_e = \sqrt{\frac{SS_1 + SS_2 + SS_3 + SS_4}{4(n-1)}} = 1.84$$

4. Determine the degrees of freedom for Duncan's Range Test, where:

$$df = N - r = 28 - 4 = 24$$

5. Assuming a 1 percent level test, enter Table 9-6 to determine the appropriate values of r_p. Since we have four means in the present example, we need to enter Table 9-6 at the columns labeled 4, 3, and 2. With 24 df we find the values of r_p to be:

Number of Groups

	2	3	4
r_p	3.96	4.14	4.24

6. Compute the least significant ranges (R_p) for comparisons between two groups, among three groups, and among four groups. The equation [Equation (9-4)] is:

$$R_p = (s_e)\,(r_p)\sqrt{1/n}$$

Making the appropriate substitutions to determine R_p for two groups (i.e., R_2), and performing the indicated operations:

$$R_2 = (1.84)\,(3.96)\,\sqrt{1/7} = 2.75$$

Similar substitutions and performance of the operations result in the values of R_p for three and for four groups:

$$R_3 = (1.84)\,(4.14)\,\sqrt{1/7} = 2.88$$

$$R_4 = (1.84)\,(4.24)\,\sqrt{1/7} = 2.95$$

The computed values of R_p may now be summarized:

	Number of Groups		
	2	3	4
R_p	2.75	2.88	2.95

7. The next step is to rank the means of the groups from lowest to highest.

	Group 1	Group 2	Group 4	Group 3
$\bar{X}$	4.00	4.57	9.29	10.00

8. We now test for significant differences among the various pairs of means. Starting with the highest (Group 3) and the lowest (Group 1), we can see that the difference between their means is 6.00. Determining the appropriate value of R_p for the comparison (that for four groups, hence $R_4 = 2.95$) we compare the mean difference with the value of R_p. In this case 6.00 exceeds 2.95. Therefore, the means of Groups 1 and 3 differ significantly. The next comparison is between the highest group (Group 3) and the second from the lowest group (Group 2). The difference between these pairs of means is 5.43. The value of R_p for comparing three groups is 2.88. Since 5.43 exceeds 2.88, the means of Groups 2 and 3 differ reliably. The next comparison is between the highest and the next to highest means. The mean difference is 0.71. This is a two-group comparison; hence $R_p = 2.75$. Since the difference between the means of Groups 3 and 4 (0.71) does not exceed 2.75, these two groups do not differ reliably. We now test for a reliable difference between the next to highest mean and the lowest mean. The mean difference is 5.29, and $R_3 = 2.88$. Therefore, Groups 4 and 1 differ reliably. In the test between Groups 4 and 2 the mean difference is 4.72 and $R_2 = 2.75$. Therefore Groups 4 and 2 differ reliably. The final comparison is between Groups 1 and 2. Their mean difference is 0.57. Since $R_2 = 2.75$, these two groups do not differ reliably.

9. We now summarize the results of our tests for statistical reliability. We found, in our example, that reliable differences did not exist between Groups 1 and 2, nor between Groups 3 and 4. All other differences were reliable. These findings are summarized as follows:

Group 1	Group 2	Group 4	Group 3
4.00	4.57	9.29	10.00

Problems

1. An experimenter was interested in assessing the relative sociability scores of different majors in his college. A random sample of students who were majoring in English, art, and chemistry was selected, and they were administered a standardized test of sociability. Assuming a 1 percent level of reliability, did the three groups differ on this measure? Can it be said that all three groups are reliably different from each other?

SOCIABILITY SCORES

English Majors	Art Majors	Chemistry Majors
1,2,2,2,3,3	5,5,5,6,6,6	9,9,9,9,10,10

2. A physical education professor is interested in the effect of practice on the frequency of making goals in hockey. After consulting a psychologist the following experiment was designed. Four groups were formed such that Group 1 received the most practice, Group 2 the second most practice, Group 3 the third most practice, and Group 4 the least amount of practice. Dependent variable values represent the number of goals made by each participant during a test period. Determine which groups are reliably different. (Criterion is $P < 0.05$.)

		Group	
I	II	III	IV
5	1	5	10
7	4	0	9
9	0	2	6
7	0	1	5
6	8	1	8
5	3	4	9
9	2	3	8
2	1	0	2

3. An experimenter was interested in testing the hypothesis that the greater the hunger drive, the more correct choices a rat would make in a certain number of runs in a maze. Five groups of rats were formed such that Group 1 had zero hours of food deprivation, Group 2 had 12 hours of food deprivation, Group 3 had 24 hours of food deprivation, Group 4 had 36 hours of food deprivation, and Group 5 had 48 hours of food deprivation. Setting a 5 percent level for rejecting the null hypothesis, was the hypothesis confirmed?

NUMBER OF CORRECT CHOICES

Group 1	Group 2	Group 3	Group 4	Group 5
0	1	0	3	4
0	1	1	3	5
1	3	1	4	6
2	3	2	5	7
3	4	4	6	7
3	4	4	7	8
4	4	5	7	9
5	4	5	8	10
6	5	7	9	11
7	6	8	10	12
7	7	9	11	14

4. An experiment is conducted to determine which of the three methods of teaching Spanish is superior. Assuming that the experiment has been adequately conducted, that a five percent level of reliability has been set, and that the higher the test score the better the performance after training on the three methods, which method is to be preferred?

Method A Participants	Method B Participants	Method C Participants
55	46	45
52	40	41
50	35	37
48	32	36
47	31	30
46	28	25
40	25	24
35	22	21
	21	21
	19	20
		19
		18
		17

10
Experimental Design
the factorial design

All the preceding designs are appropriate to the investigation of a single independent variable. If the independent variable is varied in two ways, one of the two-groups designs is used. If the independent variable is varied in more than two ways, the multigroup design is used. It is possible, however, to study more than one independent variable in a single experiment. One possible design for studying two or more independent variables in a single experiment is the *factorial design*. A complete factorial design is one where all possible combinations of the selected values of each of the independent variables are used. To illustrate a simple factorial design, let us consider some data from an experiment by W. F. Harley, Jr. and Sr. (personal communication) on learning during hypnosis. Among the independent variables they studied were: (1) whether or not the participants were hypnotized; and (2) susceptibility to be hypnotized. These are both continuous variables, but two values of each were selected for study so that the first was dichotomized as above, and the second according to high or low susceptibility of the participants. Variation of these two independent variables might be diagrammed as in Figure 10-1. But

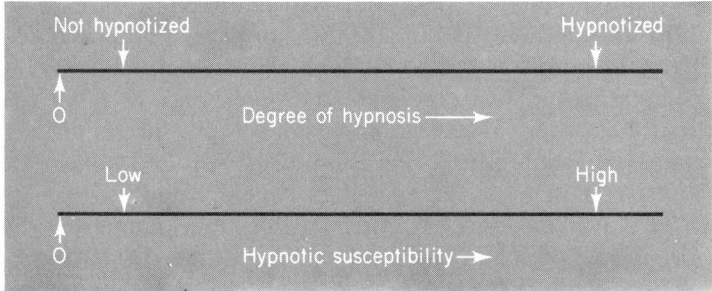

Figure 10-1. Variation of two independent variables, each in two ways.

since this was a factorial design, all possible combinations of the values of the independent variables were used, as indicated in Table 10-1.

Table 10-1 Diagram of a Factorial Design

Degree of hypnosis

		Hypnotized	Not hypnotized
Hypnotic Susceptibility	Low	(1)	(2)
	High	(3)	(4)

Table 10-1 shows that there are four possible combinations of the values of the independent variables. Each possible combination is represented by a box, a *cell:* (1) hypnotized and low susceptibility; (2) not hypnotized and low susceptibility; (3) hypnotized and high susceptibility; (4) not hypnotized and high susceptibility. With four experimental conditions there are four groups to consider in the experiment. Therefore, an equal number of participants was assigned to each of the conditions.[1]

More precisely, once the participants were tested for hypnotic susceptibility, two classes of them were formed: those high and those

[1] It is not necessary to have an equal number of participants in each cell, but the statistical analysis is more complicated with unequal n's.

low in susceptibility. Then those high in susceptibility were randomly assigned to either the hypnotic or the nonhypnotic conditions and similarly for those who tested out to be low in susceptibility.

The experiment was then conducted essentially as follows. First, all participants, while in the waking state, were presented with a paired-associate learning task, and the number of errors that they required to learn the task was tabulated. A similar count was made on a comparable paired-associate list during the experimental conditions, and the dependent variable measure was the difference in number of errors made on the two occasions. The groups were treated as follows: Group 1 consisted of participants for whom a test showed that they had low susceptibility to hypnosis, and they learned the second list while hypnotized; Group 2 was also made up of low-susceptibility subjects, but they learned the second list when in a normal awake state; Group 3 consisted of participants who were quite susceptible to hypnosis, and they learned the second list while hypnotized; Group 4 was composed of highly susceptible participants who learned the second list when not hypnotized. A statistical analysis of the dependent variable scores should then provide information concerning the following questions:

1. Does being hypnotized influence learning?
2. Does susceptibility to hypnotism influence learning?
3. Is there an interaction between degree of hypnosis and susceptibility to be hypnotized?

The procedure for answering the first two questions is straightforward, but the third will require a little more consideration. Let us examine the dependent variable values obtained for each group (Table 10-2).

Now let us place the means for the four groups in their appropriate cells (Table 10-3).

Turning to the first question first, we shall study the effect of being in a hypnotized state on learning scores. For this purpose we shall ignore the susceptibility variable. That is, we have 8 highly susceptible individuals who were hypnotized and 8 with low susceptibility who were hypnotized. Ignoring the fact that 8 were high and 8 were low in susceptibility, we have 16 participants who learned while in a state of hypnosis. Similarly, we have 16 people who learned when they were not hypnotized. We therefore have two groups who, as a whole, were treated similarly except with regard to the hypnosis variable. For the hypnosis-nonhypnosis comparison it is irrelevant that half of each group were high in susceptibility and half were low in this respect—

Table 10-2 Dependent Variable Values for the Four Groups That Compose the Factorial Design of Table 10-1

	GROUP		
1 (*Hypnotized —* *low susceptibility*)	*2* (*Not hypnotized —* *low susceptibility*)	*3* (*Hypnotized —* *high susceptibility*)	*4* (*Not hypnotized —* *high susceptibility*)
0	9	−16	−4
−8	1	0	8
1	−5	−20	−10
−20	−14	−41	9
−17	−2	−32	−10
−43	−3	−6	−23
−4	14	−42	29
−23	9	−29	−14
n: 8	8	8	8
ΣX: −114	9	−186	−15
ΣX^2: 3148	593	6002	1927
$\bar{X}$: −14.25	1.12	−23.25	−1.88

Table 10-3 Showing Means for the Experimental Conditions

Degree of hypnosis

		Hypnotized	Not hypnotized	Means
Susceptibility	Low	−14.25	1.12	−6.57
	High	−23.25	−1.88	−12.57
Means:		−18.75	−.38	−9.57

the susceptibility variable is balanced out. To make our comparison we need merely compute the mean for the 16 hypnotized individuals and for the 16 nonhypnotized participants. To do this we have computed the mean of the means for the two groups who were hypnotized (Table 10-3). (This is possible because the n's for each mean are equal.) That is, the mean of −14.25 and −23.25 is −18.75 and similarly for the nonhypnotized participants, as shown in Table 10-3. Since the two means (−18.75 and −.38) are markedly different, we suspect that being hypnotized influenced the dependent variable. We shall, however, have to await the results of a statistical test to find out if this difference is reliable.

Students who find it difficult to ignore the susceptibility variable when considering the hypnosis variable should look at the factorial design as if they are conducting only one experiment and varying only the degree of hypnosis. In this case the susceptibility variable can be temporarily considered as an extraneous variable whose effect is balanced out. Thus, the two-groups design would look like that indicated in Table 10-4.

Table 10-4 Looking at One Independent Variable of the Factorial Design as a Single Two-Groups Experiment

Value of independent variable	*Group 1* (*hypnotized*)	*Group 2* (*Not hypnotized*)
n	16	16
Mean dependent variable score	-18.75	$-.38$

We now return to question number two and compare the high versus low susceptibility classification by ignoring the hypnosis variable. The mean of the 16 participants who were low in susceptibility is -6.57 and the mean of those who were high in susceptibility is -12.57. The difference between these means is not as great as before, suggesting that perhaps this variable did not greatly, if at all, influence the learning scores. Again, however, we must await the results of a statistical test for reliability before making a final judgment.

Now that we have preliminary answers to the first two questions, let us turn to the third: Is there an interaction between the two variables? *Interaction* is one of the most important concepts discussed in this book. If you adequately understand it, you will have ample opportunity to apply it in a wide variety of situations; it will shed light on a large number of problems and considerably increase your understanding of behavior.

First, let us approach the concept of interaction from an overly simplified point of view. Assume the problem is of the following sort: Is it more efficient (timewise) for a man who is dressing to put his shirt or his trousers on first? At first glance it might seem that a suitable empirical test would yield one of two answers: (1) shirt first or (2) trousers first. However, in addition to these possibilities there is a third answer—(3) it depends. Now "it depends" embodies the basic notion of interaction. Suppose a finer analysis of the data indicates what "it depends" on. We may find that it is more efficient for tall men to put their trousers on first but for short men to put their shirts on first. In this case we may say that our answer depends on the body build of the man who is dressing. Or to put it in terms of an interaction, we may say that there is an interaction between body build and putting trousers or shirt on first. This is the basic notion of interaction. Let us take another example from everyday life before we consider the concept in a more precise manner.

The author once had to obtain the support of a senior officer in the Army to conduct an experiment. In order to control certain variables (e.g., the effect of the company commander) it was decided to use only one company. There were four methods of learning to be studied, so it was planned to divide the company into four groups. Each group (formed into a platoon) would then learn by a different method. The officer, however, objected to this design. He said that "we always train our men as a whole company. You are going to train the men in platoon sizes. Therefore, whatever results you obtain with regard to platoon-size training units may not be applicable to what we normally do with company-size units." The author had to admit this point, and it is quite a sophisticated one. It is possible that the results for platoons might be different than the results for companies — that there is an *interaction* between size of personnel training unit and the type of method used. In other words, one method might be superior if used with platoons, but another if used with companies. Actually, previous evidence suggested that such an interaction was highly unlikely in this situation, so the author didn't worry about it; he only left a slightly distressed senior officer.

An interaction exists between two independent variables if the dependent variable value that results from one independent variable is determined by the specific value assumed by the other independent variable. To illustrate, momentarily assume that there is an interaction between the two variables of degree of hypnosis (hypnotized and non-hypnotized) and susceptibility to being hypnotized (high and low). The interaction would mean that the results (learning scores) for degree of hypnosis would depend upon the degree of susceptibility of the participant. Or, more precisely, one might state the interaction as follows: whether or not being hypnotized affects amount learned depends on the degree of susceptibility of the participants.

To enlarge on our understanding of the concept of an interaction, let us temporarily assume certain fictitious sample (not population) values for the hypnosis experiment, values that indicate a lack of an interaction (Figure 10-2). On the horizontal axis we have shown the two values of the susceptibility variable. The data points represent fictitious means of the four conditions: point number one is the mean for the low-susceptibility hypnotized group; two is for the low-susceptibility nonhypnotized group; three, the high-susceptibility hypnotized group; and four, the high-susceptibility nonhypnotized group.

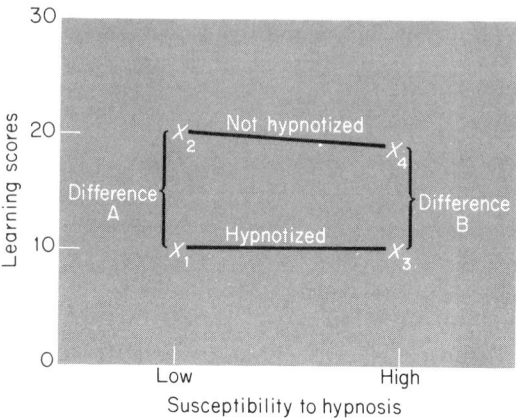

Figure 10-2. Illustration of a lack of interaction with fictitious sample means.

The line that connects points one and three represents the performance of the hypnotized participants, half of whom were low and half high in susceptibility. The line through points number two and four represents the performance of the nonhypnotized participants. If these were real data, what would be the effects of the independent variables? First, variation of the degree of susceptibility would be said not to affect learning, for both lines are essentially horizontal. Second, the nonhypnotized performed better than the hypnotized participants (the "nonhypnotized" line is higher than the "hypnotized" line). And third, the difference between the low-susceptibility hypnotized group and the low-susceptibility nonhypnotized group (Difference A) is about the same as the difference between the high-susceptibility hypnotized and the high-susceptibility nonhypnotized groups (Difference B). The performance of participants who were and were not hypnotized is thus essentially independent of their degree of susceptibility. No interaction exists between these two variables. Put another way: If the lines drawn in Figure 10-2 are approximately parallel (i.e., if Difference A is approximately the same as Difference B), it is likely that no interaction exists between the variables.[2] However, if the lines based on these sample means are clearly not parallel (i.e., if Differ-

[2] Let us emphasize that we are talking about sample values and not about population values. Thus, although this statement is true for sample values it is not true for population (true) values. Therefore, if the lines for the population values are even slightly nonparallel, there is an interaction.

ence A is distinctly different from Difference B), an interaction is present.

Another way of illustrating the same point is to compute the differences between the means of the groups. The means plotted in Figure 10-2 are specified in the cells of Table 10-5. We have computed the necessary differences so that it can be seen that the difference between the participants with low susceptibility who were hypnotized and those who were not hypnotized is −10.00 and that for the high-susceptibility participants it is −8.75. Since these are similar differences, there is probably no interaction present. The same conclusion would be reached by comparing differences in the other direction, i.e., since 0 and 1.25 are approximately the same, no interaction exists. Incidentally, the −10.00 is Difference A of Figure 10-2, and −8.75 is Difference B. Clearly, if these differences are about the same, the lines will be approximately parallel.

Table 10-5 Illustration of a Lack of an Interaction with Fictitious Means

Degree of hypnosis

		Hypnotized	Not hypnotized	Difference
Susceptibility	Low	10	20	−10.00
	High	10	18.75	−8.75
Difference		0.00	1.25	0.00

At this point you may be disappointed that we did not illustrate an interaction. This can easily be arranged by assuming for the moment that the data came out as indicated in Table 10-6. In this case our lines would look like those in Figure 10-3.

Now we note that the lines are not parallel; in fact they cross each other. Hence, if these were real data, we would make the following statements: Low-susceptibility participants who are not hypnotized are superior to low-susceptibility participants who are hypnotized; but high-susceptibility participants who are hypnotized are superior to high-susceptibility participants who are not hypnotized. Or, the logically equivalent statement is: The effect of being hypnotized depresses

Table 10-6 New Fictitious Means Designed to Show an Interaction

Degree of hypnosis

		Hypnotized	Not hypnotized	Means
Susceptibility	Low	69.1	90.0	79.55
	High	91.7	80.0	85.85
		80.40	85.00	82.70

performance for low-susceptibility participants but facilitates performance for high-susceptibility participants. Put in other words: The difference between being hypnotized and being not hypnotized depends on the susceptibility of the participants, or equally, the difference between degree of susceptibility depends on whether or not the participants are hypnotized.

This discussion should clarify the meaning of "interaction." This is a rather difficult concept, however, and the examples in the remainder of the chapter should help to illuminate it further. Note for now though, that reliably nonparallel lines indicate an interaction *but the lines do not need to intersect each other.*

To summarize, when selected values of two or more indepen-

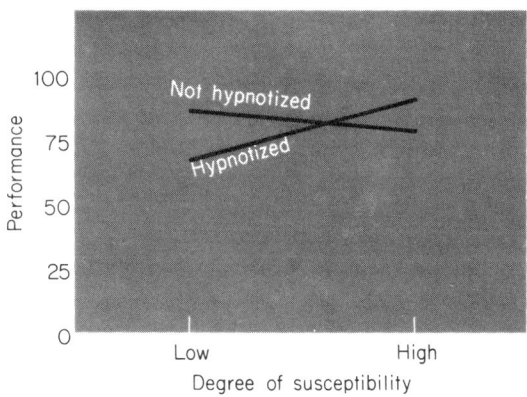

Figure 10-3. Illustration of a possible interaction with fictitious sample means.

dent variables are studied in all possible combinations, a factorial design is used. We have illustrated the factorial design by using two independent variables with two values of each. In this case participants are assigned to the four experimental conditions. Analysis of the dependent variable data yields information on: (1) the influence of each independent variable on the dependent variable; and (2) the interaction between the two independent variables.

Types of Factorial Designs

Factorial Designs with Two Independent Variables

The 2 × 2 Factorial Design. The type of factorial design that we have discussed is referred to as the 2 × 2 factorial design. In this design we study the effect of two independent variables each varied in two ways. The number of numbers used in the label indicates the number of independent variables studied in the experiment. And the size of the numbers indicates the number of values of the independent variables. Since the 2 × 2 design has two numbers (2 and 2) we can tell immediately that there are two independent variables. And since the numbers are both 2, we know that each independent variable assumed two values. From "2 × 2" we can also tell how many experimental conditions there are: 2 multiplied by 2 is 4.

The 3 × 2 Factorial Design. The 3 × 2 factorial design is one in which two independent variables are studied, one being varied in three ways, while the second assumes two values.

A 3 × 2 design was used in an experiment on programmed learning by Seidel and Rotberg (1966). The first variable concerned verbalizing the content of rules used in writing a computer program, and the three ways in which it was varied were: (1) Rules Condition—these participants periodically wrote out the content of the rules; (2) Naming Condition—they wrote down the names of the rules; and (3) Nonverbalization Condition—they wrote their programs without any verbalization of the rules. The second variable was prompting versus confirmation: (1) Prompting—under this condition participants were required to write their answers after they were given the explicit information needed to make the response; (2) Confirmation—here participants were required to write their answers prior to receiving information that confirmed the correctness of their responses. Participants were then randomly assigned to these six conditions. A representation of this design is presented in Table 10-7.

Table 10-7 A 3 × 2 Factorial Design

| | Verbalization | | |
	Write rules	Name rules	None
Prompting	17	34	368
Confirmation	24	38	389

(Information)

One set of dependent variable scores is, incidentally, included in Table 10-7. On the criterion tests the prompting versus confirmation comparison was not reliable. With regard to the verbalization variable, the participants who did not have to verbalize anything regarding the rules (the nonverbalization condition) were superior to the participants who served under the two verbalization conditions. Would you say that there is an interaction present?

The 3 × 3 Factorial Design. This design is one in which we investigate two independent variables, each varied in three ways. We therefore assign participants to nine experimental conditions. Boe's (1966) use of a 3 × 3 design involved the effect of punishment duration and punishment intensity on the extinction of an instrumental response. Rats learned a barrier-crossing response and then were punished with shocks of various durations and intensities during extinction. More specifically, the three values of the intensity of the shock were .25 ma., 2.00 ma., and 4.00 ma. (ma. = milliamp). The durations of shock were .30 seconds, 1.00 seconds, and 3.00 seconds. The design is presented in Table 10-8, and five participants were assigned to each cell. The dependent variable was a latency of response measure, but values for cells were not presented in the original article. In general, though, increases of both independent variables increased latency of responding (and therefore decreased response strength), as one would expect.

The K × L Factorial Design. Each independent variable may be varied in any number of ways. The generalized factorial design for two independent variables may be labeled the K × L factorial design, where K stands for the first independent variable and its value indicates the number of ways in which it is varied; and L similarly denotes the second independent variable. K and L might then assume

Table 10-8 Illustration of a 3 × 3 Factorial Design

Shock intensity — .25 ma. / 2.00 ma. / 4.00 ma.

Shock duration — .30 sec. / 1.00 sec. / 3.00 sec.

any value. If one independent variable is varied in four ways and the other in two ways, we would have a 4 × 2 design. If one independent variable is varied in six ways and the second in two ways, we would have a 6 × 2 design. If five values are assumed by one independent variable and three by the other we would have a 5 × 3 design, and so forth.

A 6 × 2 design was employed by Gampel (1966) in an experiment on verbal satiation. Verbal satiation for certain words is the loss of their meaning to a person when they are repeatedly presented. Briefly, this experimenter varied the duration of repetition of certain words in six ways: 0, 5, 15, 30, 60, and 120 seconds. The second variable was Word Category, and the two ways in which this variable was varied were: (1) use of the stimulus word that was "verbally satiated" or (2) use of a word that was strongly associated with that stimulus word. These two categories are referred to as "Stimulus Words" and "High Associate Words" respectively. The 6 × 2 factorial design is presented in Table 10-9.

Gampel's dependent variable was defined as the amount of time required to find a stimulus word (or its high associate) within a large group of words. It was assumed that the longer the amount of time to search out and find a certain word, the greater the verbal satiation effect for that word. Her results showed that the greater the duration of repetition, the greater the search time (and hence the greater the verbal satiation). The effect of varying the second variable was that the high associate words required a longer search time than the stimulus words.

Table 10-9 Illustration of a K × L Design Where K = 6 and L = 2

Duration of repetition (Seconds)

		0	5	15	30	60	120
Word category	Stimulus words						
	High associate words						

*Factorial Designs with
More than Two Independent Variables*

The 2 × 2 × 2 Factorial Design. The previous factorial designs have concerned two independent variables. However, in principle the number of variables that can be studied is unlimited. The 2 × 2 × 2 design is the simplest factorial for studying three independent variables. Following our preceding rule this label implies that each of the three independent variables is varied in two ways. It also follows that there are eight experimental conditions. As an illustration of a 2 × 2 × 2 factorial design, consider an experiment by Tversky and Edwards (1966). These experimenters randomly assigned 24 undergraduate students to the eight different conditions and had each student face a box that had two lights and three levers. Either the left or the right light would go on for a series of 1,000 trials. By pressing the left or the right lever, the student could guess which light was set to turn on, on the next trial. If the student did guess, no light flashed, but the response was automatically recorded. If the participant did not want to guess, he or she could depress the middle lever and observe which light came on. The three independent variables were: (1) instructional set; (2) probability value of the left versus the right light coming on; and (3) response type. The two conditions of *instructional set* were: (1) nonstationary, in which case the students were told that the probability of each light coming on would change throughout the 1,000 trials; or (2) stationary—they were told that the probabilities remained constant. *Probability value* was either (1) 60:40, or (2) 70:30, i.e., for the former the left light came on 60 percent of the time, as against 70 percent for the latter. *Response type* conditions were: (1) Forced Prediction Group—these students *had* to predict which light was going to come on; (2) Free-Choice Group—these did not have to make a prediction and could merely press the middle lever and observe which light was to come on. When a correct prediction was made, a nickel was gained; otherwise a nickel was lost. This design is diagrammed in Table 10-10 and includes means for one of the dependent variable measures used.

Table 10-10 Look versus Bet: Average Number of Observation Trials for Each Experiment Group

	Probability of left light			
	.6		.7	
	Instructional set		Instructional set	
Response type	Stationary	Nonstationary	Stationary	Nonstationary
Free choice group	505	635	410	531
Forced prediction group	165	400	135	299

Briefly considering the results, we can see the following:

1. Response Type: Free Choice versus Forced Prediction. The participants who were free to observe the next trial (by pressing the middle lever) looked at those trials more than those who were forced to give hypothetical predictions.
2. Instructional Set: Stationary versus Nonstationary Instructions. Those students who were told that the probability might change during the sequence of 1,000 trials looked more than those who received stationary instructions.
3. Probability Values: There was a tendency, though it was not reliable, for the 60:40 group to look more than the 70:30 group.

The $K \times L \times M$ Factorial Design.   It should now be apparent that any independent variable may be varied in any number of ways. The general case for the three independent variable factorial design is the $K \times L \times M$ design, where K, L, and M may assume whatever positive integer value the experimenter desires. For instance, if each independent variable assumes three values, a $3 \times 3 \times 3$ design results. If one independent variable (K) is varied in two ways, the second (L) in three ways, and the third (M) in four ways, a $2 \times 3 \times 4$ design results. A $5 \times 3 \times 3$ design was used by Johnson and Bailey (1966) in an experiment on discrimination learning. Type of stimulus was varied in five ways, age of participants in three ways (kindergarten, fourth grade, or college), and number of stimulus-response bonds in three ways (one, two, or three). This design is diagrammed in Table 10-11.

Table 10-11 Illustration of a $K \times L \times M$ Factorial Design Where $K = 5$, $L = 3$ and $M = 3$

		Stimulus type				
		I	II	III	IV	V
	Age Kindergarten					
One S-R bond	4th grade					
	College					
	Age Kindergarten					
Two S-R bonds	4th grade					
	College					
	Age Kindergarten					
Three S-R bonds	4th grade					
	College					

Statistical Analysis of Factorial Designs

We have compared the means for each of the experimental conditions in the hypnosis experiment and studied the concept of an interaction, but this has provided only tentative answers; firmer answers await the application of statistical tests to the data. Where we obtained a sizable difference between the participants who were hypnotized as against those who were not hypnotized, for example, we said that we had to find out if the apparently sizable difference in means was reliable. The statistical analysis that is most frequently applied to the factorial design is *analysis of variance*, the rudiments of which were presented in chapter 9. We shall limit our discussion to the 2×2 factorial design.

The first step in conducting an analysis of variance for the factorial design follows very closely that for any number of groups, as previously discussed. That is, we wish to compute the total SS and partition it into two major components, the among SS and the within SS. Let us return to the data in Table 10-2, which are summarized in Table 10-12.

Table 10-12 A Summary of the Necessary Ingredients for Analysis of Variance (Taken from Table 10-2)

	GROUP		
1	*2*	*3*	*4*
(*Hypnotized — Low Susceptibility*)	(*Not Hypnotized — Low Susceptibility*)	(*Hypnotized — High Susceptibility*)	(*Not Hypnotized — High Susceptibility*)
n: 8	8	8	8
ΣX: -114	9	-186	-15
ΣX^2: 3148	593	6002	1927
$\overline{X}$: -14.25	1.12	-23.25	-1.88

To compute the total SS, we substitute the appropriate values from Table 10-12 in Equation (9-9), which for four groups (always the case for the 2×2 design) is:

(10-1)
$$\text{Total } SS = (\Sigma X_1^2 + \Sigma X_2^2 + \Sigma X_3^2 + \Sigma X_4^2) - \frac{(\Sigma X_1 + \Sigma X_2 + \Sigma X_3 + \Sigma X_4)^2}{N}$$

$$= (3148 + 593 + 6002 + 1927) - \frac{(-114 + 9 - 186 - 15)^2}{32}$$

$$= 8743.88$$

Next, to compute the among groups SS, we substitute in Equation (9-10), which for four groups is:

(10-2)
$$\text{Among } SS = \frac{(\Sigma X_1)^2}{n_1} + \frac{(\Sigma X_2)^2}{n_2} + \frac{(\Sigma X_3)^2}{n_3} + \frac{(\Sigma X_4)^2}{n_4}$$
$$- \frac{(\Sigma X_1 + \Sigma X_2 + \Sigma X_3 + \Sigma X_4)^2}{N}$$

$$\text{Among } SS = \frac{(-114)^2}{8} + \frac{(9)^2}{8} + \frac{(-186)^2}{8} + \frac{(-15)^2}{8} - 2926.12$$
$$= 3061.12$$

And, as before, the within SS may be obtained by subtraction, Equation (9-11):

(10-3)
$$\text{Within } SS = \text{total } SS - \text{among } SS$$
$$= 8743.88 - 3061.12 = 5682.76$$

This completes the initial stage of the analysis of variance for a 2 × 2 factorial design, for we have now illustrated the computation of the total SS, the among SS, and the within SS. As you can see, the initial stage of the statistical computation is the same as that for the initial stage for a randomized groups design. But let us proceed further.

The among-groups SS tells us something about how all groups differ. However, we are interested not in simultaneously comparing all four groups, but only in certain comparisons. In a 2 × 2 factorial, of course, we have two independent variables, each varied in two ways. Hence we are interested in whether or not variation of each independent variable affects the dependent variable, and whether there is a significant interaction. The first step to perform is to compute the SS between groups for each independent variable. Using Table 10-1 (p. 281) as a guide, we may write our formulas for computing the between-groups SS for the specific comparisons.

The groups are as labeled in the cells. Thus to determine whether or not there is a significant difference between the two values of the first (hypnosis) variable, we need to compute the SS between these two values. For this purpose we may use Equation (10-4):

(10-4)
SS between amounts of first independent variable

$$= \frac{(\Sigma X_1 + \Sigma X_3)^2}{n_1 + n_3} + \frac{(\Sigma X_2 + \Sigma X_4)^2}{n_2 + n_4} - \frac{(\Sigma X_1 + \Sigma X_2 + \Sigma X_3 + \Sigma X_4)^2}{N}$$

For computing the SS between the conditions of the second independent variable we use Equation (10-5):

(10-5)
SS between amounts of second independent variable

$$= \frac{(\Sigma X_1 + \Sigma X_2)^2}{n_1 + n_2} + \frac{(\Sigma X_3 + \Sigma X_4)^2}{n_3 + n_4} - \frac{(\Sigma X_1 + \Sigma X_2 + \Sigma X_3 + \Sigma X_4)^2}{N}$$

Now, in our particular example we conduct tests to determine whether degree of hypnosis (our first independent variable) influences the dependent variable, whether hypnotic susceptibility (our second independent variable) influences the dependent variable, and whether there is a significant interaction. First, to determine the effect of being hypnotized we need to test the difference between the hypnotized and the nonhypnotized conditions. To make this test we shall ignore the susceptibility variable in the design.

Making the appropriate substitutions in Equation (10-4) we can compute the SS between the hypnosis conditions:

$$= \frac{(-114 - 186)^2}{8 + 8} + \frac{(9 - 15)^2}{8 + 8} - 2926.12$$

$$= \frac{(90,000)}{16} + \frac{(36)}{16} - 2926.12 = 2701.13$$

This value will be used to answer the above question. However, we shall answer all questions at once, rather than piecemeal, so let us hold it until we complete this stage of the inquiry. We have computed a sum of squares among all four groups (i.e., 3061.12), and it can be separated into parts. We have computed one of these parts above, the sum of squares between the hypnosis conditions (2701.13). There are two other parts: the sum of squares between the susceptibility conditions, and the interaction. Let us compute the former, using Equation (10-5).

Substituting the required values in Equation (10-5) we determine that the "between" SS for the susceptibility conditions is:

SS between susceptibility conditions =
$$\frac{(-114 + 9)^2}{8 + 8} + \frac{(-186 - 15)^2}{8 + 8} - 2926.12 = 288.00$$

The among SS has three parts. We have directly computed the first two parts. Hence, the difference between the sum of the first two parts and the among SS provides the third part, in this case that for the interaction:

(10-6)
> Interaction SS = among SS − between SS for first variable (hypnosis)
> − between SS for second variable (susceptibility)

Recalling that the among SS was 3061.12, the between SS for the hypnosis conditions was 2701.13, and the between SS for the susceptibility conditions was 288.00, we find that the SS for the interaction is:

$$\text{Interaction } SS = 3061.12 - 2701.13 - 288.00 = 71.99$$

This completes the computation of the sums of squares. Let us add that these values should all be positive. If your computations yield a negative SS, check your work until you discover the error. There are only several minor matters to discuss before the analysis is completed. Before we continue, however, let us summarize our findings to this point in Table 10-13.

Table 10-13 Sums of Squares for the 2 × 2 Factorial Design

Source of Variation	Sum of Squares
Among Groups	(3061.12)
Between Hypnosis (H)	2701.13
Between Susceptibility (S)	288.00
Interaction: hypnosis × susceptibility	71.99
Within groups	5682.76
Total	8743.88

We now must discuss how to determine various degrees of freedom (df) for this application of the analysis-of-variance procedure. Repeating the equations in Chapter 9, for the major components:

$$(10\text{-}7) \qquad \text{Total } df = N - 1$$

$$(10\text{-}8) \qquad \text{Among (or Between) } df = r - 1$$

$$(10\text{-}9) \qquad \text{Within } df = N - r$$

In our example, $N = 32$ and r (number of groups) $= 4$. Hence, the total df is $32 - 1 = 31$, the among df is $4 - 1 = 3$ (the among df is based on four separate groups or conditions), and the within df is $32 - 4 = 28$. The similarity between the manner in which we partition the total SS and the total df may also be continued for the among SS and the among df. The among df is 3. Since we analyzed the among SS into three parts, we may do the same for the among df, one df for each part (one df for each part is only true for a 2×2 factorial design). Take the hypnosis conditions first. Since we are temporarily ignoring the susceptibility variable, we have only two conditions of hypnosis to consider, or, if you will, two groups. Hence, the df for the between-hypnosis conditions is based on $r = 2$. Substituting this value in Equation (10-8), we see that the between-hypnosis df is $2 - 1 = 1$. The same holds true for the susceptibility variable; there are two values, hence $r = 2$ and the df for this source of variation is $2 - 1 = 1$.

Now for the interaction df. Note in Table 10-14 that the interaction is written as hypnosis × susceptibility. We may, of course, abbreviate the notation, as is frequently done, by using H × S. This is read "the interaction between hypnosis and susceptibility." The "×" sign may be used as a mnemonic device for remembering how to compute the interaction df: multiply the number of degrees of freedom for the first variable by that for the second. Since both variables have one *df,* the interaction *df* is also one, i.e., $1 \times 1 = 1$. This accounts for all three *df* that are associated with the among SS.[3] These findings are added to Table 10-13, forming Table 10-14.

Table 10-14 Sums of Squares and *df* for the 2 × 2 Factorial Design

Sources of Variation	*Sums of Squares*	*df*
Among groups	(3061.12)	(3)
Between hypnosis (H)	2701.13	1
Between susceptibility (S)	288.00	1
Interaction: hypnosis × susceptibility (H × S)	71.99	1
Within groups	5682.76	28
Total	8743.88	31

In the 2×2 factorial design there are four mean squares in which we are interested. In this experiment they are: between hypnosis conditions, between susceptibility conditions, the interaction, and within groups. To compute the mean square for the between hypnosis source of variation, we divide that sum of squares by the corresponding *df*:

$$\frac{2701.13}{1} = 2701.13$$

Similarly the within-groups mean square is computed:

$$\frac{5682.76}{28} = 202.95$$

These values are then added to our summary table of the analysis of variance, as we shall show shortly.

This completes the analysis of variance for the 2×2 design, at least in the usual form. We have analyzed the total sum of squares into its various components. In particular, we have several sources of between sums of squares to study and a term that represents the experimental error (the within-groups mean square). We said that the "between" components indicate the extent to which the various experimental conditions differ. For instance, if any given "between" component, such as that for the hypnosis conditions, is sizable, then that may be taken to indicate that hypnosis influences the dependent vari-

---

[3] If this is not clear, then you might merely remember that the *df* for the between *SS* in a 2 × 2 design is always the same, as shown in Table 10-14. That is, the *df* for the *SS* between each independent variable condition is 1, and for the interaction, 1.

able. Hence we need merely conduct the appropriate F-tests to determine whether or not the various "between" components are reliably larger than would be expected by chance. The first F for us to compute is that between the two conditions of hypnosis.[4] To do this we merely substitute the appropriate values in Equation (9-17). Since the mean square between the hypnosis conditions is 2701.13 and the mean square within groups is 202.95, we divide the former by the latter:

$$F = \frac{2701.13}{202.95} = 13.30$$

The F between the hypnosis susceptibility conditions is:

$$F = \frac{288.00}{202.95} = 1.41$$

And the F for the interaction is:

$$F = \frac{71.99}{202.95} = .35$$

These values have been entered in Table 10-15.

Table 10-15 Complete Analysis of Variance of the Performance Scores

Source of Variation	Sum of Squares	df	Mean Square	F
Between hypnosis	2701.13	1	2701.13	13.30
Between susceptibility	288.00	1	288.00	1.41
Interaction: H × S	71.99	1	71.99	.35
Within groups (error)	5682.76	28	202.95	
Total	8743.88	31		

Following the preceding discussion, let us observe that Table 10-15 is the final summary of our statistical analysis. This is the table that should be presented in the results section of an experimental write-up. All of the features of this table should be included in the results section of your report using precisely this format. We next assign a probability level to these F values. That is, we need to determine the odds that the F's could have occurred by chance. Prior to the collection of data we al-

---

[4] The factorial design offers us a good example of a point we made in Chapter 9. That is that if we are specifically interested in certain questions, then there is no need to conduct an F-test for the among-groups source of variation. With this design we are exclusively interested in whether our two independent variables are effective and whether there is an interaction. Hence we proceed directly to these questions without running an overall F-test among all four groups, although such may be easily conducted.

ways state our null hypotheses. In this design we would have previously stated three more precise null hypotheses than merely that "There is no difference between our groups."

1. There is no difference between the means of the two conditions of hypnosis.
2. There is no difference between the means of the two degrees of hypnotic susceptibility.
3. There is no interaction between the two independent variables.[5]

The general strategy is to determine the probability associated with each value of F. Assuming that we have set a reliability level of 0.05 for each F-test, we need merely confront that level with the probability associated with each F. If that probability is 0.05 or less, we can reject the appropriate null hypothesis and conclude that the independent variable in question was effective in producing the result.[6]

Let us turn to the first null hypothesis, that for the hypnosis variable. Our obtained F is 13.30. We have one df for the numerator and 28 df for the denominator. We can determine that an F of 4.20 is required for reliability at the 5 percent level with 1 and 28 df (Table A-3 in the appendix). Since our F of 13.30 exceeds this value, we may reject the first null hypothesis. The conclusion is that the two conditions of hypnosis led to significantly different performance. And since the mean for the hypnosis condition (-18.75) is lower than for the nonhypnosis condition ($-.38$), the authors concluded that hypnosis has "a strong inhibiting effect on learning".

To test the effect of varying hypnotic susceptibility, we note that the F ratio for this source of variation is 1.41. We have 1 and 28 df available for this test. The necessary F value is, as before, 4.20. Since 1.41 does not exceed 4.20, we conclude that variation of hypnotic susceptibility does not reliably influence amount learned.

To study the interaction, refer to Figure 10-4. Note that the lines do not deviate to any great extent from being parallel, suggesting that there is no reliable interaction between the variables. Incidentally, the fact that the line of the nonhypnotized condition is noticeably higher than that for the hypnotized condition is a graphic illustration of the effectiveness of the hypnosis variable.

To test the interaction we note that the F is .35. This F is considerably below 1.00. We can, therefore, conclude immediately that the

---

[5] A more precise statement of this null hypothesis should be made: There is no difference in the means of the four groups after the cell means have been adjusted for row and column effects. However, such a statement probably will only be comprehensible to you after further work in statistics.

[6] Of course, assuming adequate control procedures have been exercised.

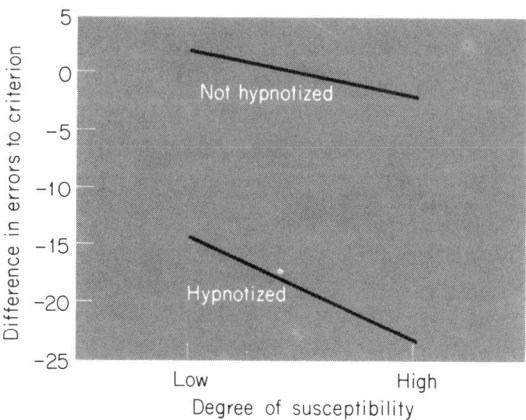

Figure 10-4. The actual data suggest that there is a lack of interaction between hypnotic susceptibility and degree of hypnosis.

interaction is *not reliable*. A check on this may be made by noting that we also have 1 and 28 *df* for this source of variation. And we know than an *F* of 4.20 is required for significance at the 5 percent level. Clearly .35 does not approach 4.20, and hence is not reliable. The third null hypothesis is not rejected.

The preceding discussion for the statistical analysis of a factorial design has been rather lengthy because of its detailed nature. Now with this background it is possible for us to breeze through the next example. This experiment, reported by Lachman, Meehan, and Bradley (1965), was an investigation of the effect of two independent variables on the learning of concepts.[7] The general procedure required the participants to face a box on which two complex stimulus pictures were presented. The pictures had borders that were black, white, or red, and within the borders were geometrical forms (circles and squares) that also varied in color. The participant's task was to guess which of the two pictures was correct by selecting a color on one of them. When correct, a light at the top of the box came on. The pairs of stimulus pictures were changed on each trial, and the participant continued making choices until the color that was always on one picture was found as indicated by 10 successive correct choices. An example of a correct concept would be always choosing the picture that had a red border. Once this initial learning was complete, a second set of stimuli was presented, and the above procedure was repeated as the participants solved their second problem.

The general question that Lachman et al. studied concerned the

[7] The very interesting theoretical questions that led to this experiment concerned the mechanisms for verbal mediation. The reader is referred to the original article for a discussion of the more theoretical issues.

relationships between the correct color on the initial learning task to the color that was correct during the later learning period. There were two such relationships. The first was the strength of the association between the names of the colors that were correct on the first and the second tasks. We may, for instance, note that the words "black" and "white" are more strongly associated than are the words "red" and "white." So we have two degrees of word association: high and low, and half of the participants served under each condition. For example, the high word association participants might have a black symbol as correct during the first learning period, and a white symbol as correct during the second. The low word association participants would first have a red symbol as the correct one, and a white symbol correct on the second task. The question, then, is: Does variation of the strength of the word association between critical symbols influence the rapidity of learning a concept?

The second question concerned the observing response. Roughly, an observing response is varied by varying the location of the critical color on the stimulus pictures. It was varied in two ways: (1) it was held constant from the first to the second learning periods, or (2) it was changed. An example of the constant observing response condition would be that if a red border of the stimulus picture was correct during initial learning, the color of the border was also correct during later learning (though the color of the border would be changed). But for the changed observing response conditions, if a red border was correct first, then the color of one of the geometrical forms within the border was correct during the second learning period. In short, the second question was whether varying the observing response (by changing the location of the critical color on the stimulus pictures) influences the rapidity of learning a new concept. The third question, of course, is whether there is an interaction between the word association variable and the observing response variable.

A diagram of the 2 × 2 factorial design is presented in Table 10-16, and the three null hypotheses that were tested are as follows:

1. There is no difference between the means for the high and low word association conditions.
2. There is no difference between the means for the observing response conditions.
3. There is no interaction between the word association and the observing response variables.

Twelve participants were randomly assigned to each cell, and the number of trials to reach criterion are presented for each participant in Table 10-17.

Table 10-16 A 2 × 2 Factorial Design with Strength of Word Association and Observing Response as the Two Variables

Word association

	Low	High
Observing response — Changed	32.25	17.08
Observing response — Constant	16.25	6.42

Table 10-17 Number of Trials to Criterion

	GROUP		
1 *Changed OR —* *Low Association*	*2* *Changed OR —* *High Association*	*3* *Constant OR —* *Low Association*	*4* *Constant OR —* *High Association*
23	12	3	33
10	3	1	2
4	32	1	2
10	18	5	1
34	12	75	1
14	10	75	4
15	17	5	5
31	28	2	12
75	59	2	4
75	4	19	10
75	6	5	2
21	4	2	1
ΣX: 387	205	195	77
ΣX^2: 20599	6367	11709	1405
$\bar{X}$: 32.25	17.08	16.25	6.42

Our first step will be to compute the total SS by substituting the values in Table 10-17 in Equation (10-1):

$$\text{Total } SS = 20599 + 6367 + 11709 + 1405 - \frac{(387 + 205 + 195 + 77)^2}{48}$$

$$= 24{,}528.00$$

Now we shall compute the among SS by appropriate substitutions in Equation (10-2).

$$\text{Among } SS = \frac{(387)^2}{12} + \frac{(205)^2}{12} + \frac{(195)^2}{12} + \frac{(77)^2}{12}$$
$$- \frac{(387 + 205 + 195 + 77)^2}{48} = 4093.60$$

The within SS is [see Equation (10-3)]

$$\text{Within } SS = 24,528.00 - 4093.60 = 20,434.40$$

As our next step we shall analyze the among SS into its three components: between the word association condition, the observing response condition, and the WA $\times$ OR interaction. Considering word association first, we substitute the appropriate values in Equation (10-4) and find that:

SS between word association conditions

$$= \frac{(387 + 195)^2}{12 + 12} + \frac{(205 + 77)^2}{12 + 12} - \frac{(387 + 205 + 195 + 77)^2}{48}$$
$$= 1875.00$$

Substituting in Equation (10-5) to compute the SS between the two conditions of observing response:

SS between observing response conditions

$$= \frac{(387 + 205)^2}{12 + 12} + \frac{(195 + 77)^2}{12 + 12} - \frac{(387 + 205 + 195 + 77)^2}{48}$$
$$= 2133.34$$

The SS for the interaction component may now be seen to be:

$$4093.60 - 1875.00 - 2133.34 = 85.26$$

The various df may now be determined.

$$\text{Total } (N - 1) = 48 - 1 = 47$$
$$\text{Over-all between } (r - 1) = 4 - 1 = 3$$
$$\text{Between word association} = 2 - 1 = 1$$
$$\text{Between observing response} = 2 - 1 = 1$$
$$\text{Interaction: WA} \times \text{OR} = 1 \times 1 = 1$$
$$\text{Within } (N - r) = 48 - 4 = 44$$

The mean squares and the F's have been computed and placed in the summary table (Table 10-18 on the following page).

Interpreting the F's

To test the F for the word association variable, let us note that we have 1 and 44 degrees of freedom available. Assuming a .05 level test, we enter

Table 10-18 Summary of the Analysis of Variance for the Concept Learning Experiment

Source of Variation	Sum of Squares	df	Mean Square	F
Between word association	1875.00	1	1875.00	4.04
Between observing response	2133.34	1	2133.34	4.59
Interactions WA × OR	85.26	1	85.26	.18
Within groups	20,434.40	44	464.42	
Total	24,528.00	47		

Table A-3 in the appendix and find that we must interpolate between 40 *df* and 60 *df*. The *F* values are 4.08 and 4.00 respectively. Consequently, an *F* with 1 and 44 *df* must exceed 4.06. The *F* for word association is 4.04; we therefore fail to reject the first null hypothesis and must conclude that variation of the word association variable did not reliably affect rapidity of concept learning.

We have the same number of *df* available for evaluating the effect of the observing response variable, and therefore the *F* for this effect must also exceed 4.06 in order to be reliable. We note that it is 4.59, and we can thus reject the second null hypothesis. The empirical conclusion is that variation of the observing response reliably influences rapidity of forming a concept.

Referring to Figure 10-5 we can visually study these findings. First, observe that the data points are lower for the high word association condition than for the low word association condition, but this decrease is not reliable. The points for the changed observing response conditions are higher than for the constant observing response condition. Since this

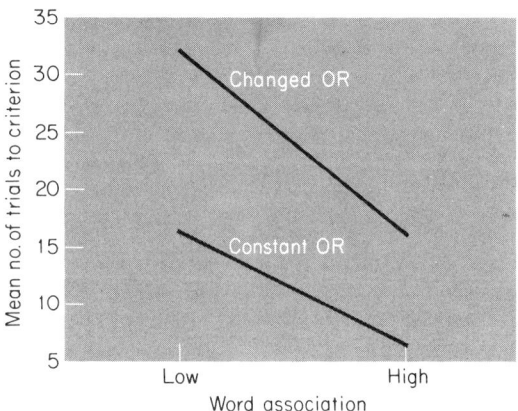

Figure 10-5. Data points for the concept learning experiment. Since the lines are approximately parallel, there probably is no interaction.

variable reliably influenced the dependent variable scores, maintaining a constant observing response facilitated the formation of a concept.

Finally, we note that the two lines are approximately parallel. The suggestion is thus that there is a lack of interaction between the independent variables, a suggestion that is confirmed by the F value for the interaction source of variation, viz., this F is well below 1.0 and we can thus immediately conclude that it is not reliable.

This completes our examples of the statistical analysis of factorial designs. We have discussed factorial designs generally, but have only illustrated the analysis for the 2×2 case. If you are interested in obtaining general principles for the analysis of any factorial design you should consult one of the references previously given or plan on taking a more advanced course. It is not likely, however, that you will get beyond the 2×2 design in your elementary work.

The Choice of a Correct Error Term

One of the most important problems of statistical analysis is the choice of the correct error term. With reference to the F-test the problem is one of choosing the correct denominator. The error term that we have used is the within-groups mean square. Although the within-groups mean square is usually the correct error term, we should be aware that sometimes it is not. To understand this, consider three types of factorial designs. The first is the case of a fixed model, the second is the case of a random model, and the third is the case of a mixed model. ("Model" in each case refers to characteristics of the independent variables of the factorial design.) We shall illustrate these three cases by means of a 2×2 design.

The Case of a Fixed Model

The 2×2 design indicates that we have two independent variables, each varied in two ways. Now, if we have some particular reason to select the two values of the two variables, it can be said that we are dealing with a fixed model. This is so because we have not arrived at the particular values of the independent variables in a random manner. In other words, we have chosen the two values of each independent variable in a premeditated way. We are interested in method A of teaching (a specific method) versus method B, for example. Or we choose to study 10 hours of training versus 20 hours. Similarly, we decide to give our rats 50 versus 100 trials, selecting these particular values for a special reason. When we intentionally select our values of the independent variables and do not select them at random, we have the case of a fixed model. *For this case the within-groups mean square is the correct error term for all F-tests being run.* If we refer to our two independent variables as K and L, and the interaction as K × L, we have the following between-groups mean squares to test: that between the two conditions of K, that between the two conditions of L, and that for K × L. For a fixed model, each of the between-groups mean squares should be divided by the within-groups mean square. This is, incidentally, the case most frequently encountered in psychological research.

The Case of a Random Model

If the values of the two independent variables have been selected at random, you are using a random model. For example, if our two variables are number of trials and IQ of participants, we would consider all possible reasonable numbers of trials and all possible reasonable IQs. Our two particular values of each independent variable would then be selected strictly at random. For instance, we might consider as reasonable possible values of the first independent variable—numbers of trials—those from 6 to 300. We would then place these 295 numbers in a hat and draw two from them. The resulting numbers would be the values that we would assign to our independent variable. The same process would be followed with regard to the IQ variable. In the case of studying various characteristics of participants such as IQ, however, we would probably do the following. If a random sample of people has been drawn, then merely grouping them into classes would satisfy our requirement. That is, we might divide all individuals into two groups, using a certain IQ score as the dividing line. This would constitute random values of this independent variable. The reason that this is so is that we have specified that our participants were selected at random from a given population. Hence, in randomly selecting participants, we also randomly select values of their characteristics, such as IQ.

The procedure for testing the between-groups mean squares for the case where both independent variables are random variables is as follows: *Test the interaction mean square by dividing it by the within-groups mean square. Then test the other mean squares by dividing them by the interaction mean square.* That is, test the K × L mean square by dividing it by the mean square within groups. Then test the mean square between the two conditions of K by dividing it by the K × L mean square, and also test the mean square between L by dividing it by the K × L mean square. We might remark that designs in which both variables are random are relatively rare in psychological research.

The Case of a Mixed Model

This is a less uncommon case than that where both variables are random, but it still does not occur as frequently as the case of a fixed model. The case of a mixed model occurs when one independent variable is fixed and the other is random. The procedure for testing the three mean squares for this case is as follows: *Divide the within-groups mean square into the interaction mean square; divide the interaction mean square into the mean square for the fixed independent variable; and divide the within mean square into the mean square for the random independent variable.*

These are the three cases that are most likely to be encountered in your research, though there are a number of variations that can occur.[8] The importance of these rules may not be immediately apparent, but a consideration of the topic of generalization (chapter 15) will correct that. Furthermore, in the appendix to this chapter we briefly present a rationale that will help you to remember these rules and to understand the reasons for them.

The Importance of Interactions

Our goal in psychology is to arrive at statements about the behavior of organisms that can be used for such purposes as explaining, predicting, and controlling behavior. For accomplishing these purposes we would like our statements to be as simple as possible. Behavior is anything but simple, however, and it would be very surprising to us if our *statements* about behavior were simple. It would seem more reasonable to expect that complex statements must be made about complex events. Those who talk about behavior in simple terms are likely to be wrong. This is illustrated by "common sense" discussions of behavior. People often say such things as "he is smart, he will do well in college," or "she is pretty, she will go far in the movies." However, such matters are not that uncomplicated; there are variables other than "smartness" that influence how well a person does in col-

[8] For a more thorough consideration of this topic you are referred to Anderson and Bancroft (1952, ch. 23), or to Wine (1964); if you prefer a psychological text, see Lindquist (1953) or Winer (1962).

lege, and there are variables other than beauty that influence job success. Furthermore, such variables do not always act on all people in the same manner. Rather, they *interact* in such a way that people with certain characteristics behave one way, but people with the same characteristics in addition to other characteristics behave another way. Let us illustrate by speculating about two variables that might influence the likelihood of a young woman succeeding in films: beauty and intelligence. Consider two values of each of these variables: beautiful and not beautiful; high intelligence and low intelligence. Were we to study these variables, we would collect data on a sample of four groups of women: beautiful women with high intelligence, beautiful women with low intelligence, not beautiful women with high intelligence, and not beautiful women with low intelligence. Suppose our dependent variable to be the frequency with which women in these four groups make film appearances and we found that beautiful women with low intelligence made film appearances most frequently, not beautiful women with high intelligence worked next most frequently, beautiful women with high intelligence appeared in films with the third greatest frequency, and low intelligence women who are not beautiful were the least frequently employed (see Figure 10-6).

Now, if these findings were actually obtained, then the simple statement, "she is pretty, she will have no trouble working in the movies" is inaccurate. Beauty is not the whole story; intelligence is also important. We cannot say that beautiful women are more likely to work

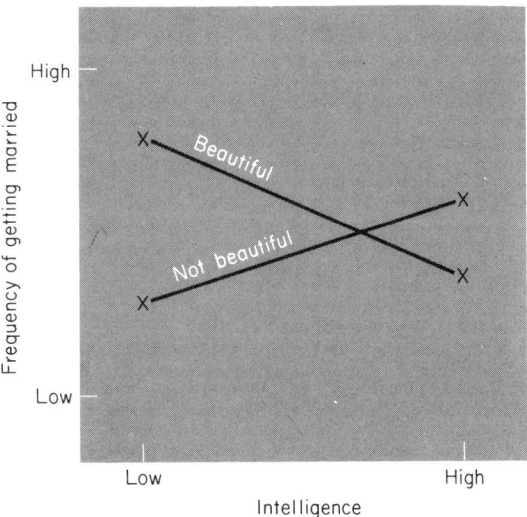

Figure 10-6. A possible interaction between beauty and intelligence.

in the movies any more than we can say that unintelligent women are the more likely. The only accurate statement is that beauty and intelligence interact; beautiful women with low intelligence are more likely to work than unbeautiful women with low intelligence; but unbeautiful women with high intelligence are more likely to work than beautiful women with high intelligence.

We have just barely begun to make completely accurate statements when we talk about interactions between two variables. Interactions of a much higher order also occur, that is, interactions among three, four, or any number of variables. To illustrate, not only might beauty and intelligence interact, but in addition such variables as motivation, social graces, and so on. Hence, for a really adequate understanding of behavior, we need to determine the large number of interactions that undoubtedly occur. In the final analysis, if such ever occurs in psychology, we will probably arrive at two general kinds of statements about behavior: those statements that tell us how everybody behaves (those ways in which people are similar), with no real exceptions; and those statements that tell us how people differ. The latter will probably involve statements about interactions. For people with certain characteristics act differently than people with other characteristics in the presence of the same stimuli. And statements that describe the varying behavior of people will probably rest on accurate determination of interactions. If such a complete determination of interactions ever comes about, we will be able to understand the behavior of what humanists call the "unique" personality.

Now let us refer the concept of interaction back to a topic considered in chapter 2. We discussed ways in which we become aware of a problem, one of them being as a result of contradictory findings in a series of experiments. For example, we considered two experiments, each using the same design and performed on the same problem, but with contradictory results. Why? One reason might be that a certain variable was not controlled in either experiment. Hence, it might assume one value in the first experiment and a different value in the second experiment. And if such an extraneous variable interacts with the independent variable(s), then the discrepant results become understandable. A new experiment could then be conducted in which that extraneous variable becomes an independent variable. As it is purposively manipulated along with the original independent variable, the nature of the interaction can be determined. In this way not only will the apparently contradictory results be understood, but a new advance in knowledge will be made.

This situation need not be limited to the case where the extraneous variable is uncontrolled. For instance, the first experimenter may hold the extraneous variable constant at a certain value while the second experimenter may also hold it constant, but at a different value. And the same result would obtain as when the variable went uncontrolled—contradictory findings in the two experiments. Let us illustrate by returning to the experiment on language suppression that was discussed on pages 23–24. Recall that the first investigator found that the experimental treatment successfully produced a suppression effect for a pronoun for the experimental group but there was no suppression for the control group. The relevant extraneous variable was the location of the experimenter, and in this study the participants could not see the experimenter. In the repetition of the experiment, however, the participant *could* see the experimenter, and the results were that there was no pronoun suppression for the experimental, as compared with the control group. The ideal solution for this problem, we said, would come by conducting a new experiment using a factorial design that incorporates experimenter location as the second variable. Hence, as shown in Table 10-19, the first variable is the original one (prior verbal stimulation), and we have varied it in two ways by using an experimental and a control group. The second variable, experimenter location, has two values: the participant cannot see the experimenter, and the participant can see him.

Table 10-19 A Design to Investigate Systematically the Effect of an Extraneous Variable

| | | Prior verbal stimulation | |
		None (control group)	Some (experimental group)
Location of experimenter	Participant cannot see him		
	Participant can see him		

In short, we are repeating the original experiment, under two conditions of the extraneous variable. A graphic illustration of the expected results is offered in Figure 10-7. We can see that the experimental group exhibits a larger suppression effect than does the con-

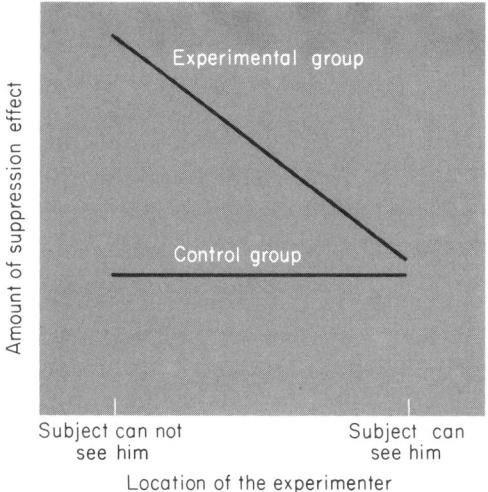

Figure 10-7 Illustration of an interaction between the independent variable and location of the experimenter. When the experimenter's location was systematically varied, the reason for conflicting results in two experiments became clear.

trol group when the participants cannot see the experimenter. But where the participants *can* see the experimenter, there is no reliable difference between the two groups. It is determined that there is an interaction between the location of the experimenter and the variable of prior verbal stimulation. What at first looked like a contradiction is resolved by isolating an interaction between the original independent variable and an extraneous variable. The problem is solved by resorting to a factorial design.

We have discussed factorial designs in which control groups are run under conditions (cells) included in those designs. Sometimes we will conduct experiments in which there are control groups that do not actually fit as one or more of the cells of the design. For instance, we might have a 2 × 2 design which consists of four experimental conditions, and we run an additional control group that doesn't "fit into the factorial design." Clearly certain design and analysis problems are entailed by this situation, and the reader might refer to the work by Hornbeck (1973), and Himmelfarb (1975) for help with experiments in which, as Himmelfarb put it, there is an "oddball" condition.

Undoubtedly these considerations hold for a wide variety of experimental findings, for the contradictions in the psychological literature are legion. By shrewd application of factorial designs to such problems their resolution should be accomplished.

Value of the Factorial Design

Not long ago the standard design in psychological research was the two-groups design. For many years, however, statisticians and researchers in such fields as agriculture and genetics, had been developing other kinds of designs. One of these was the factorial design, which, incidentally, grew with the development of analysis of variance. Slowly, psychologists started trying out these designs on their own problems. Some of them were found to be inappropriate, but the factorial design is one that has enjoyed success, and the extent of its success is still widening. It has even been held to be necessary in some areas, e.g., ". . . psychotherapy research can no longer ignore the necessity for factorial designs" (Kiesler, 1966, p. 133). Although each type of design that we have considered is appropriate for particular situations, and although we cannot say that a certain design should *always* be used where it is feasible, the factorial design is generally superior to the other designs that we discussed. Professor Ronald Fisher has made some interesting comments on this point.

> In expositions of the scientific use of experimentation it is frequent to find an excessive stress laid on the importance of varying the essential conditions *only one at a time.* The experimenter interested in the causes which contribute to a certain effect is supposed, by a process of abstraction, to isolate these causes into a number of elementary ingredients, or factors, and it is often supposed, at least for purposes of exposition, that to establish controlled conditions in which all of these factors except one can be held constant, and then to study the effects of this single factor, is the essentially scientific approach to an experimental investigation. This ideal doctrine seems to be more nearly related to expositions of elementary physical theory than to laboratory practice in any branch of research. In experiments merely designed to illustrate or demonstrate simple laws, connecting cause and effect, the relationships of which with the laws relating to other causes are already known, it provides a means by which the student may apprehend the relationship, with which he is to familiarize himself, in as simple a manner as possible. By contrast, in the state of knowledge or ignorance in which genuine research, intended to advance knowledge, has to be carried on, this simple formula is not very helpful. We are usually ignorant which, out of innumerable possible factors, may prove ultimately to be the most important, though we may have strong presuppositions that some few of them are particularly worthy of study. We have usually no knowledge that any one factor will exert its effects independently of all others that can be varied, or that its effects are particularly simply related to variations in these other factors. On the contrary, when factors are chosen for investigation, it is not because we anticipate that the laws of nature can be expressed with any particular simplicity in terms of these variables, but because they are variables which can be controlled or measured with comparative ease. If *the investigator, in these circum-*

stances, confines his attention to any single factor, we may infer either that he is the unfortunate victim of a doctrinaire theory as to how experimentation should proceed, or that the time, material or equipment at his disposal is too limited to allow him to give attention to more than one narrow aspect of his problem. The modifications possible to any complicated apparatus, machine or industrial process must always be considered as potentially interacting with one another, and must be judged by the probable effects of such interactions. If they have to be tested one at a time this is not because to do so is an ideal scientific procedure, but because to test them simultaneously would sometimes be too troublesome, or too costly. In many instances, the belief that this is so has little foundation. Indeed, in a wide class of cases an experimental investigation, at the same time as it is made more comprehensive, may also be made more efficient if by more efficient we mean that more knowledge and a higher degree of precision are obtainable by the same number of observations (Fisher, 1953, pp. 91–92; italics ours).

Let us look into this matter more thoroughly. First we may note that the amount of information obtained from a factorial design is considerably greater than that obtained from the other designs mentioned, relative to the number of participants used. For example, let us say that we have two problems: (1) does variation of independent variable K affect a given dependent variable; and (2) does variation of independent variable L affect the same dependent variable. If we investigated these two problems by the use of a two-groups design, we would obtain two values for each variable. That is, K will be varied in two ways (K_1 and K_2), and similarly for L (L_1 and L_2). With 60 participants available for each experiment, the design for the first problem would be:

Experiment # 1

| Group K_1 | Group K_2 |
| 30 participants | 30 participants |

And similarly for the second problem:

Experiment # 2

| Group L_1 | Group L_2 |
| 30 participants | 30 participants |

With a total of 120 participants we are able to evaluate the effect of the two independent variables. However, we would not be able to tell if there is an interaction between K and L if we looked at these as two separate experiments.

But what if we used a factorial design to solve our two problems? Assume that we will want 30 participants for each condition. In this case the factorial would be as in Table 10-20. We would have four groups with 15 participants per group. But for comparing the two conditions of K we would have 30 participants for condition K_1 and 30 participants for K_2. This is just what we had for Experiment #1. And the same for the second experiment: We have 30 participants for each condition of L. Here we have accomplished everything with the 2×2 factorial design that we would have accomplished with the two separate experiments with two groups. But with those two experiments we required 120 participants in order to have 30 available for each condition; however, with the factorial design we need only 60 participants to have the same number of participants for each condi-

Table 10-20 A 2×2 Design that Incorporates Two, Two-Groups Experiments (The numbers of participants for cells, conditions, and the total number in the experiment are shown)

		K		
		K_1	K_2	
L	L_1	15	15	30
	L_2	15	15	30
		30	30	60

tion. The factorial design is much more efficient because we use our participants simultaneously for testing both independent variables.[9] In addition, we can evaluate the interaction between K and L—something that we could not do for the two two-groups experiments. Although we may look at the information about the interaction as pure "gravy," we should note that some hypotheses may be constructed specifically to test for interactions. Thus, it may be that the experimenter is primarily interested in the interaction, in which case the other information may be regarded as "gravy." But whatever the case, it is obvious that the factorial design yields considerably more information than separate two-groups designs and at considerably less cost to the experimenter. Still other advantages of the factorial design could be elaborated, but they might well await your more advanced courses.

---

[9] We are assuming that the two two-groups experiments are analyzed independently. In a very atypical case, the error terms might be pooled, which would result in a larger number of degrees of freedom than for the factorial design.

Summary of an Analysis of Variance and the
Computation of an F-test for a 2 × 2 Factorial Design

Assume that the following dependent variable scores have been obtained for the four groups in a 2×2 factorial design.

		Condition A	
		A_1	A_2
Condition B:	B_1	2 3 4 4 5 6 7	3 4 5 7 9 10 13
	B_2	5 6 7 8 8 8 8	4 6 7 9 10 11 14

1. The first step is to compute ΣX, ΣX^2, and n for each condition. The values have been computed for our example:

		Condition A	
		A_1	A_2
Condition B:	B_1	$\Sigma X = 31$ $\Sigma X^2 = 155$ $n = 7$	$\Sigma X = 51$ $\Sigma X^2 = 449$ $n = 7$
	B_2	$\Sigma X = 50$ $\Sigma X^2 = 366$ $n = 7$	$\Sigma X = 61$ $\Sigma X^2 = 599$ $n = 7$

2. Using Equation (10-1), we next compute the total *SS*:

$$\text{Total } SS = (\Sigma X_1^2 + \Sigma X_2^2 + \Sigma X_3^2 + \Sigma X_4^2) - \frac{(\Sigma X_1 + \Sigma X_2 + \Sigma X_3 + \Sigma X_4)^2}{N}$$

$$= (155 + 449 + 366 + 599) - \frac{(31 + 51 + 50 + 61)^2}{28} = 238.68$$

3. The over-all among *SS* is computed by substituting in Equation (10-2):

Between *SS*

$$= \frac{(\Sigma X_1)^2}{n_1} + \frac{(\Sigma X_2)^2}{n_2} + \frac{(\Sigma X_3)^2}{n_3} + \frac{(\Sigma X_4)^2}{n_4} - \frac{(\Sigma X_1 + \Sigma X_2 + \Sigma X_3 + \Sigma X_4)^2}{N}$$

$$= \frac{(31)^2}{7} + \frac{(51)^2}{7} + \frac{(50)^2}{7} + \frac{(61)^2}{7} - \frac{(31 + 51 + 50 + 61)^2}{28}$$

$$= 67.25$$

4. The within SS is determined by subtraction, Equation (10-3):

Total SS − overall among SS = within SS

$$238.68 - 67.25 = 171.43$$

5. We now seek to analyze the over-all among SS into its components, viz., the between A SS, the between B SS, and the $A \times B$ SS. The between A SS may be computed with the use of Equation (10-4).

Between A SS

$$= \frac{(\Sigma X_1 + \Sigma X_3)^2}{n_1 + n_3} + \frac{(\Sigma X_2 + \Sigma X_4)^2}{n_2 + n_4} - \frac{(\Sigma X_1 + \Sigma X_2 + \Sigma X_3 + \Sigma X_4)^2}{N}$$

$$= \frac{(31 + 50)^2}{7 + 7} + \frac{(51 + 61)^2}{7 + 7} - \frac{(31 + 51 + 50 + 61)^2}{28} = 34.32$$

The between B SS may be computed with the use of Equation (10-5).

Between B SS

$$= \frac{(\Sigma X_1 + \Sigma X_2)^2}{n_1 + n_2} + \frac{(\Sigma X_3 + \Sigma X_4)^2}{n_3 + n_4} - \frac{(\Sigma X_1 + \Sigma X_2 + \Sigma X_3 + \Sigma X_4)^2}{N}$$

Between B SS

$$= \frac{(31 + 51)^2}{7 + 7} + \frac{(50 + 61)^2}{7 + 7} - \frac{(31 + 51 + 50 + 61)^2}{28} = 30.04$$

The sum of squares for the interaction component ($A \times B$) may be computed by subtraction:

$$A \times B \ SS = \text{overall among } SS - \text{between } A \ SS - \text{between } B \ SS$$
$$67.25 - 34.32 - 30.04 = 2.89$$

6. Compute the several degrees of freedom. In particular, determine df for the total source of variance Equation (10-7), for the overall among source Equation (10-8), and the within source Equation (10-9). Following this, allocate the overall among degrees of freedom to the components of it; namely that between A, that between B, and that for $A \times B$.

$$\text{Total } df = N - 1$$
$$= 28 - 1 = 27$$
$$\text{Over-all among } df = r - 1$$
$$= 4 - 1 = 3$$
$$\text{Within } df = N - r$$
$$= 28 - 4 = 24$$

The components of the overall among df are:

$$\text{Between } df = r - 1$$
$$\text{Between } A = 2 - 1 = 1$$
$$\text{Between } B = 2 - 1 = 1$$

$A \times B \; df = $ (number of df for between A) $\times$ (number of df for between B)
$= 1 \times 1 = 1$

7. Compute the various mean squares. This is accomplished by dividing the several sums of squares by the corresponding degrees of freedom. For our example these operations, as well as the results of the preceding ones, are summarized:

Source of Variation	Sum of Squares	df	Mean Square	F
Between A	34.32	1	34.32	4.81
Between B	30.04	1	30.04	4.21
A × B	2.89	1	2.89	.40
Within groups	171.43	24	7.14	
Total	238.68	27		

Compute an F for each "between" source of variation. In a 2×2 factorial design there are three F-tests to run. The F is computed by dividing a given mean square by the within groups mean square (assuming the case of fixed variables). These F's have been computed and entered in the above table.

9. Enter Table A-3 in the appendix to determine the probability associated with each F. To do this find the column for the number of degrees of freedom associated with the numerator and the row for the number of degrees of freedom associated with the denominator. In our example they are 1 and 24 respectively. The F of 4.81 for between A would thus be reliable beyond the 5 percent level, and accordingly we would reject the null hypothesis for this condition. The F between B (4.21) and that for the interaction (0.40), however, are not reliable at the 5 percent level; hence we would fail to reject the null hypotheses for these two sources of variation.

Problems

1. An experimenter wants to evaluate the effect of a new drug on "curing" psychotic tendencies. Two independent variables are studied—the amount of the drug administered and the type of psychotic condition. The amount of the drug administered is varied in two ways, none and 2 cc. The type of psychotic condition is also varied in

two ways, schizophrenic and manic-depressive. Diagram the factorial design used.

2. In the above experiment the psychologist used a measure of normality as the dependent variable. This measure varies between 0 and 10, where 10 is very normal and 0 is very abnormal. Seven participants were assigned to each cell. The resulting scores for the four groups were as follows. Conduct the appropriate statistical analysis and reach a conclusion about the effect of each variable and the interaction.

PSYCHOTIC CONDITION

Schizophrenics		Manic Depressives	
	Did Not		Did Not
Received Drug	Receive Drug	Received Drug	Receive Drug
6	2	5	1
6	3	6	1
6	3	6	2
7	4	7	3
8	4	8	4
8	5	8	5
9	6	9	6

3. How would the above design be diagrammed if the experimenter had varied the amount of drug in three ways (zero amount, 2 cc, and 4 cc), and the type of psychotic tendency in three ways (schizophrenic, manic-depressive, and paranoid)?

4. How would you diagram the above design if the experimenter had varied the amount of drug in four ways (zero, 2 cc, 4 cc, and 6 cc) and the type of participant in four ways (normal, schizophrenic, manic-depressive, and paranoid)?

5. A cigarette company is interested in the effect of several conditions of smoking on steadiness. They manufacture two brands, Old Zincs and Counts. Furthermore they make each brand with and without a filter. A psychologist conducts an experiment in which two independent variables are studied. The first is brand, which is varied in two ways (Old Zincs and Counts), and the second is filter, which is also varied in two ways (with a filter and without a filter). A standard steadiness test is used as the dependent variable. Diagram the resulting factorial design.

6. In the above experiment the higher the dependent variable score, the greater the steadiness. Assume that the results came out as follows (10 participants per cell). What conclusions did the experimenter reach?

OLD ZINCS		COUNTS	
With Filter	*Without Filter*	*With Filter*	*Without Filter*
7	2	2	7
7	2	3	7
8	3	3	7
8	3	3	8
9	3	3	9
9	4	4	9
10	4	4	10
10	5	5	10
11	5	5	11
11	5	6	11

7. An experiment is conducted to investigate the effect of opium and marijuana on hallucinatory activity. Both of these independent variables were varied in two ways. Seven participants were assigned to cells and amount of hallucinatory activity was scaled so that a high number indicates considerable hallucination. Assuming that adequate controls have been realized, and that a 5 percent level of reliability was set, what conclusions can be reached?

SMOKED OPIUM		DID NOT SMOKE OPIUM	
Smoked Marijuana	*Did Not Smoke Marijuana*	*Smoked Marijuana*	*Did Not Smoke Marijuana*
7	5	6	3
7	5	5	2
7	4	5	2
6	4	4	1
6	3	4	1
5	3	4	0
4	3	3	0

Appendix to Chapter 10

The Rationale for Selecting Error Terms

In a 2×2 factorial experiment we partition the total sum of squares into the sum of four component sums of squares, known as factor A, factor B, interaction AB, and within-groups sum of squares. Tests of the null hypotheses are made in terms of the four mean squares obtained from these sum of squares. With the exception of the within groups mean square, each of the other mean squares may contain more than one source of variation (or component of variance). The expected value of each mean square is shown in Table 10-21. Let us consider the fixed model first, and so we shall read down the column of Table 10-21 for the expected mean square for that model. Note,

Table 10-21 Expected Mean Squares for the Fixed, Random, and Mixed Models for a 2 × 2 Factorial Design

SOURCE OF VARIATION	EXPECTED MEAN SQUARES FOR THE			
	Fixed Model	*Random Model*	*Mixed Model (A Fixed)*	*Mixed Model (B Fixed)*
Independent Variable A:	$\sigma_W^2 + 2n\sigma_A^2$	$\sigma_W^2 + n\sigma_{AB}^2 + 2n\sigma_A^2$	$\sigma_W^2 + n\sigma_{AB}^2 + 2n\sigma_A^2$	$\sigma_W^2 + 2n\sigma_A^2$
Independent Variable B:	$\sigma_W^2 + 2n\sigma_B^2$	$\sigma_W^2 + n\sigma_{AB}^2 + 2n\sigma_B^2$	$\sigma_W^2 + 2n\sigma_B^2$	$\sigma_W^2 + n\sigma_{AB}^2 + 2n\sigma_B^2$
A × B Interaction:	$\sigma_W^2 + n\sigma_{AB}^2$	$\sigma_W^2 + n\sigma_{AB}^2$	$\sigma_W^2 + n\sigma_{AB}^2$	$\sigma_W^2 + n\sigma_{AB}^2$
Within Groups:	σ_W^2	σ_W^2	σ_W^2	σ_W^2

for instance, that the expected mean square for independent variable A has two possible components of variance: that for the within variance (σ_W^2) and $2n$ times the measure of variation due to variable A (σ_A^2). For the present discussion we shall ignore the constants, like $2n$, and focus only on the variances. Reading further down the column, note that the two possible components for independent variable B are the within variance (σ_W^2) and a measure of variation due to variable B (σ_B^2). The interaction source of variation likewise may be due to the within variance and that for the interaction (σ_{AB}^2). Finally, we note the within variance as the only component for that source of variation. In running our F-tests let us say that we start by testing the A × B interaction. If that F is reliable we conclude that there *is* an interaction between variables A and B; otherwise, there is no such interaction. Now we can more clearly see what happens when we conduct this F-test. That is, from our sample values we compute a mean square for the among source of variation for interaction, and likewise for the within groups. These mean squares are, roughly, estimates of the (population) variances for the interaction and for within groups. To compute the F ratio for interaction we divide the mean square for within groups into the mean square for interaction. Now, if the null hypothesis is true, $\sigma_{AB}^2 = 0$. Consequently, the interaction component of the mean square will be zero in the long run, so that the F ratio for interaction will consist only of an estimate for the within groups variance in the numerator, and also an estimate for the within groups variance in the denominator, yielding a value of approximately one. But if the null hypothesis is false, there is a positive value for the interaction variance. In this case the numerator of the F ratio should contain not only a within groups mean square but also a contribution for the interaction; hence F should be greater than 1.0 — it should be reliable, in fact.

In short, when the (true) variance interaction is zero, we expect an F value of 1.0, but if there is a component for interaction, F becomes greater than 1.0. Furthermore, if the ratio is sufficiently great we say that F is reliable. In summary, we can see why the within mean square is the appropriate error term for testing the interaction mean square for a fixed model: *Since the interaction mean square may contain the within variance* and *the interaction variance, we divide that mean square by the within mean square. If the resulting value is reliably greater than 1.0, we conclude that there is a reliable interaction.*

To continue to the source of variation for independent variable B we can see that precisely the same reasoning applies. Since this expected mean square contains the within variance as well as the variance due to B, we divide this mean square by the within mean square.

The case for independent variable A is precisely the same. This reasoning will become clearer as we take up the random model.

For the random model, let us read down the column for the expected mean square for the case of random variables. As before, we note that the within source of variation has only one component, viz., the within groups variance (σ_w^2). Similarly, the A $\times$ B interaction is the same for the random as for the fixed model. We therefore test the interaction for the random model as before. We divide that mean square by the within groups mean square. The difference between the two cases occurs when we note that there are *three* variance components for independent variable B: For the random model it may contain not only the within-groups variance (σ_W^2) and ($2n$ times) the variance due to B (σ_B^2), but also (n times) the interaction variance. Hence, we need to "divide out" both the within and the interaction variances for this expected mean square. Since these are the two variance components for the A $\times$ B interaction, we use the mean square for interaction as our error term for the independent variables. As before, if the variance for independent variable B is zero, as assumed by the null hypothesis, the numerator of the F ratio does not (in the long run) contain a mean square for factor B, and the expected value of F approximates 1.0. But this F value increases as σ_B^2 increases; if the sample value is sufficiently great, the F is reliable and we can reject the null hypothesis. The same reasoning applies for variable A.

We can, incidentally, now note the error that would occur should the incorrect error term be used. For example, in the random model, if you happen to divide the mean square for independent variable A by the within-groups mean square, your F value will be artificially inflated. That is, the numerator of the F-test would contain a variance component for the interaction *and* a variance component for variable B. Your interest is in whether or not variable B is effective, but what you actually would be testing is the reliability of the variance for interaction plus the variance for variable B. In this instance if your F is reliable you do not know whether it is because the interaction or variable B is reliable or whether it is a combination of these sources of variation. Experimenters have actually concluded that variable B was reliable when they have used the wrong error term. With this understanding, you can now avoid that mistake.

We now can hastily deal with the case of mixed variables, for our principles have already been established. First let us consider the case where independent variable A is fixed and independent variable B is random. In Table 10-21 we read down the column labeled "Mixed Model (A Fixed)." Here the proper error term for testing the interaction source of variation is the within mean square, because the interaction mean square may contain a within variance component and an interaction variance component. Similarly, the mean square for the random variable (variable B) may contain a component due to within and a component due to variable B; hence, to test B we use the within-groups mean square in the denominator of the F ratio. But the mean square for fixed variable A may contain sources of variation due to A and due to the within and due to interaction; consequently to test A, one uses the interaction mean square in the denominator of the F ratio.

Finally, for the mixed model where B is fixed (the last column of Table 10-21), we can apply precisely the same reasoning as for the mixed model where A is fixed. In this instance, one tests the interaction mean square by dividing that value by the within mean square. Similarly, the random variable (variable A) is tested by using the within-groups mean square as the error term, but the fixed variable (variable B) is tested by using the interaction mean square for the error term.

11

Experimental Design
within-groups designs

The two-randomized-groups design, the matched-groups design, the more-than-two-randomized-groups design, and the factorial design are all examples of "between-groups designs." This is so because two or more values of the independent variable are selected for study, and *one* value is administered to *each* group in the experiment. We then compute the mean dependent variable value for each group, compute the mean difference *between* groups, and thus assess the effect of varying the independent variable. An alternative to a between-groups design is a "within-groups design" in which two or more values of the independent variable are administered, in turn, to the same participants. A dependent variable value is then obtained for each participant's performance under each value of the independent variable; comparisons of these dependent variable values under the different experimental treatments then allow assessment of the effects of varying the independent variable.

In short, for between-groups designs we compare dependent variable values *between* groups who have been treated differently. In within-groups designs, the same participants are treated differently at different times, and we compare their scores as a function of different experimental treatments. For example, suppose that we wish to ascertain the effects of LSD on perceptual accuracy. Use of a between-groups design would dictate (essentially) that we administer LSD to an experimental group, but not to a control group. We would then compare the means of the two groups on a test of perceptual accuracy to determine possible effects of the drug. But for a within-groups design we would administer the test of perceptual accuracy to the same participants: (1) when they were under the influence of the drug; and (2) when they were in a normal condition. If the means of the same participants change as they go from one condition to the other, we ascribe the change in behavior to LSD, if controls are adequate.

One excellent example of the use of a within-groups design (when there was only one person in the "group") is the classical experiment on memory performed by Ebbinghaus (1913). This pioneer, it will be recalled, memorized several lists of nonsense syllables and then tested himself for recall at various times after learning was completed. He then calculated the percentage of each list that was forgotten after varying periods of time. For example, he found that he had forgotten about 47 percent of one list 20 minutes after he had learned it, that 66 percent of a second list was forgotten after one day, that after two days he had forgotten 72 percent of a third list, and so forth. By thus taking repeated measures on himself Ebbinghaus was able to plot amount forgotten as a function of the passage of time since learning and obtained his famous forgetting curve.[1]

A number of other classic experiments have also been conducted with the use of a within-groups design. In addition to much work on memory, for example, there has been a myriad of studies in the area of psychophysics (cf., Underwood, 1966a; Woodworth & Schlosberg, 1955). Weber's experimentation that yielded the data necessary to formulate his famous law is but one case in point; you no doubt studied Weber's Law in introductory psychology. Let us now turn, however, to several ways in which this type of design is used in contemporary psychology.

Two Conditions, Many Participants

We already have some familiarity with the t- and A-tests for matched groups, so this provides us with a good basis for studying one kind of within-groups design. In this case a measure is obtained for each of a number of participants when they perform a certain task

---

[1] It is fortunate, incidentally, that Ebbinghaus was not a professional psychologist. For if he had been, he would have known that what he accomplished was "impossible"—psychologists of his time, for example, held that the "higher mental processes" (e.g., memory) were not susceptible to experimental attack.

or under some given experimental condition. The same measure is taken again when the participants perform another task, or under another experimental condition. A mean difference between each pair of measures is computed and tested to determine whether it is reliably different from zero.[2] If this difference is not reliable, then it cannot be concluded that the variation of the independent variable resulted in behavioral changes. Otherwise, the conclusion is in the affirmative. For example, two of my students (Norm Ostrov, and Ron Savukas) sought to test a hypothesis based on the work of Lepley (1952). Lepley's findings suggested that individuals engage in covert oral behavior ("subvocal speech") when they write words. The experiment thus called for direct recording of the amplitude of speech muscle responding during handwriting; among the speech muscle measures recorded was chin electromyograms (EMG). These experimenters had their participants relax and then systematically either draw ovals or write words. It was reasoned that the motor task of drawing ovals, which does not involve the use of language, would serve as a control condition, i.e., people are generally active when they write, but is there a greater increase in covert oral behavior during writing than occurs during comparable activity that is not language in nature? To answer this question, the experimenters first determined the increase in chin EMG during writing, i.e., they subtracted amplitude of chin EMG during resting from that during writing for each person. Then they similarly determined the increase in chin EMG amplitude for each

Table 11-1 Change in Chin Electromyograms (μv) During Handwriting and While Drawing Ovals

Participant	Handwriting	Drawing Ovals	Difference
1	23.5	12.0	11.5
2	.3	5.8	− 5.5
3	86.8	52.8	34.0
4	33.3	−29.3	62.6
5	46.4	22.9	23.5
6	− 1.6	−24.1	22.5
7	26.2	−20.7	46.9
8	6.6	− 6.0	.6
9	16.9	−13.1	30.0
10	43.6	22.6	21.0
11	143.6	6.7	136.9

$$\Sigma D = 384.0$$
$$\Sigma D^2 = 28{,}578.34$$

[2] The design popularly referred to as the "Pretest-Posttest" design fits into this paradigm.

individual during the drawing of ovals. The results are presented in Table 11-1. For example, Participant 1 increased her amplitude of covert oral responding (by this measure) during writing by 23.5 μv (μv = microvolts, which is one one-millionth of a volt); the comparable increase when she drew ovals was 12.0 μv. And so on for the other participants. The question is: Is there a reliably greater increase during the writing period than during the "ovals" period? To answer this question we compute the difference in response measures; for Participant 1 we can see the difference is 11.5 μv. To conduct a statistical test, we next compute the sum of the differences and the sum of the squared differences, values that are included in Table 11-1. If the mean of the difference values is not reliably greater than zero, we will not be able to assert that variation of the experimental tasks produced a change in covert oral behavior. The appropriate test is either the matched t-test or the A-test; since the latter is easier to compute, we select it in the hope that the saved time can be used to good advantage elsewhere. Recall from page 213 that:

$$(11\text{-}1) \qquad\qquad A = \frac{\Sigma D^2}{(\Sigma D)^2}$$

Substitution of the appropriate values from Table 11-1 results in:

$$A = \frac{28{,}578.34}{(384.00)^2} = \frac{28{,}578.34}{147{,}456.00} = .194$$

Referring to Table A-2 in the appendix, we find that this A (with 10 df) indicates that the mean of the differences between the two conditions is reliably different from zero, i.e., $P < 0.05$ (P actually would have been less than 0.02 had we set this as our level of reliability). The conclusion, thus, is that the students emitted a reliably larger amplitude of covert oral responding during silent handwriting than during a comparable motor task that was nonlanguage in nature (than drawing ovals). The interpretation of this finding is that individuals engage in covert *language* behavior when receiving and processing language stimuli (words).[3]

Incidentally, the question on which we focused was: Is there a greater change in the dependent variable when the participants engaged in task A than in task B? Often, as in this case, performance in the two tasks is ascertained by comparison with some standard condition, such as during a resting state. In this event another, but related,

---

[3] These conclusions were later confirmed in a more rigorous experiment (McGuigan, 1970).

question can also be asked, viz., did performance under condition A (and B) change reliably from the standard condition? The data in Table 11-1 can also provide answers to these questions. Since the values under "Handwriting" and "Drawing Ovals" are themselves difference values, they can also be analyzed by the A-test. That is, you will recall that a measure was obtained for each person when she was resting, and then when she was writing. The score 23.5 for Participant 1 thus was obtained by subtracting the resting level from the level during writing. To determine whether there was a reliable increase in covert oral behavior when the participants changed from resting to writing, one merely needs to compute the sum of the values under the "handwriting" column, and the sum of the squares of these scores. Then substitute these values into Equation (11-1) and ascertain whether the A value is reliable. Is it? How about the values for the "ovals" condition?

Several Conditions, Many Participants

The within-groups design in which two experimental treatments are administered to the same participants can be extended indefinitely. Let us briefly illustrate one extension by considering an experiment by Underwood (1945) in which four values of the independent variable were administered to the same group of participants. First, all participants were systematically presented with the following tasks: (1) they studied no lists; (2) they studied (for four trials) two lists of paired adjectives; (3) they studied four lists of paired adjectives; and (4) they studied six such lists. Following this they completely learned another list of paired adjectives; 25 minutes later they were tested on this list, and the dependent variable was the number of paired adjectives that they could correctly recall. The results are presented in Figure 11-1, where it can be noted that the fewer the number of prior lists studied, the better the recall. As you no doubt have observed, this was an experiment on proactive inhibition (interference), i.e., when we study something and then learn some other (related) material, the first learned material inhibits the recall of the later learned material. Put another way, earlier learned material proactively interferes with the retention of later learned material, and Underwood showed that the greater the number of prior lists learned, the greater is the amount of proactive inhibition. But regardless of the subject matter findings, the point here is that participants can be administered a number of experimental treatments by means of the within-groups design.

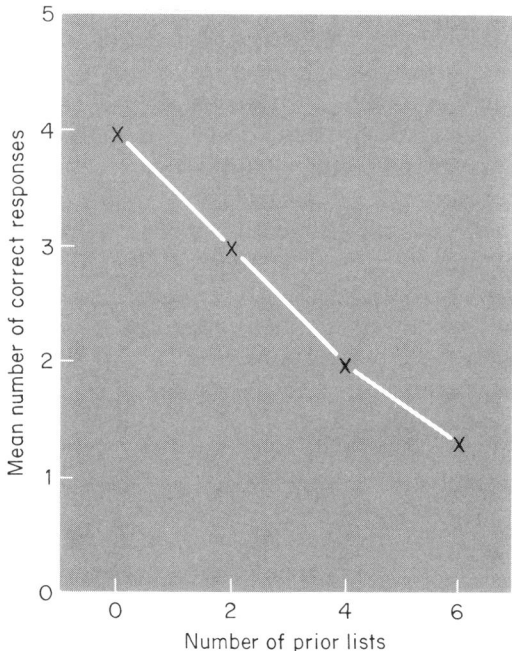

Figure 11-1. The larger the number of lists studied before learning, the greater the amount of proactive inhibition (after Underwood, 1945).

Evaluation of Within-Groups Designs

The contrast has been between within- and between-groups designs, and since we have presented both kinds it must be apparent that there are pluses and minuses for each. Let us take up some of the more straightforward points first:

1. It should be clear that the within-groups design has a great advantage as far as economy of participants is concerned. For in this design dependent variable values are available for all participants under all treatment conditions. The simplest illustration is the contrast with a two-randomized-groups design; you have two values of the independent variable and, say, 20 participants in each group for a total of 40 participants. Hence, for each condition, you have 20 dependent variable values. But in a within-groups design you would have all participants serve under both conditions; therefore you could either: (1) run a total of only 20 participants to obtain the number of values equal to that for the between groups design; or (2) you could still run 40 participants and have twice as many observations for each treatment condition.

2. The within-groups design is also relatively advantageous if your experimental procedure demands considerable time or energy in getting ready to observe your participants. For example, if you are conducting psychophysiological research, it takes a fair amount of time and patience to properly attach the necessary electrodes, etc. Or, suppose that you are conducting neuropsychological research, such as implanting brain electrodes in animals. Once you made such a sizable investment in your preparation, you probably would want to collect numerous data, probably by studying your participants under a variety of conditions.

3. The advantage of the within-groups design that is most frequently cited is that, by its means, you typically reduce your error variance. We have seen in chapter 8 that matching participants on an initial measure can result in a sizable increase in the precision of your experiment. The same logic applies here. In effect, by taking two measures on the same participant, you can reduce your error variance in proportion to the extent to which the two measures are correlated. Put another way, one reason that the error variance may be large in a between-groups design is that it includes the extent to which individuals differ. But since, in a within-groups design, you repeat your measures on the same participants, you remove individual differences from your error variance. Hence, rather than having an independent control group, you have each participant serve as his or her own control.

You are probably getting suspicious by now, wondering "what's the rub?" There are some:

4. One problem is that when one of the two (or more) treatments comes first, it may not be reasonable to then present the other. Suppose that, as in the Jacobsen et al. experiment on page 264, you wish to study the effects of injecting RNA into an organism. Further, suppose that you wish to use as a control group, animals that did not receive RNA. Clearly a between-groups design has to be used here, for you could not first administer RNA, test the animals, and then take RNA out of them and retest them. The effect of administering RNA is irreversible—an irreversible effect. An irreversible effect is one in which a given set of operations is performed in such a way that subsequent measurements are biased by the effects of the original operations. This brings us face to face with a topic that has been lurking in the background throughout this chapter, viz., the problem of the order in which the experimental treatments are presented to the same participants.

5. Order effects—this is a major problem entailed by the use of a within-groups design. The one technique that we have discussed for

systematically presenting the order of conditions is counterbalancing. If you do not adequately recall this discussion, you should restudy it now (pp. 161-64).

You have decided, let us say, to use a within-groups design and now face the question of what order to present your treatments to your participants. If you know that the order of conditions will have no effect on your dependent variable, and that there are no practice and fatigue effects, then you have no problem—whether or not you use a counterbalanced design is irrelevant. Assuming that you are in this fortunate position, you clearly should use a within-groups design. This is, however, more or less of a "thank you for nothing" answer, for unless you have a massive amount of data on your particular variables, you would never know that you are in this happy state. So we must be more realistic.

First, let us be clear about the seriousness of the problem. If you assume that one condition does not interact with another, when in fact it does, your conclusions can be drastically distorted. An excellent example of this error was pointed out by Underwood (1957a) in which Ebbinghaus' forgetting curve is reexamined. Let us emphasize here that Ebbinghaus learned a number of lists that were to be recalled at later times. Implicit in Ebbinghaus' assumptions, as we look back from our present vantage point, was that his treatments did not interact to affect his dependent variable. Put more simply, the assumption was that the learning of one list of nonsense syllables did not affect the recall of another. The results of Ebbinghaus' research indicated that the large majority of what we learn is very rapidly forgotten, e.g., after one day, according to Ebbinghaus' forgetting curve, about 66 percent is forgotten. The consequence of this research, incidentally, has been sizable and has been the source of great discouragement to educators for many decades. However, we now know that the basic assumption of Ebbinghaus' experimental design is not tenable. In fact, one of the major advances in the study of memory has been to establish the great effect of competition amongst materials that have been learned; the result has been the interference theory of forgetting. It was Underwood (1957a) who astutely demonstrated the defect in Ebbinghaus' design. For he showed that Ebbinghaus, by learning a large number of lists, created a condition in which he maximized amount of forgetting. When we take the number of previous lists learned into account, the situation is quite different. Figure 11-2 vividly makes the point, for this forgetting curve, plotted by Underwood, indicates the percent forgotten (or equally recalled) after 24 hours as a function of number of previous lists learned. There we can note that the situation is really not as bad as Ebbinghaus' results would have us believe. True, when many lists are learned, forgetting is great. But if there have been no previous lists learned, only about 25 percent is forgotten after one day. The lesson thus should be clear: By using a within-groups design Ebbinghaus gave us a highly restricted set of results that were greatly overgeneralized and that thus led to erroneous conclusions about forgetting. Had he used a between-groups design in which each participant learned only one list, he would have concluded that the amount forgotten was relatively small.

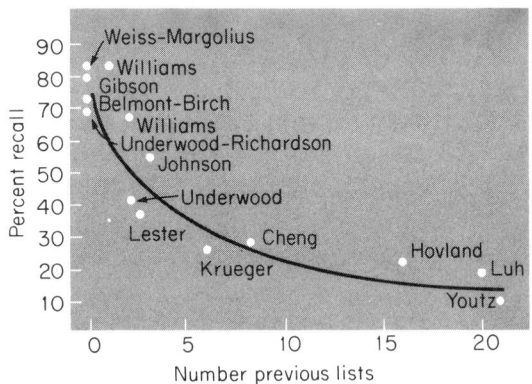

Figure 11-2. Recall as a function of number of previous lists learned (after Underwood, 1957a).

To further illustrate the difference in results that can be obtained by using the two types of designs, consider how Grice and Hunter (1964) varied the intensity of the conditional stimulus in a conditioning experiment. These experimenters used two designs. In the between-groups design one group received a low intensity conditional stimulus (soft tone), while a second group received a high intensity conditional stimulus (loud tone). They also conducted the experiment using a within-groups design in which all participants received both values of the conditional stimulus. The question was: Did variation of the intensity of the conditional stimulus affect the strength of the conditional response? The results are presented in Figure 11-3. We can see that in both experiments, there was an increase in the percentage of conditional responses made to the conditional stimulus as the intensity of the conditional stimulus increased from 50 to 100 decibels (db). But the slopes of the curves are dramatically different. The difference in percent of conditional responses as a function of stimulus intensity was not statistically reliable for the be-

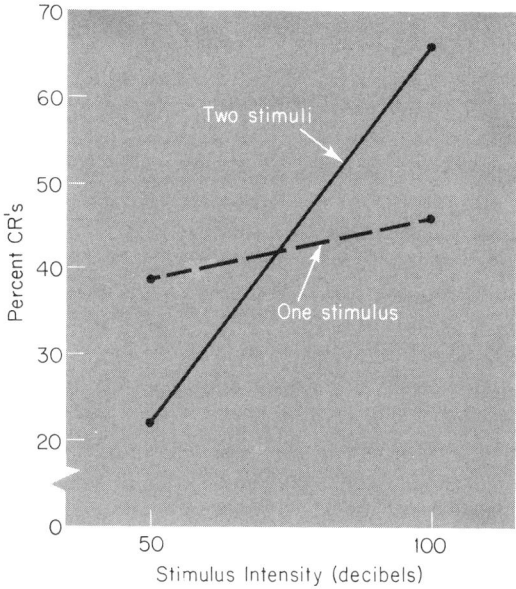

Figure 11-3. Percent of conditional responses during the last 60 trials to the loud and soft tones under the one- and two-stimulus conditions (after Grice and Hunter, 1964).

tween-groups design ("one stimulus"), while it was for the within-groups design (in which the participants received "two stimuli"). In fact, the magnitude of the intensity effect is more than five times as great for the two-stimuli condition than for the one-stimulus condition. Hence, the dependent variable values were influenced by the number of conditions in which the participants served; there was an interaction between stimulus intensity and number of presentations of stimuli.[4]

Research in other areas has also resulted in contradictory conclusions, depending on whether the researcher employed within- or between-groups designs. Pavlik and Carlton (1965) studied the effects of continuous reinforcement versus intermittent ("partial") reinforcement schedules (whether the participant is reinforced on all of the learning trials versus reinforcement on less than 100 percent of the trials). The usual intermittent reinforcement effects of greater resistance to extinction and higher terminal performance were found when using the between-groups design, but not for the within-groups design. On perhaps a more menacing dependent variable measure,

[4] This experiment was criticized by Erlebacher, (1977).

Valle (1972) found that frequency of defecation of rats was differentially affected by the type of design used (within- versus between-groups) in studying free and forced exploration.

With this appreciation of the importance of the possible interaction effects of our treatments, let us now return to the question of the order to use in a within-groups design. The purpose of counterbalancing, we have said, is to control order (practice and fatigue) effects — to distribute these extraneous variables equally over all experimental conditions. But, we pointed out, by thus controlling these variables, you might inherit problems of a different sort, viz., asymmetrical transfer effects. Hence, if you use a counterbalanced design you should demonstrate (by appropriate statistical analysis) that there was no differential transfer among your conditions. On the other hand, if you expect ("fear" might be a better word) asymmetrical transfer effects, you should seek some other method for presenting your treatments. In an excellent article by Gaito (1961), six cases for presenting repeated treatments to the same participants are examined. There are potentially serious methodological problems with five of the designs, and these should be used with caution; several methods of counterbalancing are among these five designs, as you no doubt anticipated. The sixth design is methodologically sound; it calls for the randomization of the order of the treatments. For example, if you have three treatments (A, B, and C), and all participants are to receive all treatments, then you randomly determine the order of A, B, and C for each participant. However, Gaito points out, the error variance in the design using randomization of treatments "may be quite large."

6. The problem of interaction of the different experimental conditions has also been approached from a rather different point of view, namely that concerned with the statistical analysis. When each participant is presented with more than one treatment, we have several times made the point that the dependent variable values for those conditions are not independent — a person's score under condition A is related to the individual's score under condition B, regardless of what those conditions are. As we emphasize in chapter 16 ("Assumptions Underlying the Use of Statistical Tests"), the assumption of independence of observations is critical. For instance, when we measure change in performance through gain scores, we violate the assumption of independence. O'Conner (1972) illustrated this problem of nonindependence as follows:

> The concept of a change score has considerable intuitive appeal. A person subtracts last week's weight from today's weight and talks of having gained or lost five pounds. Yet, change scores have more than their share of con-

ceptual problems. Weights are comparable—a two-hundred-pounder outweighs a one-hundred-pounder regardless of his other traits; but changes are not necessarily comparable—a loss of twenty-five pounds may be a godsend for one individual but a disaster for another. Even in cases where changes in one direction are preferred, certain comparisons of changes appear inappropriate. For example, an instructor may grade physical education students on their improvement in running the mile. All of the students running an eight-minute mile at the beginning of the course may cut more than a minute out of their times; none of the four-minute milers are likely to improve by more than a few seconds. Clearly, the eight-minute milers "improved" their time by more seconds than did the four-minute milers. Yet no instructor would give A's to the slowest runners and F's to the fastest, regardless of his commitment to the concept of grading on improvement. Somehow these "improvements" are not comparable for the purposes of evaluation. This inability to directly compare changes at different points of the scale, *even with ratio scales,* is the fundamental problem of the measurement of change (p. 73).

This problem of nonindependence of the repeated measures of your participant's behavior in within-groups designs is obviously extremely complex. While we probably should not consider the issue in much greater detail here, we might briefly mention some published thoughts that could be valuable for you should you become seriously involved with the problem in your later research. For instance, Gaito (1973), in commenting on this problem of nonindependence, emphasized a different apsect when he pointed out that the computed probability levels for the usual (univariate) F test applied in this situation may be (erroneously) three to five times as great as it should be. In analyzing the issue Gaito argued that "multivariate procedures" (which use the statistic known as Hotelling's T^2, among other things) is to be preferred, but since those procedures are computationally laborious many researchers, he held, will not employ them. "Thus to be practical one would suggest that either the Box approximate F test or the Geisser-Greenhouse approach would be appropriate. The Greenhouse-Geisser three step procedure might be most useful. However, an important point is that univariate analyses should not be used in the usual manner unless ρ is constant, or relatively constant, over treatments" (p. 75).

Huck and McLean (1975) criticized the popular procedure of using a repeated measures analysis of variance (such as the Lindquist Type I analysis of variance, or the split plot factorial design) on the grounds that incorrect conclusions may result or that there may be completely redundant reanalyses of the same data. These authors then argued that an analysis of gain scores as an alternative to the repeated analysis of variance procedure yields the same amount of useful information in a more straightforward and parsimonious (see chapter 3) manner. Finally, the analysis of covariance in which the pretest score is the covariant (mentioned later in chapter 16) is contrasted with a gain score analysis. The analysis of covariance, they conclude, is to be preferred to a gain score analysis because it is more powerful and has greater versatility.

After a thoughtful analysis of the problem, Cronbach and Furby (1970) argued that we should not even attempt to measure change (gain) except possibly under very restricted conditions. They reasoned that raw change scores formed by subtracting pretest scores from posttest scores leads to fallacious conclusions. Rather, they suggest, we would be better advised to frame our questions in different ways. For other statistical analyses and consequent recommendations on this problem of non-independence, you might also consult Namboodiri (1972), Poor (1973), and Dickinson and Wolins (1974).

7. There are yet other difficulties with within-groups designs. For instance, Labouvie, Bartsch, Nesselroade, and Baltes (1974) studied characteristics of participants in a longitudinal design (in which repeated measures are taken on the same individuals). They found that a sample of adolescents who volunteered for the study were different intellectually from other potential participants, so that what was found for the volunteers could not be safely generalized (see chapter 15) to those who did not volunteer. Finally, we may note that Poulton (1973) analyzed a large number of results from within-groups designs and concluded that this type of design leads to what he referred to as "unwanted range effects" of a wide variety. Such range effects, he pointed out in some detail, have led researchers to numerous unwarranted and conflicting conclusions, and even led them to develop theories that explained these erroneous conclusions: "At present, widely used statistical texts (Edwards, 1968; Lindquist, 1953; Winer, 1971) fail to warn students that within-subject designs produce unwanted range effects. Clearly a warning needs to be given, both in texts and in courses on experimental design. The day should come then when no reputable psychologist will use a within-subject design, except for a special purpose, without combining it with a separate-groups design" (p. 119). For further discussion of some of the pros and cons of this line of argument on the use of within-groups designs see Poulton (1974), Rothstein (1974), Greenwald (1976), and Erlebacher (1977).

In summary, it is quite clear that there are several advantages of the within-groups design over the between-groups design, and vice versa. If you do proceed with a within-groups design, but cannot effectively handle the control problems entailed by counterbalancing, then you can present your treatments to your participants in a random order. However, if you are not satisfied with your counterbalancing, you probably should use a between-groups design. If you select a between-groups design, it is interesting to note that Gaito suggested serious consideration of the matched-groups design (chapter 8).

The problem of how to analyze statistically various kinds of within-groups designs (instances of which are variously called "pretest-posttest designs," "repeated measures designs," "repeated treatments designs," "longitudinal designs," or "developmental designs") has long constituted a major stumbling block to their proper employment. Many years ago on a very pleasant walk with Mr. Snedecor (see the item on p. 267), the author enjoyed listening to him consider this problem out loud. He admitted that we did not have a good solution, but that we did have to use within-groups designs under some conditions. Consequently we might just as well do the best we can "for now," hoping that with continued contrasts of within- and between-groups designs, a good solution will eventually evolve. I think we *are* making some progress in better understanding the problem.

Single-Participant ($n = 1$) Designs with Replication

The designs on which we have thus far concentrated have, in general, been based on the assumption that many participants would be studied for a short period of time. The effects of varying the independent variable were studied by computing group means and evaluating them relative to the amount of error in the experiment. If changes in group means were sufficiently larger than experimental error, it was concluded that there was a relationship between the independent and the dependent variables. For example, in designs analyzed by analysis of variance the value of the numerator of the F ratio is an indication of the effects of varying the independent variable, (the among-mean square) while the denominator (the within-mean square) is the error variance. The F-test yields a significant value if the numerator is sufficiently larger than the denominator. In short, the strategy has been to attempt to determine whether changes in behavior produced by the independent variable were sufficiently great to show through the "noise" in the experiment. This strategy has been vigorously criticized by Skinner (e.g., 1959) and his associates (especially Sidman, 1960). Although we shall take up the topic of error variance in greater detail in chapter 16, this brief introduction serves to explain the rationale of the present design. Skinner's methodology, which is referred to as "The Experimental Analysis of Behavior," is an effort to sizably reduce the error variance in the experiment. From the present point of view, experimental error has two major components: (1) that due to individual differences among the participants; and (2) that due to ineffective control procedures. Briefly, the former is eliminated in this design, simply put, by using only one or a few participants in each experiment; the latter is reduced by establishing highly controlled conditions in the experimental situation. Rather than studying a relatively large number of participants for a short period of time, Skinner has proposed that we study a few participants over an extended period. There is, then, replication such that the same experimental conditions are repeated with several participants.

The participant is placed in a well-controlled environment, typically in an operant conditioning chamber[5] if it is an animal subject. In the operant conditioning chamber an effort is made to eliminate or hold constant all extraneous stimuli so that external noises are minimized, lighting is constant, no unique olfactory cues are present, and so forth. Then the animal performs for a lengthy period during which time a baseline level of performance is established. For example, a hungry rat might be conditioned to press a lever and receive food. Conditioning procedures are continued until the rat displays a stable rate of bar pressing, as shown by a cumulative record.

The cumulative record is established as follows. The writing pen on an ink recorder is automatically activated each time the rat presses the bar. The pen writes on a continuously moving piece of paper and is elevated one unit for each bar press. Figure 11-4 shows this process. Imagine that the paper is moving from right to left. Then each bar press moves the pen up one unit. When no response is made, the pen indicates this by continuing to move horizontally. Hence, we can note that after one minute the rat made a response, that it did not make another response until two minutes had elapsed, that a third response was made after two and one half minutes, and so forth. If we wish to know the total number of responses made after any given time in the experimental situation, we merely read up to the curve from that point in time and over to the vertical axis. For example, we can see that after five minutes the rat had made five responses, as read off of the vertical axis. Incidentally, the cumulative response curve is a summation of the total number of responses made since time zero; this means that the curve can never decrease, i.e., after the rat has made a response, as indicated by an upward mark, that response can never be

---

[5] The operant conditioning chamber, also referred to as a "free experimental space," has more popularly been called a "Skinner Box." Professor Skinner has expressed the hope that he will be remembered more for his scientific work than as the inventor of a "box." In deference to Skinner, we wish to help to counter this tendency.

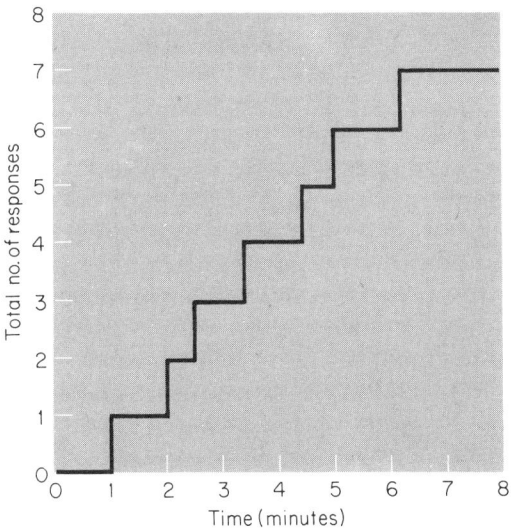

Figure 11-4. A cumulative response curve shown in detail.

unmade; the pen can never move down. Think about this point, if the cumulative response curve is new to you.

We have shown in Figure 11-4 a short portion of the cumulative response curve. The experimenter allows the participant to continue responding so that a considerable amount of the organism's experimental history is recorded. The animal eventually performs in a very stable fashion so that the response rate is quite constant. Now, once this steady state has been established, it is reasonable to extrapolate the curve indefinitely, so long as the conditions remain unchanged. It is at this time that the experimenter introduces some unique treatment. The logic is quite straightforward — if the response curve changes, that change can be ascribed to the effects of the new stimulus condition. Once it has been established that the curve changes, the experimental condition can be removed and, providing there are no lasting or irreversible effects, the curve should return to its previously stable rate. Additional conditions can then be presented, as the experimenter wishes. It can thus be seen that this is a within-groups design in which repeated treatments are administered to the same participant, forming "groups" by replicating the experiment with several participants.

One of the interesting aspects of Skinner's research has been his concern for technological matters. Although his laboratory work yielded a number of valuable principles of behavior, he and his followers have made a strong effort to explore applications of the principles to the problems of everyday life. Let us illustrate the Skinnerian methodology more completely by considering a portion of an applied study conducted by Harris, Wolf, and Baer (1964). These experimenters studied a four-year-old boy who cried a great deal after he experienced minor frustrations. In fact, it was determined that he cried about eight times during each school morning. The cumulative number of crying episodes can be studied for the first 10 days of the experiment in Figure 11-5. The question was: What is the reinforcing event that maintains this crying behavior? The experimenters hypothesized that it was the special attention from the teacher that the crying brought. The paradigm is thus the same as that for the rat in the operant chamber: When the response is made (the bar is pressed or the child cries), reinforcement occurs (food is delivered or the teacher comes to the child). After 10 days when the response rate was stabilized, the experimental treatment was effected: For the next 10 days the teacher ignored the child's crying episodes, but she did reinforce more constructive responses (verbal and self-help behaviors) by giving the child approving attention. As can be seen in Figure 11-5, the number of crying episodes sharply decreased with the withdrawal of the teacher's reinforcement for crying and during the last 5 of these 10 days only one crying response was recorded. During the next 10 days, reinforcement was reinstated — whenever the child cried, the

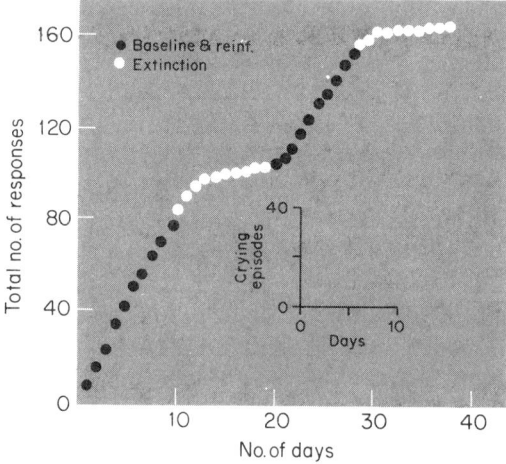

Figure 11-5. Cumulative record of the daily number of crying episodes. The teacher reinforced crying during the first 10 days (dark circles) and withdrew reinforcement during the second 10 days (light circles). Reinforcement was reinstituted during the third period of 10 days (dark circles) and withdrawn again during the last 10 days (after Harris et al., 1964).

teacher attended to the boy as she had originally done. Approximately the original rate of responding was reinstituted. Then, for the last 10 days of the experiment, reinforcement was again withdrawn and the response rate returned to a near zero level, as you can note by the last series of light circles. Furthermore, it remained there after the experiment was terminated. The experiment was repeated with another four-year-old boy, with the same general results.

Let us emphasize the importance of this last point in Skinner's methodology. Once it has been determined in an experiment with a single participant that some given treatment affects rate of responding, the experiment is replicated. When, under highly controlled conditions, it is ascertained that several other participants behave in the same way to the change in stimulus conditions, the results are generalized to the population of organisms sampled. The point we made several times earlier in the book applies here too, i.e., the extent to which the results can be generalized to the population of organisms depends on the extent to which that population has been sampled. Unfortunately, researchers who use the present methodology often do not include a large enough sample of organisms in their experiment.

In considering the method illustrated in Figure 11-5, we may well ask what is the likelihood that the changes in the response curve actually are due to the variations of the independent variable.[6] That is, might the response changes have occurred regardless of whether or not schedules of reinforcement and extinction were introduced as they were? This is essentially the question of a control condition. A variety of procedures has been introduced to increase the likelihood of the conclusion that any response change actually is a function of the introduction of the experimental treatment. One type is known as a "withdrawal design." Basically, the withdrawal design, which is used primarily in operant-conditioning studies, involves running a pre-experimental baseline, then introducing the experimental treatment. If the effects of the independent variable are reversible, the removal of the independent variable should return the response curve back to baseline level. This is sometimes referred to as the "ABA paradigm" in which behavior is studied to see whether it changes from A (the baseline or control period) to B, the treatment condition, and whether it returns back to baseline (A) when the independent variable or treatment is withdrawn. If behavior actually does increase and then decrease again during the ABA treatment series, then the likelihood that the response change is a function of the independent variable is in-

---

[6] My appreciation to Randy Flory for his good suggestions regarding this section.

creased. To further increase the likelihood of the conclusion that behavior is functionally related to the independent variable, one may introduce an ABAB sequence such that the experimental effect is produced twice with reference to changes from baseline. One could even further increase the likelihood of the conclusion by requiring additional changes such as ABABA or even ABABABA.

Another way in which the relationship between a behavioral change and the introduction of the independent variable may be further confirmed is by introducing the experimental treatment at a random point in the experimental session rather than at a predetermined time. That is, rather than introducing the experimental treatment after 10 days, as in Figure 11-5, the day on which the schedule change is effected could be randomly determined as any day — perhaps between day 5 and day 15. If, then, the same behavioral change always occurs in several participants immediately after the introduction of the experimental treatment, even though the treatment appeared randomly at different times, one could more firmly conclude that the two are functionally related.

The "reversal design" is another popular design used in the experimental analysis of behavior. Typically, two incompatible behaviors are initially selected for experimental study. Baselines are run on both of these classes of behavior, following which one behavior is subjected to one given treatment while the other behavior receives another type of treatment (or perhaps no treatment at all). For instance, the two incompatible behaviors might be talking and crying. Baselines would be established to ascertain the operant level for both. Then each time the child talked (the first behavior) reinforcement would be administered, and crying is not reinforced (the second behavior). Then, after sufficient data are accumulated, the treatment conditions are reversed so that the first behavior receives the treatment initially given the second behavior and the second behavior receives the treatment (or no treatment at all) that was associated with the first behavior initially. In other words, there is literally a reversal of the treatment conditions for the two behaviors. For instance, in this second phase crying would be reinforced and talking is ignored. Usually there is a final condition in which the desired treatment is reinstated such that talking (and therefore not crying) would be reinforced.

There are various other applications of basic operant designs in the experimental analysis of behavior, such as the multiple schedule design in which behavior is brought under the control of a discriminative stimulus. A variety of good sources are available for these other designs such as Sidman (1960) or Leitenberg (1973). Still, single-participant designs are not free from the same problems that occur for other types of within-groups designs. For one, there can be order effects such that practice on one treatment improves performance under a later treatment condition (or the first may lead to fatigue for the second); the effects of treatment may *not* be reversible so that a second treatment may have to be evaluated from quite a different baseline than a first, leading to potentially ambiguous conclusions. With regard to the question of whether or not a change in behavior actually (reliably) occurred, while there may be some change in behavior following the introduction of the independent variable, the change may not have been great enough to allow us to believe that it was truly a reliable change. One must evaluate the likelihood that any such change really is a function of the introduction of the independent variable. The withdrawal and reversal designs, the random introduction of the independent variable, etc. are techniques that attempt to increase the likelihood of that conclusion. Yet, it still is possible to statistically evaluate the effects of an experimental treatment even with single-participant designs. As was pointed out in the 1960 edition of the book, one can easily conduct statistical tests to remove any reasonable doubt that an apparent change in behavior due to the introduction of an experimental treatment is, in fact, statistically reliable. It is heartening to note the small avalanche of articles in recent years devoted to this topic. For those who are interested in the statistical analyses of single-participant designs please consult Shine and Bower (1971), Edgington (1967, 1972c), Revusky (1967), etc.

The methods used in the experimental analysis of behavior have much to recommend them. Skinner's work, and that inspired by him, has had a major influence on contemporary psychology. In addition to his contributions to pure science, it is important to emphasize that the results of using this methodology have had a sizable impact in such technological areas as education (e.g., programmed learning), social control, clinical psychology (through behavior modification), and so forth. Some of these practical issues will be taken up in chapter 12. It is likely that should you continue to progress in psychology, you will find that you can make good use of this type of design. Particularly at this stage in our development we should encourage a variety of approaches in psychology, for we have many questions that appear difficult to answer. No single methodological approach can seriously claim that it will be universally successful, and we should maintain as large an arsenal as possible. Sometimes a given problem can be most effectively attacked by one kind of design, while another is more likely to yield to a different design.

Review of Chapter 11

1. Distinguish between within-groups and between-groups designs. Identify examples of each category. Are these designs necessarily experimental designs, or might the method of systematic observation be used with either type?

2. Summarize some of the arguments for and against the use of within-groups designs. You might include a discussion of the efficiency as far as number of participants is concerned, the relevance of the concept of error variance, the irreversibility issue, order effects, and interaction among experimental treatments.

3. If you were limited to the t-test, which form of it would you employ for a within-groups design — the randomized (chapter 5) or the paired t (chapter 8)? (Naturally, implied here is the question of *why* you would use which form.)

4. Discuss operant conditioning paradigms within the context of within-groups designs. Would you apply a statistical test to determine whether or not an experimental effect with an $n = 1$ design is reliable?

12
Quasi-Experimental Design
seeking solutions to society's problems

Applied *Versus?* Pure Science

The spirit in which this book was originally written (1960), from one point of view, was that various of our endeavors are not mutually exclusive. Rather, they can facilitate each other. For instance, we have held that there should be no controversy between science and technology, or between (classical) experimental and clinical psychology. The fruits of pure science can often be applied for the solution of society's problems just as research on technological (applied, practical) problems may provide foundations for scientific ("basic," "pure") advances. The existence of practical problems may make gaps in our scientific knowledge apparent, and technological research can demand the development of new methods and principles in science. It is, furthermore, common for a researcher to engage in both scientific and in applied research at different times, or a research project may be astutely designed to yield scientific knowledge while curing some of society's ills. The issue, then, is not whether we are to favor pure science to the exclusion of applied matters or vice versa—we can do both. One need not be an experimentalist *or* a clinician—one could be an experimental clinician. Contrary to much popular opinion, there is no need for us to choose up sides on such issues.

B. F. Skinner is a good example of a scientist-technologist, as we mentioned in the last chapter when discussing single-participant research designs. Much of his life has been dedicated to the acquisition of knowledge for its own sake, but probably most of it has been devoted to the application of the principles of behavior for the solution of society's problems.

Historically, in our attempts to solve practical problems, as we noted in the introduction of chapter 6, society has made minimal use of control conditions. Skinner characterized it this way:

> So far, men have designed their cultures largely by guesswork, including some very lucky hits; but we are not far from a stage of knowledge in which this can be changed. The change does not require that we be able to describe some distant state of mankind toward which we are moving or "deciding" to move. Early physical technology could not have foreseen the modern world, though it led to it. Progress and improvement are local changes. We better ourselves and our world as we go (Skinner, 1961, pp. 545–46).

The guesswork has, much to Skinner's dismay, often involved punitive techniques for controlling behavior—principles of the Old Testament ("An eye for an eye and a tooth for a tooth") are to a very large extent applied to our everyday dealings with others. How often do we observe parents beating their children in efforts to control behavior?[1]

Skinner has studied various aspects of our culture to see how it might be improved through the application of psychological principles:

> Governmental, religious, economic, educational, and therapeutic institutions have been analyzed in many ways—for example, as systems which exalt such entities as sovereignty, virtue, utility, wisdom, and health. There is a considerable advantage in considering these institutions simply as behavioral technologies. Each one uses an identifiable set of techniques for the control of human behavior, distin-

---

[1] There is an apparent increase in the frequency of child abuse throughout the country. From a rational and long term point of view, one wonders why some parents bear children when they so terribly abuse them. We have licenses for so many things—to show our proficiency in driving an automobile, to practice psychology, etc. Some have advocated that potential parents should demonstrate at least a minimal level of child rearing proficiency before they are granted the privilege of having children (like they really want children, or that they probably would not injure their bodies). The argument continues that ineptly raised children become society's problems, so society should have *something* to say about who is born and how they are raised.

guished by the variables manipulated. The discovery and invention of such techniques and their later abandonment or continued use — in short, their evolution — are, or should be, a part of the history of technology (Skinner, 1961, pp. 543–44).

In place of the application of punishment, with all of its unfortunate consequences, Skinner has advocated selective reinforcement of behavior to improve our culture. In simplest form, the principle is to reinforce culturally desirable responses and to not reinforce undesirable behavior, though punishment still can play an effective if minor role. In his popular *Walden Two* (1948) Skinner illustrated in detail how he would design the ideal culture. The key is to arrange effective and desirable contingencies of behavior — one should wisely reinforce (and occasionally punish) social responses (see especially Skinner, 1953).

Skinner thus principally advocated the application of existing scientific knowledge to solve our practical problems. Certainly the wise and effective application of behavioral principles to such mounting problems as those of crime, drugs, auto accidents, and child-rearing abuses, would be far better than mere guesswork. In conducting business at our various governmental levels, we are constantly changing policies and introducing reforms. A new president or mayor is elected with campaign promises to change this or that — to abolish welfare, to extend it, to modify the penal system, etc. Unfortunately, however, society seldom systematically evaluates the effects of reforms, and we have little in the way of an objective basis for ascertaining whether or not a new policy has actually improved matters. The same can be said for many aspects of our society other than levels of government, such as, for instance, in our universities and colleges. We are constantly changing our educational practices, the character of our curricula, the nature of our graduation requirements. The pendulum continues to swing between extremes of decreasing and increasing course requirements for students. The essence of this chapter is that we are often in a position to systematically evaluate cultural changes and thus to gradually develop more beneficial practices. Just what are beneficial practices is difficult to say, but few would disagree that ills of society such as traffic in drugs should be cured. Systematic research cannot guarantee our happiness, but as we are promised by our Declaration of Independence, perhaps it can at least help to provide us with the opportunity to *pursue* happiness more effectively.

Some may deny that current societal reforms are only guesswork and say that data *are* presently collected on various of our cultural practices. Certainly we agree that there are acres and acres of storehouses of government records that constitute data of sorts. However, they are seldom if ever employed for the purpose of improving our governmental practices — they simply are not systematically related to independent variable conditions under which they were gathered. Our question is whether or not the fruits of systematic research can replace unused data gathered under conditions of chaotically changing policies? If so, we might be able to confirm that the application of a scientific principle does solve a technological problem. Where a problem exists for which science does not have a suggested answer, such systematic research "in the field" would have to "start from scratch;" this would be an instance of a "gap in our knowledge," as discussed in chapter 2. In either event "field" research would be suggested.

Unfortunately, however, we often cannot conduct true experiments with proper control conditions in everyday life. Consequently, we are presented with a serious dilemma, for a major theme of this book is that we countenance sound research and shun shoddy research. You will recall the section in chapter 6 entitled "When to Abandon the Experiment," the point there being that an unsolvable confound may entail the decision to abandon the study. These statements are easy to make when we are talking about the acquisition of knowledge for knowledge's sake ("pure science") — it is hard to conceive of a situation in which poorly designed *scientific* research can be tolerated. But many technological issues pose another question. To solve an important problem of society, the researcher may simply not be able to properly conduct a ("true") experiment. Consider a study of the effects of welfare programs on unemployment, or the effects of capital punishment for deterring crime. One can imagine the national furor if we attempted to randomly assign half of the present welfare population to a control condition in which their welfare checks were discontinued, or if we randomly assigned convicted murderers to experimental conditions either of death or life imprisonment. The Nazis in World War II conducted atrocious medical experiments with little regard for human life, but in a civilized society such extremes for the sake of research are simply not tolerated — the kind of research cited in the introduction in chapter 6 in which half of the "participants" were administered a potentially effective antidote to prevent death due to poisoning may have been allowable in ancient times, but not today. In our discussion of research ethics in chapter 16 it is not necessary to caution against Nazi-like mutilation of the human body.

Since, then, it is often not feasible to conduct research that satisfies the highest standards, the question is whether or not compromises in rigorous methodology are justifiable. If the problem is sufficiently important, one that *demands* solution, it may be better to compromise research standards than to not attempt a solution at all. Society is replete with examples where some research was better than none. For instance, research that fell short of high laboratory standards provided a reasonably good solution to the problem of airplane hijackings (Dailey & Pickrel, 1975); the psychological procedures thereby developed have contributed to the elimination of air highjackings. The quasi-experimental designs presented by Campbell and Stanley (1963) have been widely studied for these purposes.

Quasi-Experimental Designs

The defining feature of a quasi-experimental design is that participants are *not* randomly assigned to different conditions. The method of systematic observation (chapter 4) is a quasi-experimental design in which participants are classified according to some characteristic, such as high versus low intelligence; their performance is then compared on a dependent variable measure. The shortcoming of such a quasi-experimental design is that the independent variable is confounded with extraneous variables so that we do not know whether or not any change in the dependent variable is actually due to variation of the independent variable (see chapter 6). That is, the probability of a conclusion that the independent variable produced a given behavioral change (reduced dependence on welfare, decreased drug traffic, etc.) when using a quasi-experimental design is lower than when it results from an experiment. From chapter 3, we know that we never know anything about the empirical world with certainty, but we do seek the highest probability for our conclusions that is reasonable. The best of our experiments may yield faulty conclusions, as in rejecting the null hypothesis 5 times out of 100 ("by chance") when it should not be rejected. Consequently, the empirical probability of the conclusion from a well-designed experiment may be, say, only 0.92. If we *must* settle for less than a rigorous experiment, as use of one of the better quasi-experimental designs, perhaps the probability of a conclusion may drop to 0.70. Even less rigorous quasi-experimental designs may yield lower probabilities (perhaps 0.50, or 0.40). The probability of a conclusion from a clinical or a case history study (see chapter 4) would be yet lower (perhaps 0.25), but still may be the best information that we have. Certainly it is preferable for us to operate on the basis of low probability knowledge than on no knowledge (0.00 probability relationships) whatsoever. As Campbell has developed this theme:

The general ethic, here advocated for public administrators as well as social scientists, is to use the very best method possible, aiming at "true experiments" with random control groups. But where randomized treatments are not possible, a self-critical use of quasi-experimental designs is advocated. We must do the best we can with what is available to us (Campbell, 1969, p. 411).

In short, to design the most desirable of cultures, as Skinner would have us do, we should accumulate as much knowledge of as high degree of probability as we can. For such a purpose we may need quasi-experimental designs.

Types of Quasi-Experimental Designs

In their excellent monograph, Campbell and Stanley (1963) presented 16 quasi-experimental designs, in addition to "correlational" designs, "ex post facto" designs, and other nonexperimental designs. Perhaps the most commonly used of the quasi-experimental designs is what they referred to as the Nonequivalent Comparison Group Design (which is similar to The Method of Systematic Observation except that it employs a pretest).

The Nonequivalent Comparison Group Design. A control group is one of the groups *in an experiment* to which participants are randomly assigned. A comparison group, however, is one that is already formed and is used in a nonexperimental design in place of a control group (chapters 6 and 7). This distinction between control and comparison groups is an important labeling difference because it immediately alerts the researcher to expect confounding. That is, the term "comparison group" implies confounding with an attendant reduction in the confidence that one may place in the empirical conclusion. With The Nonequivalent Comparison Group Design, two or more groups that have already been naturally assembled are studied, e.g., two different fifth grade classes in an elementary school. The participants thus have *not* been randomly assigned to the two groups, so that neither is a control group (one may be a comparison group).

Both groups are administered a pretest, which provides some information as to their "equality" prior to the administration of the experimental treatment. However, even if the two groups are shown to be equivalent with regard to the pretest, they no doubt differ in many other ways—even with identical pretest scores we have no reason to consider them as equivalent groups. Regardless of whether or not the groups are equivalent on the pretest, the experimental treatment is administered to one of the groups, following which both groups receive posttests on the dependent variable. The researcher should, preferably, randomly determine which of the two or more groups receive the experimental treatment.

Campbell and Stanley illustrated this design with a study that was conducted by Sanford and Hemphill at the United States Naval Academy at Annapolis. The question was whether midshipmen who took a psychology course developed greater confidence in social situations. The second-year class was chosen to take the psychology course while the third-year class constituted the comparison group. The second-year class reliably increased confidence scores on a social situations questionnaire from 43.26 to 51.42, but the third-year class only increased their scores from 55.80 to 56.78. From these data one might conclude that taking the psychology course *did* result in greater confidence in social situations. However, although this conclusion is possible, alternative explanations are obvious. For instance, the greater gains made by the second-year class could have been due to some general sophistication process that occurs maximally in the second year and only minimally in the third year. If this were so, the sizable increase in scores for the second-year class would have occurred regardless of whether or not the midshipmen took the psychology course. This alternative conclusion is further strengthened by noting that the second-year class had substantially lower pretest scores and, although their gain score was greater, their posttest score was still not as high as the pretest score of the third-year class.[2]

The statistical analysis of this type of design is probably obvious to you when you relate it to chapter 8. You would probably wish to evaluate gain scores for each group separately, so that you could determine whether or not there was a reliable change in the dependent variable measure for each of your groups. For this purpose you would employ the A-test (or matched t-test). Finally you may wish to determine whether or not any change from pre- to posttest was greater for one of the groups than for the other. For this purpose, you would probably wish to conduct an independent groups t-test (chapter 5) between the two groups, employing a gain score for each of the participants in the study. However, be sure to recall our discussion of problems in measuring gain (chapter 11).

Although some extraneous variables are controlled with this design (e.g., both groups receive the pretest and the posttest) there are numerous differences in how the groups are treated during the conduct of the research. For instance, the two classes probably have two

---

[2] A preferable design would have been to form two groups out of the second-year class and to have given the psychology course to only one (randomly chosen).

different teachers, perhaps they meet at different times of the day and are influenced by different characteristics in the separate classrooms. Such confounded extraneous variables are the reason that ambiguous conclusions are always possible with quasi-experimental designs. Finally, we may note that Campbell (1969) cautioned that one should not match participants of the two groups on pretest scores, because this matching procedure results in regression artifacts which is, incidentally, a shortcoming of matched-groups designs in general (chapter 8).

Interrupted Time Series Designs. For this type of design, periodic measurements are made on a group or individual in an effort to establish a baseline. Eventually, an experimental change is introduced into the time series of measurements, and the researcher seeks to determine whether or not a change in the dependent variable occurs. If so, one would hope that the change in the time series (the dependent variable) was systematically related to the experimental treatment. This design is thus similar to the Single Participant Design of chapter 11, the major difference being that much less control is possible in the use of this design in "the field" situation.

As an illustration of the interrupted time series design, Campbell (1969) presented some data on the 1955 Connecticut crackdown on speeding. After a record high of traffic fatalities in 1955 a severe crackdown on speeding was initiated. As can be noted in Figure 12-1, a year after the crackdown, the number of fatalities decreased from 324 to 284. The conclusion the governor offered was that "With the saving of 40 lives in 1956, a reduction of 12.3% from the 1955 motor vehicle death toll, we can say that the program is definitely worthwhile" (Campbell, 1969, p. 412). In Figure 12-2 the data of Figure 12-1 are presented as part of an extended time series. There we may note that the baseline actually is quite unstable, which illustrates one of the difficulties in employing this design in the field situation — quite in contrast to the single participant design of chapter 11 where the operant methodology calls for greater control to establish a stable baseline. With such an unstable baseline, it is difficult to evaluate the effect of a treatment, regardless of where in the time series the treatment is introduced. In Figure 12-2 the "experimental treatment" (the crackdown) was initiated at the highest point of the time series. Consequently, the number of fatalities in 1956 would on the average be less than in 1955, regardless of whether or not the crackdown had been initiated at that point. Campbell attributes this feature to the instability of the time series curve and refers to the reduction in fatalities from 1955 to 1956 as at least in part due to a "regression artifact":

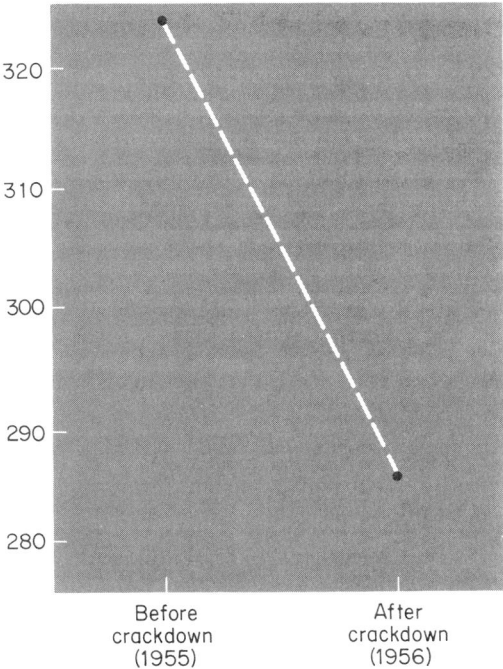

Figure 12-1. Connecticut traffic fatalities (after Campbell, 1969). Copyright (1969) by the American Psychological Association. Reprinted by permission.

Regression artifacts are probably the most recurrent form of self-deceptions in the experimental social reform literature. It is hard to make them intuitively obvious. . . . Take any time series with variability, including one generated of pure error. Move along it as in a time dimension. Pick a point that is the "highest so far." Look then at the next point. On the average this next point will be lower, or nearer the general trend (Cambell, 1969, p. 414).

In short, we could expect the time series to have decreased after the high point regardless of any treatment effect.

Another reason that we cannot firmly reach a conclusion about a causal relationship in this study was that the death rates already were going down year after year, relative to miles driven or population of automobiles, regardless of the crackdown. Consequently, other variables may have operated to produce the decrease after 1955 and these were thus confounded with the independent variable. To further illustrate how one may attempt to reason with the use of the interrupted time series design (and with quasi-experimental designs more generally) we may note that Campbell did argue against this latter interpretation. He pointed out that in Figure 12-2 the general slope

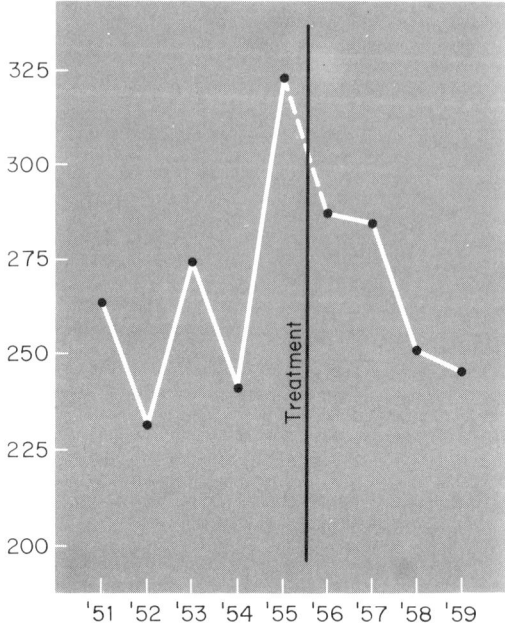

Figure 12-2. Connecticut traffic fatalities. (Same data as in Figure 12-1 presented as part of an extended time series; after Campbell, 1969.) Copyright (1969) by the American Psychological Association. Reprinted by permission.

prior to the crackdown is an increasing one, whereas it is a decreasing slope thereafter. If the national trend toward a reduction in fatalities had been present in Connecticut prior to 1955, one would have expected a decreasing slope prior to the crackdown. Although this reasoning does help to increase the likelihood of the conclusion that the crackdown was beneficial, the argument is certainly not definitive.

The interrupted time series design would typically be used when no control group is possible and where the total governmental unit has received the experimental treatment (that which is designed as the social reform). Because of the serious confounding with this design, Campbell argued for the inclusion of comparison groups wherever possible, even though they may be poor substitutes for control groups. The next design is an effort to improve on the Interrupted Time Series Design by adding a comparison series of data measurements from a similar institution, group, or individual not undergoing the experimental change.

The Multiple Time Series Design. This design is basically that of The Nonequivalent Comparison Group Design with the exception that multiple time series measures of the dependent variable are taken. For instance, the time series data for Connecticut in Figure 12-2 might be compared with similar data from some neighboring state such as Massachusetts. If the decreasing slope of the curve of Figure 12-2 after the crackdown is in contrast to values for Massachusetts, the conclusion that the reduction in traffic fatalities was produced by the crackdown would gain strength. With the Multiple Time Series Design then, any possible dependent variable change may be evaluated relative to a baseline value (as in the preceding design) and also relative to a change or lack of change in a comparison series for another governmental unit. One further method of increasing the likelihood of a valid conclusion is to introduce the experimental treatment randomly at some point in the series. We noted this strategy in the Single Participant Design in chapter 11. For a further attempt to reduce confounding and for a discussion of statistical analyses, see Gottman, McFall, and Barnett (1969). Campbell and Stanley regard this design as "an excellent quasi-experimental design, perhaps the best of the more feasible designs" (1963, p. 57).

Techniques of Naturalistic Observation

In chapter 4 we discussed the clinical or case study methods which closely resemble what are known as techniques of naturalistic observation. It may be well to briefly contrast these techniques here with those of experimentation. In techniques of naturalistic observation there is no intervention or treatment condition involved, only the gathering of systematic data protocols on behavior in naturally existing groups (families, pre-schoolers, school classes). These techniques are preferably made in unobtrusive ways so that natural patterns of behavior are preserved. Unfortunately behavioral research has often lacked unobtrusive naturalistic observation methodology, one possible solution being in the use of radio telemetry in which voice or other data may be detected from the participants and "radioed" through transmitters to the receiver of the researcher (Miklich, Purcell, and Weiss, 1974). A sizable amount of naturalistic observation research is being conducted in education, developmental, clinical, and social areas, together with discussion of lively methodological issues. These procedures have their own distinct problems of design and analysis. Ethologists too have highly developed techniques for observing animal behavior in their natural habitat and under various special conditions.

While this approach is clearly not experimental, sometimes it could be argued that it does follow within the genre of quasi-experimental research since the group being studied is identified for reason of some naturally existing "treatment" such as being disadvantaged, divorced, chronically ill, and a comparison group without these afflictions is also often used. This approach is thus included for mention because it falls within the concerns of this chapter, yet differs from what we have discussed so far.[3]

Conclusion

These examples illustrate for us the nature of quasi-experimental designs and, for further information about the large number of other designs you might refer to Kenny (1975), Campbell (1975), Campbell and Boruch (1975), and Cook and Campbell (1976). Some of the difficulties in carrying out true experiments in everyday life are obvious, but the shortcomings of the quasi-experimental designs make it clear that true experiments are to be preferred if at all possible. Other difficulties in conducting quasi-experimental designs are discussed in the above cited references. As one final illustration, Campbell (1969) pointed out that when a reform is introduced, the administrator often improves the recordkeeping too. Consequently, there is a confounding of the reform treatment with the accuracy of keeping records. For instance, in one situation where a reform was introduced in order to decrease larcenies, the records after the reform actually indicated an *increase* in larcenies due apparently to increased accuracy of the records. Maybe the reform *was* actually effective, but because of poor recordkeeping prior to the reform (perhaps there was inattention to the importance of that particular crime) the effectiveness of the reform was obscured.

While our study of quasi-experimental designs may be profitable in learning how to solve some technological problems, it can also provide us with an opportunity to better appreciate experimentation for, by recognizing the shortcomings of quasi-experimental designs, we might thereby improve our ability to plan and to conduct well-designed experiments.

Review of Chapter 12

1. Distinguish between applied and pure science. Must a scientist be always one or the other?
2. If you are a clinical psychologist, does this mean that you cannot also be an experimental psychologist?

---

[3] Thanks to Meredith Richards for bringing some of these points to my attention.

3. How would *you* attempt to solve what you regard as some of society's most pressing problems? Can a public administrator, well educated in the everyday wisdom of life, adequately solve our problems if merely given the power to do so? Or must governmental authorities rely on systematic technological research over the long run?
4. If you were given complete power over the penal system or the welfare system in this country what would you do? Would you attempt to change the system? If so, precisely how would you proceed?
5. Distinguish between experimental and quasi-experimental designs.
6. Confounding is always present in a quasi-experimental design. True or false?
7. No doubt you would want to review and summarize well for yourself the various types of quasi-experimental designs presented, including especially the method of systematic observation discussed in previous chapters.
8. Consider some instances in which you would advocate the use of naturalistic observation.

13

The Logical Bases of Experimental Inferences

We have said that an experimenter should state the hypothesis explicitly. An experiment has as its purpose the gathering of data that are relevant to the hypothesis. These data are summarized in the form of an evidence report, and the experimenter confronts the hypothesis with the evidence report. If the two are in agreement, it is concluded that the hypothesis is confirmed. Otherwise it is not. Let us analyze the relationship between the hypothesis and the evidence report.

The hypothesis, preferably, is stated in the conditional form; that is, as an if–then relation. To pursue the example of chapter 3, recall that the hypothesis that industrial work groups in great inner conflict have low production levels was stated as follows: *If* an industrial work group is in great inner conflict, *then* that work group will have a low production level. To test this hypothesis we might form two groups of workers, one in great conflict and the other quite harmonious. We would then collect data on the production output of the two groups. Assume that the statistical analysis of the data shows that the in-conflict-group has a reliably lower production level than the harmonious group.

Forming the Evidence Report[1]

An evidence report is a summary statement of the results of an empirical investigation; it is a sentence that precisely summarizes what was found. In addition, the evidence report states that the antecedent conditions of the hypothesis were realized. Hence, the evidence report consists of two parts: a statement that the antecedent conditions of the hypothesis held, and that the consequent conditions were found to be either true or false. The general form for stating the evidence report is thus that of a conjunction. Recalling the general form of the hypothesis is "If a, then b," a denotes the antecedent conditions of the hypothesis and b the consequent conditions. Hence, the possible evidence reports would be "a and b," or "a and not b," for the cases where the consequent conditions were found to be (probably) true and false respectively. The former is a positive evidence report and the latter is a negative one.

We shall illustrate by continuing the example. Let a stand for "an industrial work group is in great inner conflict" and b for "that work group has a lowered production level." In our hypothetical research we had an industrial work group that was in great inner conflict, and so we may assert that the antecedent conditions of our hypothesis were realized, that they were present in the research situation. Since our finding was that that work group had a lower production level than a control group, we may also assert that the consequent conditions were found to be true. Thus, our evidence report is: "An industrial work group was in great inner conflict *and* that work group had a lowered production level."

At this point, let us examine how we tell whether the consequent conditions of the hypothesis are true or false. In this example the group with inner conflict had a lower production level than did a control group. We needed the control group as a basis of comparison. For without such a basis "lower production level" does not mean anything—it must be lower than something. And so it is for all experiments. The way to tell whether or not consequent conditions are true is by comparing the results obtained under an experimental condition with those for some other condition, usually by means of a control group. That the hypothesis implicitly assumes the existence of a control or other group may be clarified by stating the hypothesis in the following manner: "If an industrial work group is in great inner conflict, then that work group will have a lower production level *than that*

---

[1] The term "evidence report" is taken from Hempel (1945) and is used because of its descriptive nature. Similar terms are "observational sentence," "protocol sentence," and "concept by inspection."

of a group that is not in inner conflict." The direction of the comparison is determined by the consequent conditions of the hypothesis. In this example, the hypothesis states that the group with inner conflict should have a *lower* production level than a control group. If the statistical analysis indicates that it is reliably lower than the control group, we conclude that the consequent conditions are probably true. If the statistical analysis indicates that the group with the inner conflict has a production level that is reliably higher than the control group, however, or if there is no reliable difference between the two levels, we conclude that the consequent conditions are probably false. The evidence report would then be: "An industrial work group was in great inner conflict and that work group did not have a lowered production level."

With this format for forming the evidence report before us, we shall now consider the nature of the inferences made from it to the hypothesis. Before we do this, however, it will be necessary to discuss the general topic of inferences and how they are made.

Inductive and Deductive Inferences

Let us say that we have a set of statements that we shall denote by A. These statements contain information on the basis of which we can reach another statement, B. Now, when we proceed from A to B, we make an *inference*. An inference, then, is a process by which we reach a conclusion on the basis of certain preceding statements — it is a process of reasoning whereby we start with A and arrive at B. We then have some belief that B is true on the basis of A. There are two kinds of inferences, inductive and deductive. In both types, our belief in the truth of B is based on the assumption that A is true. The essential difference between the two is that in an inductive inference, we reach the conclusion that, if A is true, B is true with some degree of probability; with a deductive inference, however, we can conclude that B is *necessarily* true if A is true.

The statement "Every morning that I have arisen, I have seen the sun rise" might be A. On the basis of this statement, we may infer the statement B: "The sun will always rise each morning." Now, does B necessarily follow from A? It *does not*, for while you may have always observed the rise of the sun in the past, it does not follow that it will *always* rise in the future. B is not *necessarily* true on the basis of A. Although it may seem unlikely to you now, it is entirely possible that one day, regardless of what you have observed in the past, the sun will not rise. We can only say, then, that B has some degree of probability (is probably true) on the basis of the information contained in A. When we make an inference that may be in error, we say that the result of the inference (the conclusion) can only have a certain probability of being true. Thus, the probability that statement B can be (inductively) inferred to be true on the basis of statement A may be high, medium, or low. The fact that we make inferences from one statement to another with a certain degree of probability sometimes leads us to use the term *probability inference* as a synonym for *inductive inference*.

It is possible (at least in principle) to determine the probability of an inference precisely as a specific number, rather than simply to say "high" or "low." Conventionally, the probability of an inductive inference may be expressed by any number from zero to one.[2] Thus, we may say that the probability (P) of the inference from A to B is 0.40, or 0.65, or whatever. Furthermore, the closer P is to 1.0, the higher the probability that the inference will result in a true conclusion (again, assuming that A is true). By analogy, the closer P is to 0.0, the lower the probability that the inference will result in a true conclusion, or, if you will, the higher the probability that the inference will result in a false conclusion. At this point we may note that if the probability of an inference is 1.0, the conclusion is necessarily true if the statements on which it is based (i.e., A) are true; and if P is 0.0, the conclusion is necessarily false. As we have previously noted, however, neither of these situations obtain when an inductive inference is made.

We may thus say that if the probability that B follows from A is 0.99, it is rather sure that B is true. The previous example of the inference from "Every morning that I have arisen, I have seen the sun rise" to "the sun will always rise each morning" is an example of an inference with a high probability. The probability of this inference is, in fact, extremely close to, but still not quite, 1.0. On the other hand, if the probability of the inference from A to B is 0.03, we know that this probability is extremely low and thus we are not likely to accept B as true on the basis of A. For example, the probability of the inference from the statement "a person has red hair" to the conclusion that "that person is very temperamental," would be one with a very low probability.

In short, then, an inductive inference is made when one passes from one statement to another with a lack of certainty; we infer that one statement probably implies another. And we may express our degree of belief in the truth of the results of such an inference by the

---

[2] To emphasize that 0 to 1 is an arbitrary range we may note that some authors allow P to assume any value between -1 and $+1$.

use of a number that varies from 0.0 to 1.0. The higher the number, the greater the probability that the inference results in a true conclusion.

We said that a deductive inference is made when the truth of one statement is necessary, based on another one or set of statements, if statement A necessarily implies B. In this case the inference is strict. Consider the following statements as an example of a deductive inference. We might know that "all anxious people bite their nails" and further that "John Jones is anxious." We may, therefore, deductively infer that "John Jones bites his nails." In this example, if the first two statements are true (they are called premises), the final statement (the conclusion) is necessarily true.

The determination of whether or not a given inference is deductive or inductive lies in the realm of logic. In both deductive and inductive (probability) logic certain rules (known as rules of inference or transformation) have been developed that indicate how to proceed from one set of statements to another. If an inference conforms to the rules of deductive logic, it is deductive. If it conforms to the rules of inductive logic, it is inductive. However, since this is not a book in logic, we shall not explore the various rules to any great extent. We shall simply indicate the rules that are used in making inferences from evidence reports to various types of hypotheses. In doing this, we shall indicate which are valid inferences, and whether they are inductive or deductive.

Direct Versus Indirect Statements

The statements with which science deals may be divided into two categories, direct or indirect. A direct statement is one that refers to limited phenomena that are immediately observable, that is, phenomena that can all be observed directly with the senses. For example, the statement "that bird is red" is direct. Of course the use of various kinds of auxiliary apparatus (e.g., microscopes, telescopes) to aid the senses may be used in forming direct statements. Hence, the statements "there is a sun spot," "there is an amoeba," or "that is a covert response" are also direct statements, since a person's sensory apparatus can be extended by various types of equipment. The procedure for testing a direct statement is straightforward: Compare the statement with an observation of the phenomenon with which it is concerned. More precisely, compare the statement with the evidence report. If the two are in agreement, the statement may be regarded as true; otherwise it is false. For example, in testing the direct statement

"That door is open," we observe the door. If we find that it is open, our observation agrees with the direct statement, and we conclude that the statement is true. If we observe the door to be closed, we conclude that the direct statement is false.

An indirect statement is one that cannot be directly tested. Such statements usually deal with phenomena that cannot be directly observed (logical constructs such as atoms, electricity, or habits) or that are so numerous or extended in time that it is impossible to view them all. A universal hypothesis is of this type—"All men are anxious." It is certainly impossible to observe all men (living, dead, and as yet unborn) to see if the statement is true. The universal hypothesis is the type in which scientists are most interested, since it is an attempt to say something about one or several variables for all time, at all places.[3]

Clearly, then, indirect statements cannot be directly tested. In order to test indirect statements it is necessary to reduce them to direct statements s_1, s_2, etc., we expect to find at least some that are direct in deductive inferences. Consider an indirect statement S. By drawing deductive inferences from S we may arrive at certain logical consequences, which we shall denote s_1, s_2, and so forth. Now among the statements s_1, s_2, etc., we expect to find at least some that are direct in nature. Such statements may be tested by comparing them with appropriate evidence reports. Now, if these directly testable consequences of our indirect statement are found to be true, we may make a further (inductive) inference that the indirect statement itself is probably true. That is, although we cannot directly test an indirect statement, we can derive deductive inferences from such a statement and directly test those inferences. If such directly testable statements turn out to be true, we may inductively infer that the indirect statement is probably true (see note 1 in the appendix to this chapter). But if its consequences turn out to be false, we must infer that the indirect statement is also false. In short, indirect statements that have true consequences are themselves probably true, but indirect statements that have false consequences are themselves false. This procedure is represented in Figure 13-1, although it will be necessary to analyze it more thoroughly later.

To illustrate this procedure let us consider the universal hypothesis, "All men are anxious." Assume we know that "John Jones is a man," and "Harry Smith is a man." From these statements (premises)

---

[3] Don't get too universal though, as one student did who defined a universal statement as a "relationship between *all* variables for all time and for all places."

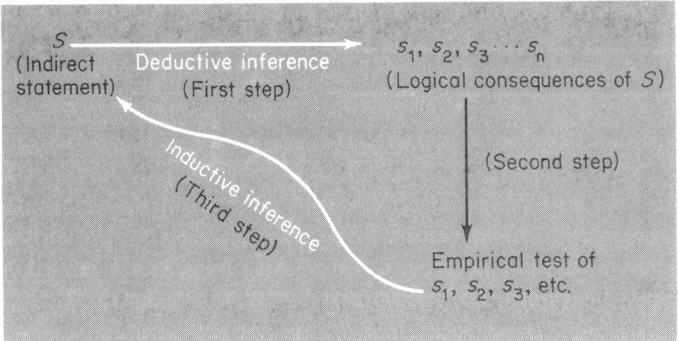

Figure 13-1. Representation of the procedure for testing indirect statements. First, deductive inferences result in consequences of general statements that are empirically testable. If those specific statements are confirmed in empirical tests, those confirmed consequences form the basis for an inductive inference that the indirect sentence is probably true.

we can state (by deductive inference) that "John Jones is anxious" and "Harry Smith is anxious." Since the universal hypothesis is an indirect statement, it cannot be directly tested. However, the deductive inferences derived from this indirect statement are directly testable. We only need to determine the truth or falsity of these direct statements. If we perform suitable empirical operations and thereby conclude that the several direct statements are true, we may now conclude, by way of an *inductive inference,* that the indirect statement is confirmed.

The class of variables with which the indirect statement deals is of infinite number. For this reason it is impossible to test all the logically possible consequences of that indirect statement (e.g., we cannot test the hypothesis for all men). Further, it is impossible to make a deductive inference from the direct statements back to the indirect statement—rather, we must be satisfied with an inductive inference. And we know that an inductive inference is liable to error; its probability is less than 1.0. As long as we seek to test indirect statements, we must be satisfied with a probability estimate of their truth. We will never know absolutely that they are true. We can never know for sure that anything is absolutely true—"truths" are held only until further notice.

Confirmation Versus Verification

Our goal is to determine whether a given universal statement is true or false. To accomplish this goal we reason thusly: *If* the hypothesis is true, *then* the direct statements that are the result of deductive inferences are also true. Now, if we find that the evidence reports are in accord with the logical consequences (the direct statements), we conclude that the logical consequences are true. And if the logical consequences are true, we inductively infer that the hypothesis itself is probably true.

Note that we have been cautious and limited in our statements about concluding that a universal hypothesis is false. Under certain circumstances it is possible to conclude that a universal hypothesis is strictly false (not merely improbably or probably false) on the assumption that the evidence report is reliable. More generally (i.e., with regard to any type of hypothesis), it can be shown that under certain circumstances it is possible strictly to determine that a hypothesis is true or false, rather than probable or improbable, but always on the assumption that the evidence report is true. We will here distinguish between the processes of *verification* and *confirmation*. By verification we mean a process of attempting to determine that a hypothesis is strictly true or strictly false; confirmation is an attempt to determine whether a hypothesis is probable or improbable. This ties in with the distinction between inductive and deductive inferences. Under certain conditions it is possible to make a deductive inference from the consequence of a hypothesis (which has been determined to be true or false) back to that hypothesis. Thus, where it is possible to make such a deductive inference, we are able to engage in the process of verification. Where we must be restricted to inductive inferences, the process of confirmation is used. To enlarge on this matter, let us now turn to a consideration of the ways in which the various types of hypotheses are tested (see note 2 in the appendix to this chapter).

Inferences from the Evidence Report to the Hypothesis

Universal Hypotheses. Recall that the universal hypothesis specifies that all things referred to in the hypothesis have a certain characteristic. For such hypotheses we have indicated a preference for the "If a, then b" form. With such a form it is understood that we are referring to all a and all b. For example, if a stands for "rats are reinforced at the end of their maze runs" and b for "those rats will learn to run that maze with no errors," it is understood that we are talking about all rats and all mazes on which those rats might be trained. The general procedure for testing this hypothesis may be represented as follows (see note 3 in the appendix to this chapter):

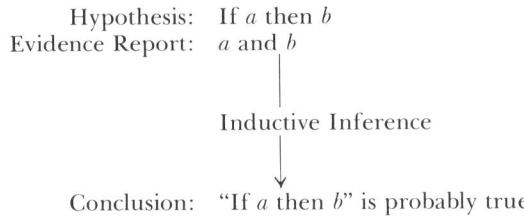

Hypothesis: If a then b
Evidence Report: a and b

Inductive Inference

Conclusion: "If a then b" is probably true.

To illustrate this procedure by means of our example we might form two groups of rats; Group E is reinforced at the end of each maze run, but Group C is not. Let us say that at the end of a certain number of trials Group E is able to run the maze with no errors, but Group C is still making errors. A t-test indicates that Group E's performance is reliably superior to that of Group C. We are thus able to assert that the antecedent conditions of the hypothesis were realized and that the data were in accord with the consequent conditions. The evidence report is positive. The inferences involved in the test of this hypothesis may be illustrated as follows:

Universal Hypothesis: If rats are reinforced at the end of their maze runs, then those rats will learn to run that maze with no errors.

Positive Evidence Report: A (specific) group of rats was reinforced at the end of their maze runs and those rats learned to run the maze with no errors.

Conclusion: The hypothesis is probably true.

We have attempted to show some specific steps in testing a hypothesis, to give you some insight into the various inferences that must be made for this purpose. In your actual work, however, you need not specify each step, for that would become cumbersome. Rather, you should simply rely on the brief rules that we present for testing each type of hypothesis. To summarize the rule for testing a universal hypothesis for the case of a positive evidence report, we shall merely say that when the evidence report agrees with the hypothesis, that hypothesis shall be said to be confirmed.

To understand the test of the universal hypothesis when the evidence report is negative, we refer to the distinction between confirmation and verification. Clearly, the case of confronting the universal hypothesis with a positive evidence report is an example of confirmation. When the evidence report is negative, however, we are able to apply the procedure of verification. This is possible because the rules of deductive logic tell us that a deductive inference may be made from a negative evidence report to a universal hypothesis. The procedure for this situation may be illustrated as follows:

> Universal Hypothesis: If a then b
> Evidence Report: a and not b
>
> |
>
> Deductive Inference
>
> ↓
>
> Conclusion: "If a then b" is false

For example:

Universal Hypothesis: If rats are reinforced at the end of their maze runs, then those rats will learn to run that maze with no errors.

Negative Evidence Report: A group of rats was reinforced at the end of their maze runs and those rats did not learn to run that maze without any errors.

Conclusion: The hypothesis is false (see note 4 in appendix to this chapter).

We may thus see that it is possible to determine that a universal hypothesis is (strictly) false (through verification) if the evidence report is negative. But if the evidence report is positive, we cannot determine that the hypothesis is (strictly) true; rather, we can only say that it is probable (through confirmation). This characteristic of being able to verify a hypothesis in only one direction (for example being able to determine that it is strictly false, but not being able to determine that it is strictly true) has been called by Reichenbach (1949) *unilateral verifiability*. And we shall see that it is characteristic of all universal and existential hypotheses.

Existential Hypotheses. This type of hypothesis says that there is at least one thing that has a certain characteristic. Our example, stated as a positive existential hypothesis would be: "There is a (at least one) rat that, if it is reinforced at the end of its maze runs, then it will learn to run that maze with no errors." If this type of hypothesis is strictly true, then it can be verified by observing a series of appropriate events until we come upon a positive instance; and a single positive case is sufficient to determine that the hypothesis is true. On the other hand, if the class of variables specified in the hypothesis is infinite in size, or at least indefinitely large, we shall never be able to determine that the hypothesis is strictly false. We can only observe a finite number of events. If, in that finite number of events, we do not observe a positive instance of our hypothesis, we may well expect this state of affairs to continue for future observations. But we will never be sure that a negative instance of the hypothesis will not eventually occur. Hence, we can appreciate that this type of hypothesis is also unilaterally verifiable; it is possible to determine that it is strictly true but not possible to determine that it is strictly false. Let us illustrate the inferences involved in testing the existential hypothesis. For the case of a positive evidence report:

Existential Hypothesis: There is an a such that if a then b
Positive Evidence Report: a and b

Deductive Inference

Conclusion: Therefore, the hypothesis is (strictly) true.

For the case of a negative evidence report:

Existential Hypothesis: There is an a such that if a then b
Negative Evidence Report: a and not b

Inductive Inference

Conclusion: Therefore, the hypothesis is not confirmed.

To illustrate by means of our previous example:

Existential Hypothesis: There is a rat that, if it is reinforced at the end of its maze runs, will learn to run that maze with no errors.
Positive Evidence Report: A group of rats was reinforced at the end of their maze runs and at least one of those rats learned to run that maze with no errors.
Conclusion: The hypothesis is (strictly) true.

Existential Hypothesis: There is a rat that, if it is reinforced at the end of its maze runs, will learn to run that maze with no errors.
Negative Evidence Report: A group of rats was reinforced at the end of their maze runs and none of those rats learned to run that maze with no errors.
Conclusion: The hypothesis is not confirmed.

Limited Hypotheses. Limited hypotheses (offered as direct statements) are always verifiable because it is possible to make a deductive inference from the evidence report (whether positive or negative) to the hypotheses. Thus, a limited hypothesis, such as "if rat number 3 is reinforced at the end of its maze runs, then that rat will learn to run that maze with no errors" is completely (bilaterally) verifiable. For instance, if we find that rat number 3 does learn to run the maze with no errors, we may conclude that the hypothesis is true. But if he does not, we may conclude that the hypothesis is false.

Irrelevant Evidence Reports. One final matter needs to be emphasized: the importance of satisfying the antecedent conditions of the hypothesis in the experimental situation. If this requirement is not satisfied, no inference can be made from the evidence report to the hypothesis. Instead, we may only say that the evidence report is irrelevant to the hypothesis and thus does not constitute a test of the hypothesis. For example, if the experimental group of rats in our hypothetical experiment was not reinforced at the end of its maze runs, then whatever the results with regard to the number of errors made, we cannot say that a test of the hypothesis has occurred.

The Reason for the "If . . . Then. . . ." Form

In chapter 3, we said that universal hypotheses can be viewed as assuming the "If . . . then. . . ." form. After reading this chapter, the reason for our desire to state them in this form, or at least to know that we could state them in this form if we wanted to, should be apparent. To be more explicit, however, let us note that various inferences can only be made validly if the statements (hypotheses and evidence reports) assume certain forms. For the logical rules that we have borrowed are applicable only to certain forms of statements. By using the conditional form for stating hypotheses and the conjunctional form for stating evidence reports, we are able to satisfy these rules.

Hence the inferences that we discuss in this chapter are valid because they conform to the rules of logic. Let us emphasize, however, that strictly speaking, it is not necessary actually to state your hypothesis in the if-then form. All that is required is that, whatever the form in which you state your hypothesis, it is *possible* to restate it in the if-then form. If this is not possible, then it is not possible to make valid inferences from the evidence report to your hypothesis. Similar considerations hold true for the evidence report; in order to perform a valid inference from it to the hypothesis, that evidence report must assume the conjunctional form. If you do not actually state your evidence report in the conjunctional form, however, you need not be concerned *as long as it would be possible to restate it in that form.*

Appendix to Chapter 13

Note 1 (from page 365). It is necessary to point out an error that we are perpetuating in this chapter. It concerns the inductive inference that is made from the evidence report to the hypothesis. Let us consider the universal hypothesis. For this case we said that an inductive inference may be made from a positive evidence report to the hypothesis, an inference that results in the conclusion that the hypothesis is confirmed. But this is not quite right. Rather, it is a procedure that is universally used. To understand the error, let us emphasize that the valid inferences of inductive logic are specified by the rules of probability logic (cf. Reichenbach, 1949), and it is not possible to find a rule of this sort in the calculus of probability.

To make the difficulty more specific, let us say that we seek to test the hypothesis that "If a then b." Further, assume a positive evidence report was obtained: a and b. Now, we make an inductive inference from that evidence report to the hypothesis and conclude that the hypothesis is confirmed. However, our hypothesis is not the only hypothesis that will also predict the consequent condition b. Rather, there is an unspecifiable number of additional hypotheses which will also have b as a consequent condition—if a' then b, if a'' then b, if a''' then b, and so on. When we actually find that b is true, then, we do not know which of the numerous hypotheses that imply b is confirmed. Although it is customary to assume that it is our hypothesis, it may just as well be one of the other possible hypotheses that is confirmed. About the best we can say at this point is that any hypothesis that implies b may be confirmed, providing that its antecedent conditions were present in the experimental situation. And since there are numerous unspecified antecedent conditions present in the experimental situation, we cannot say that only a was present, for a', a'', a''', etc., may also have been present. To avoid prolonging our discussion of this problem, let us simply note that there are certain inferences in probability logic that would, in principle, satisfy our needs. However, much additional work needs to be accomplished before we can properly make these more complicated inferences in our everyday research. For more information on the nature of the difficulty and a proposed solution you might refer to McGuigan (1956).

Note 2 (from page 367). At this point, an apparent contradiction in what we have said might occur to you. We previously stated that it is impossible to determine strictly that a hypothesis is true or false. Rather, we must always settle for a probability statement—that the hypothesis has some degree of probability. Here, however, we have said that we can sometimes determine that a hypothesis is false, not merely improbable. How may we reconcile these two statements?

The answer may be stated thus: The deductive inference from the evidence report to the universal hypothesis in effect states that if it is true that the evidence report is negative, then the universal hypothesis is necessarily false. But the evidence report itself rests on a probability statement, for we have formed it by failing to reject the null hypothesis. Hence, even though we have every reason to believe that the evidence report *really is* negative, we may actually be in error. For this reason we may erroneously conclude that the universal hypothesis is false, even though we are making a deductive inference. Remember that a deductive inference does not lead to an absolutely irrefutable statement about knowledge. Rather, it allows us to say that *if* the premises are true, then it is necessarily the case that the conclusion is true. But the determination of the truth or falsity of the premises (here the evidence report) is still an empirical matter and thus liable to error. It may thus be seen that confirmation and verification are really only different in degree with regard to the final conclusion. Although inductive inferences are used in the former and deductive inferences in the latter, the conclusion regarding the hypothesis still must be evaluated in terms of probability. But the reasons that the conclusion in the two cases is limited to a probability statement are different. In confirmation, the probability character of the conclusion results from the probability of the evidence report *and* the fact that an inductive inference must be made. In verification, however, the probability character of the conclusion rests *only* on the probability character of the evidence report since a deductive (not an inductive) inference is made. In general then, we can say that the conclusion in the case of verification is considerably more probable than that in the case of confirmation. This is so because of the high probability (frequently extremely high) of the evidence report.

Note 3 (from page 367). But this procedure is not quite right. To understand this let us observe that the hypothesis is universal in nature—as we said above, it refers to *all* rats, etc. But we are certainly in no position to test it on all rats; we must be content with a particular sample of rats that we have drawn from a population. We draw a deductive inference that the hypothesis is applicable to our particular sample of rats, i.e., if the hypothesis covers all rats, it certainly covers the particular rats with which we are dealing. By the same token, our evidence report is limited to results obtained on our particular group of rats. But as we have stated, the evidence report (i.e., *a* and *b*) is universal in nature. Thus, "*a* and *b*" states, in effect, that "*all* rats were reinforced at the end of their maze runs and *all* those rats learned to run the maze with no error." But this is not true. We are able to advance our evidence report for *only* those rats that we studied. Hence, the evidence report is much more specialized or restricted in scope.

We should indicate this restriction in some way. This may be accomplished by placing a subscript to the variables contained in the evidence report: a_1 and b_1, which shall then be read "a (certain) group of rats was reinforced at the end of its maze runs and that group of rats learned to run the maze with no errors." Similarly, the hypothesis that is the result of the deductive inference from the general hypothesis must be stated in specialized form: If a_1 then b_1. We then confront the specialized hypothesis with the (specialized) evidence report and reach a conclusion (by way of a deductive inference) as to its truth or falsity. And from this conclusion we go back to our universal hypothesis, again by way of an inductive inference, to determine its probability. This general procedure may thus be represented as follows:

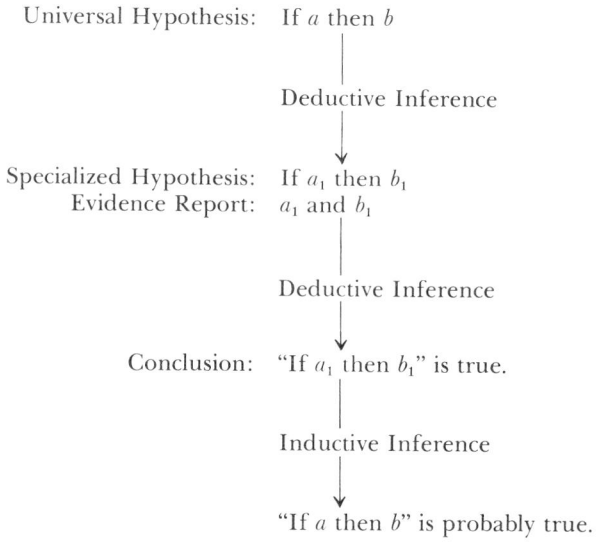

Universal Hypothesis: If a then b

Deductive Inference

Specialized Hypothesis: If a_1 then b_1
Evidence Report: a_1 and b_1

Deductive Inference

Conclusion: "If a_1 then b_1" is true.

Inductive Inference

"If a then b" is probably true.

In terms of our example, the steps are:

Universal Hypothesis: If (all) rats are reinforced at the end of their maze runs, then those rats will learn to run that maze with no errors.

Specialized Hypothesis: If a (specific) group of rats is reinforced at the end of its maze runs, then *those* rats will learn to run that maze with no errors.

Evidence Report: A (specific) group of rats was reinforced at the end of its maze runs and *those* rats learned to run that maze with no errors.

Conclusion: (a) The specialized hypothesis is true.
(b) The universal hypothesis is probably true.

We can now see that there are two reasons that it is an inductive inference that is made from the evidence report to the hypothesis: (1) because the evidence report states what was found for a sample of participants and the hypothesis is concerned with populations, hence one proceeds from "the specific to the general"; (2) because it is inferred that it is the hypothesis "if a then b" that is confirmed rather than, as discussed in note 1 above, the hypotheses "if a' then b," "if a'' then b," and so forth.

Note 4 (from p. 369). The probability character of empirical hypotheses exerts itself in still other ways, even though they are verifiable, as discussed above. For instance, we may still ask how many trials a rat should run before we are convinced that he would never learn to run the maze perfectly. Say that after a goodly number of trials the rat has still failed to learn. Yet it may be on just the next trial (no matter where it stopped running) that it would demonstrate flawless performance. One answer is to be more precise in the statement of our antecedent conditions, i.e., to specify a certain number of trials. For example, "If rats are reinforced at the end of each of 30 maze runs, then. . . ." In this case, if they demonstrate the required performance or if they don't after 30 trials, our conclusion is "certain" and the verifiable nature of the hypothesis is preserved.

Review of Chapter 13

1. You probably will want to make sure that you can define the basic terms used throughout the book, and especially in this chapter. Perhaps the glossary at the end of the book will also be useful for your review. Some basic terms that you should not forget are: evidence report, various kinds of hypotheses, the difference between inductive and deductive inferences, direct versus indirect statements, confirmation versus verification, and the conditional relationship (the "if . . . then . . ." proposition).

2. You might outline for yourself the basic procedures by which universal and existential hypotheses are subjected to empirical tests. Basic here of course is the use of inductive and deductive inferences.

14

The Inductive Schema

an overview of some characteristics of science

"Dr. Watson, Sherlock Holmes," said Stamford introducing us.

"How are you?" he said cordially, gripping my hand with a strength for which I should hardly have given him credit. "You have been in Afghanistan, I perceive."

"How on earth did you know that?" I asked in astonishment . . . "You were told, no doubt."

"Nothing of the sort. I knew you came from Afghanistan. From long habit the train of thoughts ran so swiftly through my mind that I arrived at the conclusion without being conscious of intermediate steps. There were such steps, however. The train of reasoning ran, 'Here is a gentleman of a medical type, but with the air of a military man. Clearly an army doctor, then. He has just come from the tropics, for his face is dark, and that is not the natural tint of his skin, for his wrists are fair. He has undergone hardship and sickness, as his haggard face says clearly. His left arm has been injured. He holds it in a stiff and unnatural manner. Where in the tropics could an English army doctor have seen so much hardship and had his arm wounded? Clearly in Afghanistan.' The whole train of thought did not occupy a second. I then remarked that you came from Afghanistan, and you were astonished" (Doyle, 1938, pp. 6,14).[1]

[1] Reprinted by permission of the Estate of Sir Arthur Conan Doyle.

This, the first meeting between Holmes and Watson, is a relatively simple demonstration of Holmes' ability to reach conclusions that confound and amaze Watson. It serves well to illustrate what Reichenbach has called the *inductive schema*. The reconstruction of Holmes' reasoning is presented in the inductive schema shown in Figure 14-1. The observational information available to Holmes is at the bottom. On the basis of this information Holmes infers certain intermediate conclusions. For example, he observed that Watson's face was dark, but that his wrists were fair. These two bits of information immediately led to the conclusion that Watson's skin is not naturally dark. He must therefore have recently been in an area where there was considerable sun; Watson had probably "just come from the tropics." From the several intermediate conclusions it was then possible for Holmes to infer the final conclusion, that Watson had just recently been in Afghanistan. You should trace through each step of Holmes' reasoning process, as represented in the inductive schema, to make sure that you understand how it was constructed. You might even want to construct such a schema for yourself from any of Holmes' other amazing processes of reasoning.

Let us now turn to an example from physics, an inductive schema that represents the process of scientific reasoning. In Figure 14-2 we have partially reconstructed the evolution of this science. In the bottom row are some of the basic data (evidence reports of investigation) from which more general statements were made.

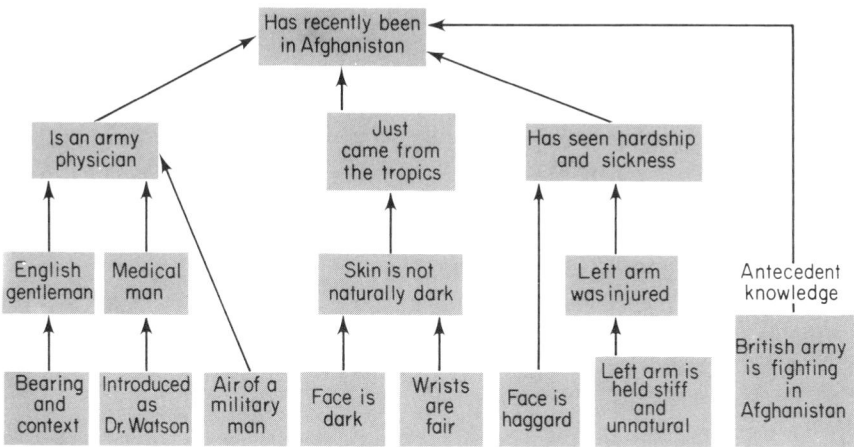

Figure 14-1. An inductive schema based on Sherlock Holmes' first meeting with Dr. Watson.

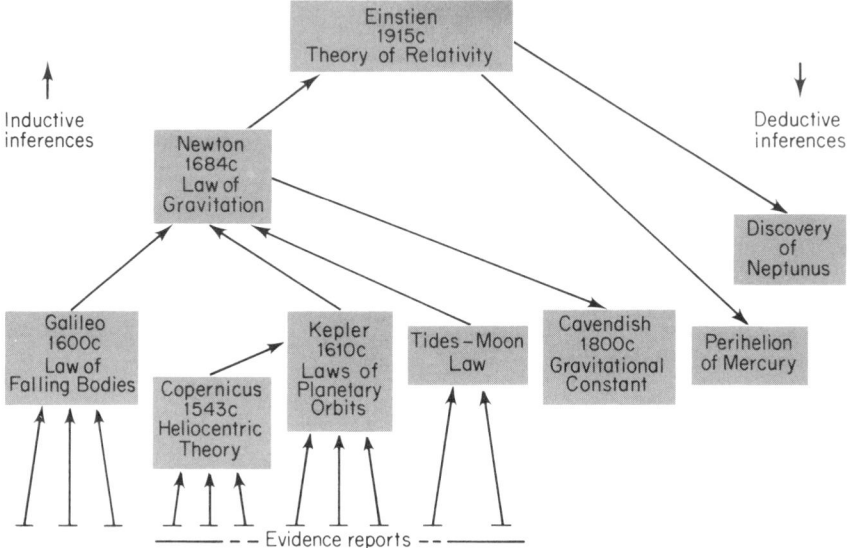

Figure 14-2. An inductive schema which partially represents the development of physics (after Reichenbach).

For instance, Galileo conducted some experiments in which he rolled balls down inclined planes. He measured two variables, the time that the bodies were in motion and the distance covered at the end of various periods of time. He found that the distance traveled was related in a specific manner to the amount of time that the bodies were in motion. This relationship is known as the Law of Falling Bodies.[2]

Copernicus was dissatisfied with the Ptolemaic theory that the sun rotated around the earth, and on the basis of extensive observations and considerable reasoning advanced the Heliocentric (Copernican) Theory of Planetary Motion that the planets rotate around the sun. Kepler based his laws on his own meticulous observations, the observations of others, and on Copernicus' theory. The statement of his three laws of Planetary Orbits (among which was the statement that the earth's orbit is an ellipse) was a considerable advance in our knowledge.

There has always been interest in the height of the tides at various localities, and it is natural that precise recordings of this phenom-

---

[2] More precisely, the Law of Falling Bodies is that $S = 1/2\ gt^2$ where S is the distance the body falls, g the gravitational constant, and t the time that it is in motion. History is somewhat unclear about whether Galileo conducted similar experiments in other situations, but it is said that he also dropped various objects off the Leaning Tower of Pisa and obtained similar measurements.

enon would have been made at various times during the day. Similar observations were made of the location of the moon. Then, on the basis of these two sets of observations, it was possible to state a relationship, known as the Tides-Moon Law. This relationship states that high tides occur only on the parts of the earth that are nearest to, and farthest from, the moon, respectively. It follows that as the moon moves about the earth, the location of high tides shifts accordingly.

Largely on the basis of the preceding relationships, Newton was able to formulate his law of gravitation. Briefly, this law states that the force of attraction between two bodies varies inversely with the square of the distance between them. As an example of a prediction from a general law, we illustrate (the first downward arrow of Figure 14-2) the prediction of the gravitational constant from Newton's law, the precise determination of which was made by Cavendish.

The crowning achievement of the portion of physics that we are considering came with Einstein's statement of his general theory of relativity. One particularly interesting prediction made from the theory of relativity concerned the perihelion of Mercury. Newton's equations failed to account for a slight discrepancy in Mercury's perihelion. This discrepancy between Newton's equations and the observational findings was accounted for precisely by Einstein's theory. The interrelationship of discoveries in science is illustrated by the fact that research on the movement of Mercury's perihelion was associated with the discovery of the planet Neptunus by Leverrier.

This brief discussion of the evolution of a portion of physics is, of course, inadequate for a proper understanding of the subject matter involved. Each step in the story constitutes an exciting tale that you might wish to follow up in detail. And where does the story go from here? One of the problems that has been bothering physicists and philosophers is how to reconcile the area of physics depicted in Figure 14-2 with a similar area known as quantum mechanics. To this end physicists such as Einstein and Schrodinger attempted to develop a "unified field" theory to encompass Einstein's theory of relativity as well as the principles of quantum mechanics. Consideration of such higher-level principles is beyond our present scope. Our purpose is fulfilled by presenting the general *form* of scientific progress, as shown in Figure 14-2, which is an evolution that may easily be our crowning intellectual achievement.

We shall now use these two schemata as a basis for considering a number of characteristics of science. Specifically, they will help us (1) to discuss the process of scientific generalization; (2) to elaborate our distinction between inductive and deductive inferences; (3) to understand Reichenbach's concept of concatenation; (4) to understand the nature of explanation; and (5) to understand the nature of prediction.

Generalization

Galileo conducted a number of specific experiments. Each experiment resulted in a statement that there was a relationship between the distance traveled by balls rolling down an inclined plane and the time that they were in motion. From these specific statements he then advanced to a more general statement: The relationship between distance and time obtained for the bodies in motion was true for *all* falling bodies, at all locations, and at all times.

Copernicus observed the position of the planets relative to the sun. After making a number of specific observations, he was willing to generalize to positions of the planets that he had not observed. The observations that he made fitted the heliocentric theory, that the planets revolved around the sun. He then made the statement that the heliocentric theory held for positions of the planets that he had not observed. And so it is for Kepler's laws and for the Tides-Moon law. In each case a number of specific statements based on observation (evidence reports) were made. Then from these specific statements came a more general statement. It is this process of proceeding from a set of specific statements to a more general statement that is referred to as *generalization*. The general statement, then, includes not only the specific statements that led to it but also a wide variety of other phenomena that have not been observed.

This process of increasing generalization continues as we read up the inductive schema. Thus Newton's law of gravitation is more general than any of those that are lower in the schema. We may say that it generalizes Galileo's, Copernicus', Kepler's, and the Tides-Moon Laws. Newton's law is more general in the sense that it includes these more specific laws and that it makes statements about phenomena other than the ones on which it was based. In turn, Einstein formulated principles that were more general than Newton's, principles that included Newton's and therefore all of those lower in the schema.

More on Inductive and Deductive Inferences

In the inductive schema of Figure 14-2, we may observe that inductive inferences are represented when arrows point up, deductive inferences by arrows that point down. In Figure 14-1 we have only inductive inferences, and these are liable to error. For instance, Watson was introduced as *"Dr.* Watson"; on the basis of this information Holmes concluded that Watson was a medical man. Is this necessarily the case? Obviously not, for Watson may have been some other kind of doctor, such as a Doctor of Philosophy. Similarly, consider the observational information, "left hand held stiff and unnatural," on the basis of which Holmes concluded that "the left arm was injured." This conclusion does not necessarily follow, since there could be other reasons for the condition (Watson might have been organically deformed at birth). In fact, was it necessarily the case that Watson had just come from Afghanistan? The story may well have gone something like this: Holmes: "You have been in Afghanistan, I perceive." Watson: "Certainly not. I have not been out of London for forty years. Are you out of your mind?"

In a similar vein we may note that Galileo's law was advanced as a general law, asserting that *any* falling body *anywhere at any time* obeyed his law. Is this necessarily true? Obviously not, for perhaps a stone falling off Mount Everest or a hat falling off of a man's head in New York may fall according to a different law than that offered for a set of balls rolling down an inclined plane in Italy many years ago. (We would assume that Galileo's limiting conditions such as that concerning the resistance of air would not be ignored.)

And so it is with the other statements in Figure 14-2. Each conclusion may be in error. As long as inductive inferences are used in such situations, the conclusion will only have a certain degree of probability of being true. Yet, we must continue using inductive inferences. You have no doubt noted that each time a generalization is made, inductive inferences have been used to arrive at that generalization. Since the making of a generalization necessitates saying something about as yet unobserved phenomena, the generalization *must* be susceptible to error.

Let us illustrate deductive logic by referring to our inductive schema for physics. Since Galileo's and Kepler's laws were generalized by Newton's, it follows that they may be deduced from them. In this case, it may be said "If Newton's laws are true, then it is necessarily the case that Galileo's is true, and also that Kepler's are true." Similarly, on the basis of Newton's laws, the gravitational constant was deduced and empirically verified by Cavendish. This deductive inference takes the form: "If Newton's laws are true, then the gravitational constant is such and such." Furthermore, concerning Einstein's principles, we may say: "If Einstein's theory is true, then the previous discrepancy in the perihelion of Mercury may be accounted for."

This discussion allows us to emphasize a very important matter that we have previously mentioned. That is, that a deductive inference does not guarantee that the conclusion is true. The deductive inference discussed above, for example, does not say that Galileo's law is true. It does say that *if* Newton's laws are true, Galileo's law is true. One may well ask, at this point, how we determine that Newton's laws are true. Or, more generally, how do we determine that the premises of a deductive inference are true. The answer, of course, is through the use of inductive logic. For example, we have determined by empirical investigation that the probability of Newton's laws being true is very high, It is, in fact, sufficiently high that we wish to say that it *is* true (in an approximate sense, of course).

Concatenation

As we move up the inductive schema we arrive at statements that are increasingly general. And as the generality of a statement increases in the manner depicted in our inductive schema, there is a certain increase in the probability of the statement being true. This increase in probability is the result of two factors. *First,* since the more general statement rests on more numerous and more varied evidence, it usually has been confirmed to a greater degree than has a less general statement. For example, there is a certain addition to the probability of Newton's law of gravitation that is not present for Galileo's law of falling bodies, since the former is based on inductions from more numerous data of wider scope. *Second,* the more general statement is *concatenated* with other general statements. By concatenated we mean that the statement is "chained together" with other statements and is thus consistent with these other statements. For example, Galileo's law of falling bodies is not concatenated with other statements, and Newton's is. The fact that Newton's law is linked with other statements gives it an increment of probability that cannot be said of Galileo's. We may say that the probability of the whole system in Figure 14-2 being true is greater than the sum of the probabilities of each statement taken separately. It is the compatibility of the whole system and the support gained from the concatenation that provide the added likelihood.

It also follows that when each individual generalization in the system is confirmed, the entire system gains increased credence. For instance, if Einstein's theory was based entirely on his own observations, and those which it stimulated, its probability would be much lower than it actually is, considering that it is also based on all of the lower generalizations in Figure 14-2. Or let us put the matter another way. Suppose that a new and extensive test determined that Galileo's laws were false. This would mean the complete "downfall" of Galileo's laws, but it would only slightly reduce the probability of Einstein's theory since there is a wide variety of additional confirming data for the latter.

Explanation

The concept of explanation as used in science is sometimes difficult for students to understand, probably because of the common sense use of the term to which they have previously been exposed. One of the common sense "meanings" of the term concerns familiarity. Suppose that you learn about a scientific phenomenon that is new to you. You want it explained; you want to know "why" it is so. This desire on your part is a psychological phenomenon, a motive. When somebody can relate the scientific phenomenon to something that is already familiar to you, your psychological motive is satisfied. You feel as if you understand the phenomenon because of its association with knowledge that is familiar to you. A metaphor is frequently used for this purpose. At a very elementary level, for example, it might be said that the splitting of an atom is like shooting an incendiary bullet into a bag of gunpowder.

Any satisfaction of your motive to relate a new phenomenon to a familiar phenomenon is far from an explanation of it. Explanation is the placing of a statement within the context of a more general statement. If we are able to show that a specific statement belongs within the category of a more general statement, we may say that the specific statement has been explained. To establish this relationship we must show that the specific statement may be logically deduced from the more general statement. To return to a previous example, we might ask how to explain the statement that "John Jones is anxious." The answer is that this statement can be logically deduced from the more general statement that "All men who bite their fingernails are anxious," and that "John Jones bites his fingernails." This immediately brings us face to face with a matter that we have approached previously from different angles: that such an explanation is accomplished on the assumption that the more general statement is true. Hence, to be more complete, we need to say: "If it is true that 'all men who bite their fingernails are anxious,' and if it is true that 'John Jones is a man who bites his fingernails,' then it is true that 'John Jones is anxious.' " By so deductively inferring this conclusion, we have explained why John Jones is anxious; we have logically deduced that specific statement from the more general statement. Of course, we immediately want to go on: Why are all men anxious? But this is outside our scope, except that we may note that such an explanation would be accomplished by deducing that statement from a still more general one.

Referring to Figure 14-2 we can see that Kepler's laws are more general than the Copernican theory. And since the latter is included in the former, it may be logically deduced from it—Kepler's laws explain the Copernican theory. In turn, Newton's law, being more general than Galileo's, Kepler's, and the Tides-Moon laws, explains these more specific laws; they may all be logically deduced from Newton's law. And finally, all of the lower generalizations may be deduced from Einstein's theory, and we may therefore say that Einstein's theory explains all of the lower generalizations.

Prediction

To make a prediction we apply a generalization to a situation that has not yet been studied. The generalization says that all of something has a certain characteristic. When we extend the generalization to the new situation, we are simply saying that the new situation should have the characteristic specified in the generalization. In its simplest form this is what a prediction is, and we have illustrated three predictions in Figure 14-2, the gravitational constant, the perihelion of Mercury, and the discovery of Neptunus. Whether or not the prediction is confirmed, of course, is quite important for the generalization. For if the prediction is confirmed, the probability of the generalization is considerably increased. If it is not confirmed, however, (assuming the evidence report is true and that the deduction is valid), then either the probability of the generalization is decreased, or the generalization must be restricted so that it does not apply to the type of phenomena with which the prediction was concerned.

As an illustration of a prediction let us say that a certain hypothesis was formulated about the behavior of school children in the fourth grade. It was then tested on those children and found to be probably true. The experimenter may wonder whether this hypothesis is also true for all school children. If so, the hypothesis might be generalized to make it applicable to all school children. From such a generalized hypothesis it is possible to derive specific statements concerning any given school grade. For example, the experimenter could deductively derive the conclusion that the hypothesis is applicable to the behavior of school children in the fifth grade, in which case prediction is being made about as yet unobserved children of that grade level. Consequently a prediction is made that the hypothesis is applicable to a novel situation.

We have, in this chapter, attempted to present some of the important characteristics and processes of science in general. With an understanding of these principles we can now turn to a consideration of their role in psychology.

Review of Chapter 14

1. To help yourself better understand the nature of the inductive schema, you might construct a schema from one of Sherlock Holmes' other stories.

2. Review basic definitions such as generalization, prediction, explanation, concatenation.

3. Did Sherlock Holmes misuse the word "deduction?" Do you notice the word being misused in your newspaper and other aspects of everyday life?

4. Perhaps you would also want to consider developing an inductive schema for some area of scientific inquiry that is especially interesting for you. For instance, you might develop one for a given theory of vision, for the "big bang" theory of the origin of the universe, or for a theory of how life developed. In so doing, you would then emphasize the basic terms of explanation, generalization, and prediction in a new context.

15

Generalization, Explanation, and Prediction in Experimentation

In the last chapter we discussed generalization, explanation, and prediction, as well as several other topics, in a rather general way. It now remains for us to consider these topics as they fit into the day-to-day work of the experimenter. We shall want to discuss some of the mechanics that one might use in these, the final phases of experimentation. The three questions to which we now turn are: How and what does the experimenter generalize? How do we explain our results? And how do we predict to other situations?

Generalization

A distinction is frequently made between what is known as applied science (technology) and basic or pure science. In applied science the investigator attempts to solve some relatively limited problem, whereas in basic science one attempts to arrive at a general principle. The answer that the applied scientist obtains will usually be applicable only under the specific conditions of the experiment conducted to solve the problems. The basic scientist's results, however, are likely to be more widely applicable.

An applied psychologist might be called in by the Burpo Company to find out why their soft drink sales in Atlanta, Georgia, were below normal for the month of December. The basic scientist, on the other hand, would be more likely to study the problem of the general relationship between temperature and consumption of liquids. The applied psychologist, in this example, might find that Atlanta was unseasonably cold during December and hypothesize that it was for that reason that sales declined. The basic scientist however, might reach the more general conclusion that the amount of liquid consumed by humans depends on the temperature—the lower the temperature, the less they consume. Thus, the latter finding would account for the specific phenomenon in Atlanta, as well as a wide variety of additional phenomena; it is by far the more general in scope.

Although both kinds of research are important, we shall limit our considerations here to basic research. Our immediate goal shall be to see how the experimenter arrives at general statements, rather than only specific statements about the results of research. We shall start our discussion with the assumption that the experimenter wants to generalize results as widely as is reasonable. Hence we shall consider two questions in detail: By what procedure does one generalize results; and how wide is "reasonable"?

The Mechanics of Generalization

Let us say that an experimenter has selected 20 participants for an experiment. It should be apparent that the interest is not in these 20 participants in and for themselves but only insofar as they are typical of a larger group. Whatever the researcher finds out about these participants will be assumed to be true for the larger group. In short, the wish is to *generalize* from the 20 participants to the larger group of individuals. The terms we have previously used in this connection are "sample" and "population." As we said, an experimenter defines a population of participants to make statements about. This population is usually quite large, such as all the students in the university, all dogs of a certain species, or perhaps even all humans. Since it is not feasible to study all the members of such large populations, the experimenter randomly selects a sample therefrom. And since that sample has been randomly selected from the population, it should be representative of that population. Therefore, the experimenter is able to say that what is probably true for the sample is also probably true for the population; a generalization is made from the sample of participants to the entire population from which they came.[1]

The *most important* requirement for generalizing from a sample is that the sample must be *representative* of the population. The technique that we are using for obtaining representativeness is randomization; if the sample has been randomly drawn from the population, it is reasonable to assume that it is representative of the population. *Only when the sample is representative of the population are we able to generalize from the sample to the population.* We are emphasizing this point to a great extent for two reasons: because of its great importance in generalizing to populations, and because we ourselves want to state a generalization. We want to generalize from what we have said about populations of organisms to a wide variety of other populations.[2] For when you conduct an experiment you actually have a number of populations, in addition to a population of people, dogs, etc., to which you might generalize.

To illustrate, suppose you are conducting an experiment on knowledge of results. You take two groups of participants and assign them to two conditions: One group receives knowledge of results, and the second (control) group doesn't. The classic task used in studying this problem is line drawing. Participants are blindfolded and asked to draw 5-inch lines. The knowledge-of-results group would be told whether their lines were too long, too short, or correct, while the control group would be given no knowledge about the lengths of their lines. We are dealing with several populations: people, experimenters, tasks, and various stimulus conditions. Since we wish to generalize to a population of people, we randomly draw our sample from that population and randomly assign them to the two groups. If we find, as we certainly should, that the knowledge-of-results group performs better than the control group, we can safely say that this is probably also true for the entire population of people sampled.

But what about the experimenter? We have controlled this variable, presumably, by having a single experimenter handle all the par-

---

[1] Even though this statement offers the general idea, it is not quite accurate. If we were to follow this procedure, we would determine that the mean of a sample is, say, 10.32, and generalize to the population, inferring that its mean is also 10.32. Strictly speaking, this procedure is not reasonable, for it could be shown that the probability of such an inference is 0.00. A more suitable procedure is known as "confidence interval estimation," whereby one infers that the mean of the population is "close to" that for the sample. Hence, the more appropriate inference might be that, on the basis of a sample mean of 10.32, the population mean is between 10.10 and 10.54.

[2] For an elaboration of matters relating to generalization to nonsubject populations you might refer to the classic work of Brunswick (1956) on representative design of psychological research, also elaborated on by Hammond (1948, 1954).

ticipants. But can we say that the knowledge-of-results group will always be superior to the control group *regardless of who is the experimenter?* In short, can we generalize from the results obtained by our single experimenter to all experimenters? This question is difficult to answer. Let us imagine a population of experimenters, made up of all psychologists who conduct experiments. Strictly speaking, then, we should take a random sample from that population of experimenters and have each member of our sample conduct a separate experiment. Suppose that we define our population of experimenters in such a way that it includes 500 psychologists and that we randomly select a sample of 10 experimenters from that population. Further assume that we have selected a sample of 100 participants. We would randomly assign the 100 participants to two groups; then we would randomly assign 5 participants in each group to each experimenter. In effect, then, we will repeat our experiment 10 times. We have now not only controlled the experimenter variable by balancing, but we have also sampled from a population of experiments. Assume that the results come out approximately the same for each experimenter—that the performance of the knowledge-of-results participants is about equally superior to their corresponding controls for all 10 experimenters. In this case we are able to generalize the results to the population of experimenters as follows: For the population of experimenters sampled (and also for the population of participants sampled), providing knowledge of results under the conditions of this experiment leads to performance that is superior to that derived from not providing knowledge of results.

By "under the conditions of this experiment" we mean two things: with the specific task used, and under the specific stimulus conditions that were present for the participants. Concerning the first, our question is this: Since we found that the knowledge-of-results group was superior to the control group on a line-drawing task, would that group also be superior in learning other tasks? Of course, the answer is that we do not know from this experiment. Consider a population of *all* the tasks that humans could learn, such as drawing lines, learning Morse code, hitting a golf ball, assembling parts of a radio, and so forth. If we wish to make a statement about the effectiveness of knowledge of results for all tasks, then, as before, we must obtain a representative sample from that population of tasks. By selecting only a line-drawing task we did not do this and therefore cannot generalize back to the larger population of tasks. The proper procedure to generalize to all tasks would be randomly to select a number of tasks from that population. We would then conduct the same experiment for each of those tasks. If we find that on each task studied the knowledge-of-results group is superior to the control group, then we can say that for all tasks, knowledge of results leads to performance that is superior to that gained from a lack of knowledge of results.

Now what about the various stimulus conditions that were present for our participants? For one, they were blindfolded. But there are different techniques for "blindfolding" people. One experimenter might use a large handkerchief, another might use opaque glasses, and still another might place a large screen between the participant's eyes and hands so that although the participant would be able to see, the length of the lines drawn could not be seen. Would the knowledge-of-results condition be superior to the control condition regardless of the technique of blindfolding? What about other stimulus conditions? Would the specific temperature be relevant? How about the noise level? And so on — one can conceive of a number of populations of these stimulus conditions. Strictly speaking, if an experimenter wishes to generalize to the populations of stimuli present, random samples should be drawn from those populations. Take temperature as an example. If one wishes to generalize results to all reasonable values of this variable, then a number of temperatures should be randomly selected. The experiment would then be repeated for each temperature value studied. If it is found that regardless of the temperature value studied the knowledge-of-results condition is always superior, one can generalize those findings to the population of temperatures sampled. Only by systematically sampling the various stimulus populations can the experimenter, strictly speaking, generalize results to those populations.

At this point it might appear that the successful conduct of psychological experimentation is hopelessly complicated. One of the most discouraging features of psychological research is the difficulty encountered in confirming the results of previous experiments. When one experimenter (Jones) finds that variable A affects variable B, all too frequently another experimenter (Smith) achieves different results. The reason for this lack of repeatability was discussed in chapter 6, on control, and in chapter 10, on factorial designs. Looking at it from the present point of view, we might explain the differences in findings by the fact that in the experiment conducted by Jones a number of conditions were held constant and a generalization to the populations of these conditions was made. For example, Jones may have held the experimenter variable constant, and at least implicitly generalized to a population of experimenters. Strictly speaking, that should not have been done, for Jones did not randomly sample from a population of experimenters. Let us then assume that Jones' generalization was in error and that the results obtained are valid only for that experimenter. If this is the case, then a different set of results may very well be obtained with a different experimenter.

Psychological research (or *any* research for that matter) frequently does become discouraging. After all, if it were easy there would be little joy in it. The toughest nut to crack yields the tastiest meat. Psychologists, however, are beginning to investigate systematically experimental situations more thoroughly than in the past. They are increasingly seeking to account for conflicting results in independent experiments. This is one of the reasons that factorial designs are being more widely used, for they are wonderful devices for sampling a number of populations simultaneously. To illustrate, suppose that we wish to generalize our results to populations of people, experimenters, tasks, and temperature conditions. We could conduct several experiments here, but let us say that we conduct only one experiment using four independent variables, each varied in the following ways: (1) knowledge of results, two ways (knowledge and no knowledge); (2) experimenters varied in six ways; (3) tasks varied in five ways; and (4) temperature varied in four ways. Assume that we have chosen the values of the last three variables at random. The resulting factorial is presented in Table 15-1.

It can thus be seen that we have a $6 \times 5 \times 4 \times 2$ factorial design. What if we find a significant difference for the knowledge of results variable, but no significant interactions? In this case we could rather safely generalize about knowledge of results to our experimenter population, to our task population, to our temperature population, and also, of course, to our population of humans.

At this point it is well to recall our discussion from chapter 10 on factorial designs. There we distinguished between the case of a fixed model and the case of a random model. For the case of a fixed model, we said that the experimenter selects the values of the independent variables for some specific reason; one does not randomly select them from a population. For the case of a random model, however, we define a population and then randomly select values from that population. The relevance of that distinction should be apparent, for only in the case of random variables can you safely generalize to the population. If you select the values of your variables in a nonrandom fashion, any conclusions must be restricted to those values. Let us illustrate by considering the temperature variable again. Suppose that we are particularly interested in three specific values of this variable: 60 degrees, 70 degrees, and 80 degrees. Now, whatever our results, they will be limited to those particular temperature values. On the other hand, if we are interested in generalizing to all temperatures between 40 and 105, we would write each number between 40 and 105 on a

Table 15-1 A 6 × 5 × 4 × 2 Factorial Design for Studying the Effect of Knowledge of Results When Randomly Sampling from Populations of Experimenters, Tasks, Temperatures, and People

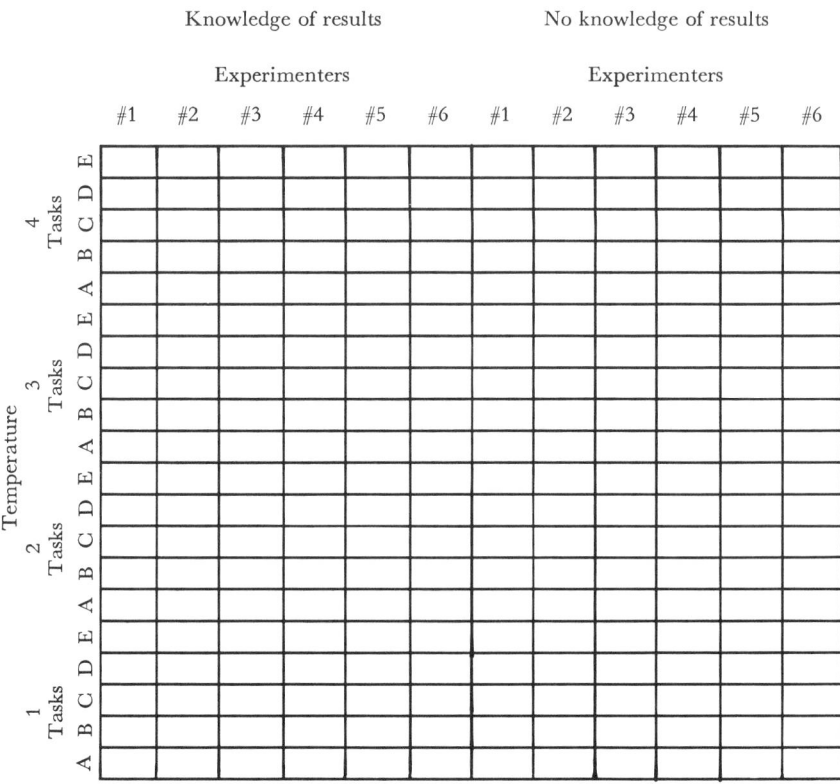

piece of paper, place all these numbers in a hat, and draw several values from the hat. Then whatever the experimental results we obtain, we can safely generalize back to that population of values, for we have randomly selected our values from it.[3]

The Limitation of Generalizations

How widely is it reasonable to generalize? Let's say that we are interested in whether Method A or Method B of learning leads to superior performance. Assume that one experimenter tested these methods on

---

[3] Assuming, of course, that we select enough values to study. Just as with sampling from a population of people, the larger the number of values selected, the more likely that the sample is representative of the population.

a sample of college students and found Method A to be superior. Unhesitatingly a generalization of the results was made to all college students. Another experimenter becomes interested in the problem and repeats the experiment. It is found that Method B is superior. We wish to resolve the contradiction. After studying the two experiments we may find that the first experimenter was in a women's college, whereas the second was in a men's college. A possible reason for the different results is now apparent. The first experimenter generalized to a population of male and female students without randomly sampling from the former (as also did the second, but without sampling females). To determine whether we have correctly ascertained the reason for the conflicting results we design a 2×2 factorial experiment in which our first variable is methods of learning, varied in two ways, and our second is gender, varied, of course, in two ways. We randomly draw a sample of males and females from a college population. Assume that our results come out with the following mean values, where the higher the score, the better the performance (Table 15-2).

Table 15-2 A 2×2 Factorial Design with Fictitious Means

Methods

		A	B
Sex	Males	10	20
	Females	20	12

Graphing these results, we can clearly see that an interaction exists between gender and methods such that females are superior with Method A and males are superior with Method B (Figure 15-1). We have thus confirmed the results of the first experimenter in that we found, as did he, that Method A is superior for females; similarly, we have confirmed the results of the second experimenter since we found that Method B is superior for males. We have, therefore, established the reason for conflicting results. But we cannot make a simple statement about the superiority of a method that generally applies to everybody; the discovery of this interaction limits the extent to which a simple generalization can be offered.

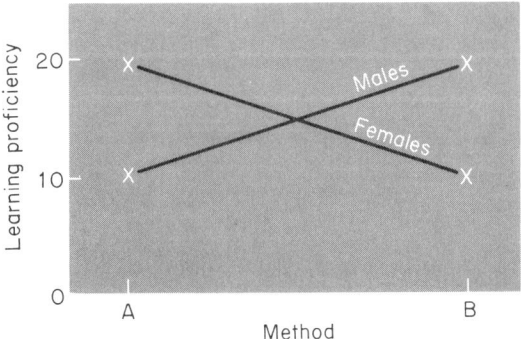

Figure 15-1. Indicating an interaction between methods of learning and gender.

This example provides us with the opportunity to consider the matter of limitations of our generalizations in a broader fashion. Our goal, we have said, is to attempt to make statements of as great a degree of generality as we can. And we would like those statements to be as simple (parsimonious) as possible (see p. 62). Unfortunately, though, nature does not always oblige us and, to make general statements, we often must complicate them in order to accurately describe events that we study. We can, in short, expect to find a large number of interactions between our experimental treatments, and we should explicitly design our experiments so that we can establish interactions where in fact they exist. The alternative of failing to look for interactions amounts to blinding ourselves to truth, with the consequence that we have difficulty in confirming previous experimentation.

In general, then, the experimenter should systematically study variables that might interact with the variables of primary interest. It is often very easy to construct an experimental design so that such interactions can be studied. The example given above is a case in point, for one can conveniently analyze the results as a function of gender, or other participant characteristic such as anxiety. We have emphasized the importance of controlling the experimenter variable and have shown that this variable may exert subtle influences that may affect dependent variable measures. Where more than one experimenter collects data in a given experiment (and this happens in about half of published experiments) it is "a natural" to analyze the results as a function of experimenters in order to see if this variable interacts with that of primary interest. Similar variables that may be built into a factorial design for this purpose might be environmental temperature, type of task, nature of equipment used (e.g., memory drum X as against memory drum Y), and so forth.

Let us now examine more closely the possible outcomes with regard to the variable of secondary interest. We shall consider three possible cases. Assume that we vary the independent variable of primary interest in two ways so that we have what we conventionally call an experimental and a control group. The variable of secondary interest may be varied in several ways but, for the moment, let us vary it in only two ways. For instance, let us say that two experimenters collect data in a two-randomized groups design so that we can analyze the data as a 2×2 factorial design (see Table 15-3). The three possible outcomes are discussed as the following cases.

Table 15-3 A Two-Groups Design in Which the Data Are Analyzed as a Function of Two Experimenters

Independent variable

		Experimental	Control
Experimenter	#1		
	#2		

Case I. This case occurs when Experimenters 1 and 2 obtain precisely the same results. The results are graphed in Figure 15-2 where we can note that the lines are parallel. In this instance the variable of secondary interest does not influence the dependent variable measure and we would conclude that it does not interact with our variable of primary interest. In this case a difference between our experimental and control groups can be generalized with regard to the variable of secondary interest. There is but one remaining point: We could not possibly have known this unless we had designed and analyzed our experiment to find it out.

As an empirical illustration of Case I, consider an experiment involving two methods of learning and three data collectors (McGuigan, Hutchens, Eason, & Reynolds, 1964). An analysis of variance indicated that there was a reliable difference between methods but that the differences among experimenters, and the experimenter $\times$ methods interaction were not reliable. A graph of the dependent variable values for the two methods as a function of experimenters is presented in Figure 15-3. Lines of best fit for these sample points do not deviate significantly from horizontal lines. Hence, we have some reason for generalizing the methods results to a population of experimenters, though of course a larger sample of this population would be preferred.

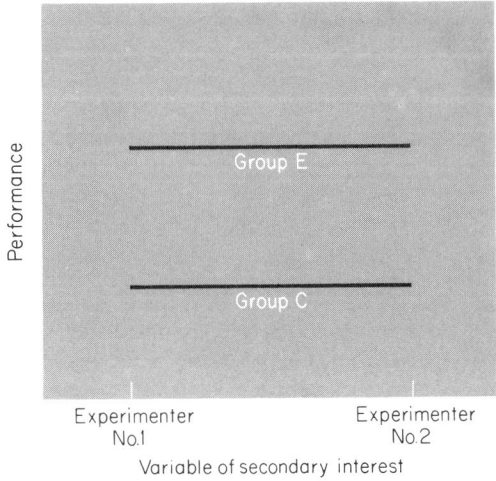

Performance

Group E

Group C

Experimenter No.1 — Experimenter No.2

Variable of secondary interest

Figure 15-2. Population values for Case I showing no interaction between the variable of secondary interest (e.g., experimenters) and the primary independent variable. Variation of the variable of secondary interest does not differentially affect the dependent variable values.

Case II. The second general possibility is that variation of the variable of secondary interest *does* affect the dependent variable, but it affects all participants in the same way, regardless of the experimental condition to which those participants were assigned. For example, we might suppose that participants assigned to Experimenter 1 (or temperature A, or task X) perform at a higher level on the average than do those assigned to Experimenter 2 (or temperature B, or task Y). But the experimental group is equally superior to the control group for both experimenters, or what have you. This case is illustrated in Figure 15-4.

In Figure 15-5 we have selected values from an experiment in which there was a significant difference between experimenters, but lack of an interaction between experimenters and methods as an empirical illustration of Case II (McGuigan, 1959). Since in Case II we are able to reach the same conclusion with regard to our hypothesis regardless of which experimenter conducted the experiment, we are not immediately interested in the experimenter difference and have a basis for generalizing the results with regard to methods to that population. There is, in short, a lack of interaction that could limit our generalization. As an adjunct to this case, however, we note that a particular kind of behavior *is* influenced by this secondary variable, information that may be valuable for further experimentation.

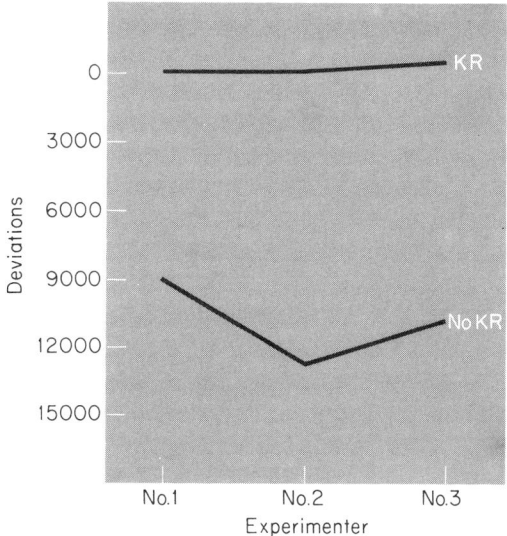

Figure 15-3. Sample values illustrating Case I. Three experimenters and two methods (knowledge of results) were used. The interaction between experimenters and methods is not reliable.

Case III. In Cases I and II we have justification for generalizing to the population of the secondary variable to the extent to which that population has been sampled. In Case III, however, we must deal with an interaction. To take an extreme example, suppose that the control group is superior to the experimental group for one experimenter but that the reverse is the case for the second experimenter (Figure 15-6). In this event the extent to which we can generalize to a population is sharply restricted, particularly since we probably don't know the precise ways in which the two experimenters differ. To understate the matter, the discovery of an interaction of this sort tells us to proceed with caution. It is, however, heartening to note that psychologists have increased the extent to which they investigate the effects of the experimenter (and other secondary variables) on their data. Let us briefly look at two interesting studies in which interactions with experimenters have been established.

The first was a verbal conditioning study using the response class of hostile words emitted in sentences (Binder, McConnell, & Sjoholm, 1957). Whenever the participant used a hostile word in a sentence the experimenter reinforced that response by saying "good." Two groups were used, a different experimenter for each group. The

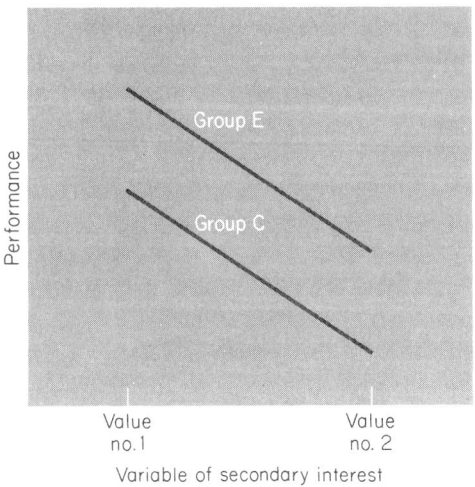

Value no.1 Value no. 2

Variable of secondary interest

Figure 15-4. Population values for Case II showing no interaction between methods and the variable of secondary interest. But the variable of secondary interest does differentially affect the dependent variable values.

two experimenters differed in gender, height, weight, age, appearance, and personality:

> The first ... was ... an attractive, soft-spoken, reserved young lady ... 5'1/2" in height, and 90 pounds in weight. The ... second ... was very masculine, 6'5" tall, 220 pounds in weight, and had many of the unrestrained personality characteristics which might be expected of a former marine captain — perhaps more important than their actual age difference of about 12 years was the difference in their age appearance: The young lady could have passed for a high school sophomore while the male experimenter was often mistaken for a faculty member (Binder et al., 1957, p. 309).

The results of this experiment are presented in Figure 15-7. First we may note that since the number of hostile words emitted by both groups increases as number of trials increase, the participants of both experiments were successfully conditioned. During the first two blocks of learning trials, however, the participants of the female experimenter were inferior to those of the male experimenter. On succeeding blocks the reverse is the case, and the two curves intersect. In short, there is an interaction between experimenters and learning trials such that the slope of the learning curve for the female experimenter is steeper than that for the male experimenter. If we, therefore, wish to offer a generalization about the characteristics of the learning curve, it must be tempered by considering the nature of the experimenter. Exactly why this difference occurred is not clear, but

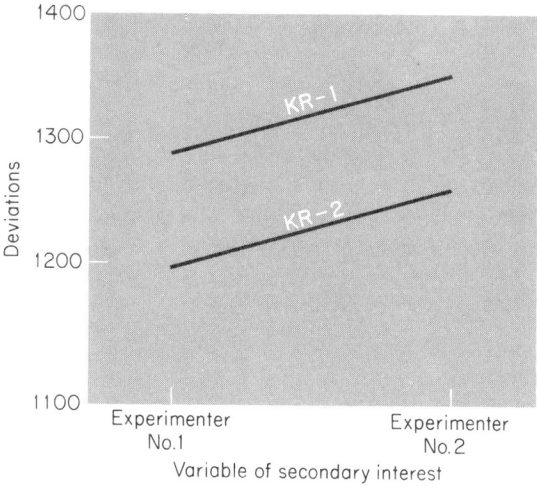

Figure 15-5. Sample values illustrating Case II. The results are for two methods of presenting knowledge of results.

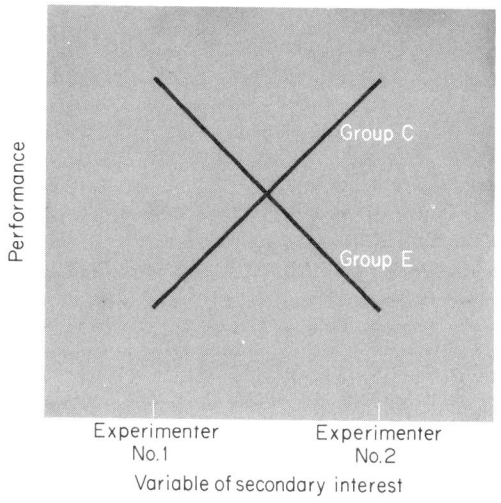

Figure 15-6. Population values for Case III showing one possible interaction between a variable of secondary interest (here experimenters) and the primary independent variable.

we may speculate with the authors that the female experimenter provided a less threatening environment, and the participants consequently were less inhibited in the tendency to increase their frequency of usage of hostile words. Presumably some reverse effect was present early in learning.

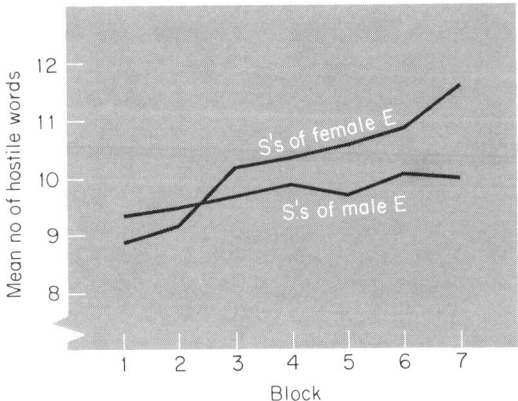

Figure 15-7. Learning curves for two groups treated the same, but with different experimenters. The steeper slope for the participants of the female experimenter illustrates an interaction between experimenters and stage of learning (after Binder et al., 1957).

In the second example of Case III, Spires (1960) selected a group of people who scored high on the Hysteria Scale of the Minnesota Multiphasic Personality Inventory and a second who scored high on the Psychasthenic Scale. The participants were then given one of two sets when they entered the experimental situation: for the positive set the participant was told that the experimenter was a "warm, friendly person, and you should get along very well"; for the negative set the participant was told that the experimenter may "irritate him a bit, that he's not very friendly, in fact kind of cold." The experimenter was the same person in both cases, however.

The participants were then conditioned to emit a class of pronouns that was reinforced by saying "good." An analysis of variance of Spire's results indicated that there was a reliable difference between positive and negative sets for the experimenter such that participants with the positive set conditioned better than those with a negative set. Furthermore, and this is the point of present interest, there was a significant interaction between set for the experimenter and personality of the participant (whether the individual was an hysteric or a psychasthenic). To illustrate this interaction we have plotted the terminal conditioning scores under these four conditions in Figure 15-8. We can thus see that the hysterics who were given a positive set had higher scores than those given a negative set. There is, though, little difference between the two groups of psychasthenics.

This type of research is especially valuable to us because of its analytic nature; it suggests, for instance, that we can generalize condi-

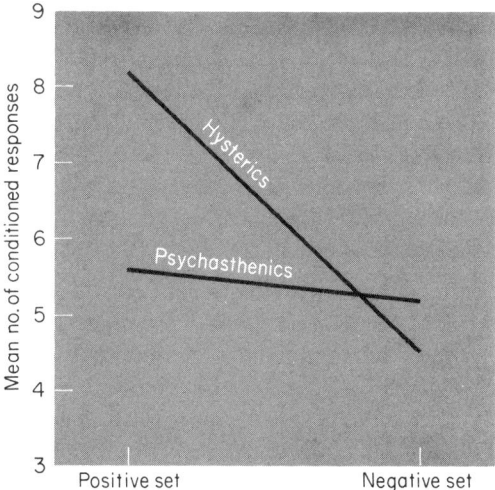

Figure 15-8. An interaction between set for the experimenter and personality characteristic of participants. The effect of set depends on whether people are hysterics or psychasthenics (after Spires, 1960).

tioning results with regard to this particular experimenter characteristic for one kind of person, but not for another kind. Continuous and sustained analysis of the various secondary variables in an experimental situation will eventually allow us to advance our generalizations so that we can have great confidence in their confirmability.

Having now specified these three cases, let us attempt to summarize where we stand.

First, if you have not sampled from a population of some variable, you should, strictly speaking, not generalize to that population. If, for instance, there is but one data collector for your experiment the best that you can do is to attempt to hold his or her influence on the participants constant. If, however, you have previous knowledge that no interaction has been found between your independent variable and the populations for other variables to which you wish to generalize, then your generalization to those populations will probably be valid.

Second, if you have systematically varied some variable of secondary interest to you, then you should investigate the possibility that an interaction exists between it and your variable of primary interest. If, for instance, more than one data collector has been used you should: (1) specify techniques for controlling this variable (see p. 390); (2) analyze and report your data as a function of experimenters; and (3) test for interactions between experimenters and treatments. Should your analysis indicate that the experiment is an instance of Cases I or II, the results are generalizable to a population to the extent to which that population has been sampled.[4] We grant that completely satisfactory sampling of secondary variables can seldom occur, but at least some sampling is better than none. And it is beneficial to *know* and be able to *state* that, within those limitations, the results appear to be instances of Cases I or II.

Third, if you find that your data are an instance of Case III, then you cannot offer a simple generalization for them. If your variable of secondary interest is an operationally defined one, then your generalization can be quite precise, if a bit complicated. Consider as an example the experiment by Reynolds and Meeker (1966) in which the effects of shock and thiosemicarbazide were studied on the development of gastric ulcers in rats. Their generalization is quite exact, and incidentally theoretically very interesting, precisely *because* the variables interact. They found that rats who received thiosemicarbazide and shock developed fewer ulcers than those who received shock only, thiosemicarbazide only, or nothing. Hence, there is an interaction such that the combination of the drug and shock led to an increased resistance to stress (ulcers). This is a generalization incorporating two variables and it is therefore not as simple as those involving only one variable. If, on the other hand, you cannot adequately specify the ways in which your secondary variable differ (as in the case of different data collectors), the extent to which you can generalize is sharply limited. You can only say, for instance, that Method A will be superior to Method B when experimenters similar to Experimenter 1 are used, but that the reverse is the case when experimenters similar to Experimenter 2 are used. This knowledge is of course valuable, but only in a negative sense since we do not know what the different characteristics of the two experimenters are; an interaction of this kind tells us to proceed with considerable caution (cf. McGuigan, 1963).

This all may sound rather demanding, and rather than conclude this topic on such a note let us return to the most typical situation that you are likely to face in your elementary work, namely that specified in the first point above.

If you have no knowledge about interactions between your independent variable and the populations to which you wish to generalize, then it is possible to tentatively offer your generalization. Other experimenters may then repeat your experiment in their own laboratories. This implies that the various extraneous variables will assume different

---

[4] We are assuming that a random model is used (see p. 308).

values from those that occurred in your experiment (either as the result of intentional control or because they were allowed to randomly vary). If, in the repetitions of your experiment your results are confirmed, it is likely that the populations to which you have generalized do not interact with your independent variable. On the other hand, if repetitions of your experiment by others, with differences in tasks, stimulus conditions, and other factors do not confirm your findings, then there is probably at least one interaction that needs to be discovered. At this point thorough and piecemeal analysis of the differences between your experiment and the repetitions of it needs to take place in order to discover the interactions. Such an analysis might assume the form of a factorial design such as that diagrammed in Table 15-2 and illustrated by Figure 15-1.

This last point leads us to consider an interesting proposal. Some experimenters have suggested that we should use a highly standardized experimental situation for studying particular types of behavior. All experimenters who are studying a given type of behavior should use the same values of a number of extraneous variables—lighting, temperature, noise, and so forth. In this way we can exercise better control and be more likely to confirm each other's findings. The Skinner Box is a good example of an attempt to introduce standardized values of extraneous variables into an experiment, for in the Skinner Box the lighting is controlled (the box is opaque), the noise level is controlled (it is sound deadened), and a variety of other external stimuli are prevented from entering into the experimental space. On the other hand, under such highly standardized conditions the extent to which we can generalize our findings is sharply limited. If we continue to proceed in our present direction with extraneous variables assuming different values in different experiments, then when experimental findings are confirmed, we can be rather sure that interactions do not exist. And when findings are not confirmed, we suspect that we have interactions present that limit our generalizations, and hence we should initiate experimentation in order to discover them. Regardless of your opinion on these two positions, that in favor of standardization or that opposed, the matter is probably only academic. It is unlikely that much in the way of standardization will occur in the foreseeable future.

The final matters for us to discuss in this chapter are those of explanation and prediction, particularly as these are relevant to the day-to-day work of the experimenter. First we shall consider explanation, for after that it will be possible to cover the topic of prediction very briefly.

Explanation[5]

The question "why" lies at the heart of scientific investigation.[6] In the last chapter we saw that in science "why" amounts to asking: according to what general statements (laws) does a particular event occur? Thus by placing the particular event within the context of a more general statement we can say that the particular event is explained. Let us now consider this process in greater detail.

In a classic paper, Hempel and Oppenheim (1948) offered an example wherein a mercury thermometer is rapidly immersed in hot water. A temporary drop of the mercury column occurs, after which the column rises swiftly. Why does this occur? That is, how might we explain it? Since the increase in temperature affects at first only the glass tube of the thermometer, the tube expands and thus provides a larger space for the mercury inside. To fill this larger space, of course, the mercury level drops. But as soon as the increase in heat is conducted through the glass tube and reaches the mercury, the mercury also expands. And since mercury expands more than does glass (i.e., the coefficient of expansion of mercury is greater than that of glass), the level of the mercury rises.

Now this account, as Hemple and Oppenheim pointed out, consists of two kinds of statements. Some statements indicate certain conditions that exist before the phenomenon to be explained occurs. These conditions may be referred to as *antecedent conditions*, and they include the fact that the thermometer consists of a glass tube that is partly filled with mercury, that it is immersed in hot water, and so on. The second kind of statement expresses general laws, an example of which would be a statement about the thermal conductivity of glass. Now the fact that the phenomenon to be explained can be logically deduced from the general laws with the help of the antecedent conditions constitutes an explanation of that phenomenon. That is, the way in which we may find out that a given phenomenon can be subsumed under a general law is by determining that the former can be deduced (deductively inferred) from the latter. The schema for accomplishing an explanation can be indicated as follows:

Deductive inference: $\begin{cases} \text{Statement of the general law(s)} \\ \text{Statement of the antecedent conditions} \end{cases}$

$\hookrightarrow$ Description of the phenomenon to be explained

[5] Consult Scriven (1959) for alternative approaches to explanation and prediction.
[6] Some authorities hold that we never ask "why" in science, but rather "what" and "how." We do not mean to quibble about these words, for our actual positions probably would not differ to any great extent. "What" and "how" could easily replace "why."

In this example it can be seen that the phenomenon to be explained (the immediate drop of the mercury level, followed by its swift rise) may be logically deduced according to the above schema. To illustrate further the nature of explanation, we might develop an analogy using the familiar syllogism concerning Socrates. Say that the phenomenon to be explained is Socrates' death. In the syllogism the two kinds of statements that we require for an explanation are offered. First, the antecedent condition is that "Socrates is a man." And second, the general law is that "All men are mortal." From these two statements, of course, we can deductively infer that Socrates is mortal. With this general understanding of the nature of explanation, let us now ask where the procedure enters the work of the experimental psychologists.

Deductive inference: $\begin{cases}\text{General law: All men are mortal}\\ \text{Antecedent condition: Socrates is a man}\end{cases}$
$\quad\quad\quad$ ↳Phenomenon to be explained
$\quad\quad\quad\quad$ (i.e., Why did Socrates die?): Socrates is mortal

Assume that a researcher wishes to test the hypothesis that the higher the anxiety the better the performance on a relatively simple task. The Manifest Anxiety Scale (Taylor, 1953) is used to measure anxiety. Say that the researcher decides to vary anxiety in two ways. Two groups of participants are selected such that one group is composed of individuals who have considerable anxiety, a second group of those with little anxiety. A relatively simple task is constructed for the participants to learn. The evidence report states, in effect, that the high-anxiety group performed better than did the low-anxiety group. The evidence report is thus positive, and since it is in accord with the hypothesis, we may say that the hypothesis is confirmed. The investigation is completed, the problem is solved. But is it really? Although this may be said of the limited problem for which the study was conducted, there is still a nagging question—why is the hypothesis "true"? How might it be explained? To answer this question, of course, one must refer to a principle that is more general than that hypothesis. Let us say that the psychologist appeals to a principle, from stimulus-response theory, that says that performance is equal to the amount learned times the drive level present. Letting E stand for performance, H (habit strength) for the learning factor, and D for drive, the principle may be stated as $E = H \times D$.

Assuming that anxiety is a specific drive, our psychologist's hypothesis can be established as a specific case of this more general principle. For instance, let us say that the high-anxiety group exhibits a drive factor of 80 units. To simplify matters, assume that both groups learned the task equally well, thus causing the learning factor to be the same for both groups. For instance, H might be 0.50 units. In this event the performance (E) factor for the high-drive group is:

$$E = 0.50 \times 80 = 40.00$$

Assume that the low-anxiety group exhibits a drive factor of 20 units, in which case its performance factor would be:

$$E = 0.50 \times 20 = 10.00$$

Clearly, then, the performance of the high-drive (high-anxiety) group should be superior to the low-drive group, according to this principle. And the principle is quite general in that it ostensibly covers all drives in addition to including a consideration of the learning factor, H. Following our previous schema, then, we have the following situation:

Deductive Inference:
General Law: The higher the drive, the better the performance (i.e., $E = H \times D$).
Antecedent Conditions: Participants had two levels of drive, they performed a simple task, anxiety is a drive, etc.

Phenomenon to be explained:
High-anxiety participants performed a simple task better than did low-anxiety participants.

Since it would be possible logically to deduce the hypothesis (stated as "the phenomenon to be explained") from the stimulus-response principle together with the necessary antecedent conditions, we may say that the hypothesis is explained.

In chapter 14 we illustrated the ever-continuing search for a higher-level explanation for general statements. We have shown how a relatively specific hypothesis about anxiety and performance can be explained by a more general principle about (1) drives in general and (2) a learning factor (which we ignored because it was not relevant to the present discussion). The next question, obviously, is how to explain this stimulus-response principle. At this point, however, our immediate purpose is accomplished so that we shall conveniently slip off to other topics.

It is important to emphasize that the logical deduction is made on the assumption that the general principle and the statement of the antecedent conditions were actually true. Hence, a more cautious statement about our explanation would be this: Assuming that (1) the general law is true, and (2) the antecedent conditions obtained, then the phenomenon of interest is explained. But how can we be sure that the general principle is, indeed, true? We can never be *absolutely* sure, for it must always assume a probability value. It might someday turn out that the general principle used to explain a particular phenomenon was actually false. In this case what we accepted as a "true" explanation was in reality no explanation at all. Unfortunately, we can do nothing more with this situation — our explanations must always be of a tentative sort. As Feigl has put it, "scientific truths are held only until further notice." We must, therefore, always realize that when we explain a phenomenon it is on the assumption that the general principle used in the explanation is true. If the probability of the general principle is high, then we can feel rather safe. We can, however, never feel absolutely secure. This is merely another indication that we have been given a "probabilistic universe" in which to live. And the sooner we learn to accept this fact (in the present context, the sooner we learn to accept the probabilistic nature of our explanations) the better adjusted to reality we will be.

One final thought on the topic of explanation. We have indicated that an explanation is accomplished by logical deduction. But how frequently do psychologists actually explain their phenomena in such a formal manner? How frequently do they actually cite a general law, state their antecedent conditions, and deductively infer their phenomena from them? The answer, clearly, is that this is done very infrequently. Almost never will you find such a formal process being used in the actual report of scientific investigations. Rather, much more informal methods of reasoning are substituted. One need not set out on a scientific career armed with books of logical formulae and the like. But we should be familiar with the basic logical processes that one could go through in order to accomplish an explanation. As with several matters that we have previously discussed, such as stating hypotheses and evidence reports, it is not necessary that you rigidly follow the procedures that we have set down. What *is* important, and what we hope you have gained from this discussion, is that you *could* explain a phenomenon in a formal, logical manner if you wanted or needed to.

Prediction

The *processes* of making predictions and offering explanations are precisely the same, so that everything we have said about explanation is applicable to prediction. The only difference between explanation and prediction is that a prediction is made before the phenomenon is observed, whereas explanation occurs after the phenomenon has been recorded. In explanation, then, we start with the phenomenon and logically deduce it from a general law and the attendant antecedent conditions. In prediction, on the other hand, we start with the general law and antecedent conditions and derive our logical consequences. That is, from the general law we infer that a certain phenomenon should occur. We then conduct our experiment to see if it does occur. If it does, then our prediction has been successful. And as we have previously pointed out, this considerably increases the probability of the general law (unless of course the general law already has a very high degree of probability).

To illustrate a possible prediction briefly, say that we are in possession of the general stimulus-response principle that we previously discussed. We might reason thusly: This principle asserts that the higher the drive, the better the performance; anxiety is a specific drive; therefore, we would predict that high-anxiety individuals would perform a given task better than low-anxiety individuals. The conduct of such an experiment would then inform us of the success (or lack of success) of our prediction. Actually, this is precisely what has been done (cf. Spence, Farber, & McFann, 1956).

Review of Chapter 15

1. Summarize the procedures by which we generalize from a sample to a population. Consider how you assure that you have a representative sample.
2. Does the sample from which you generalize have to be restricted to organisms? What other populations exist in an experiment from which you should obtain random samples?
3. Review Table 15-1. You might consider an experiment that you have conducted or would like to conduct in this course and diagram a similar factorial design for your experiment for the purpose of generalizing to various populations.
4. Specify the ways in which interactions limit the extent to which you can advance your generalization.
5. Define the three cases relevant to the advancing of generalizations. Can you find instances of these three cases in the scientific literature? You might look for them when you read over the journals in your library. In the experiments that you are involved in, how might you explain your findings, and how might you make predictions to new situations?

16
Miscellany

We have attempted to logically develop the major aspects of experimentation throughout the preceding chapters. Starting with the nature of the problem and concluding with prediction of behavior, we have attempted to weave into a coherent pattern the problems and procedures that are important to the experimental psychologist. However, it obviously was not feasible to include there *all* matters of importance. Many of these will simply have to await your future study, but in this chapter we will take up several topics that did not conveniently fit into our general plan of development, though they deserve emphasis.

Concerning Accuracy of the Data Analysis

In one sense, we would like to place this particular section at the beginning of the book, in the boldest type possible. For no matter how much care you give to the other aspects of experimentation, if you are not accurate in your records and statistical analysis, the experiment is worthless. Unfortunately, there are no set rules that anybody can give you to guarantee accuracy. The best that we can do is to offer you some suggestions which, if followed, will reduce the number of errors, and if you are sufficently vigilant, eliminate them completely.

The first important point concerns "attitude." Students frequently feel that they must record their data and run their statistical analysis only once, and in so doing, they have amazing confidence in the accuracy of their results. Checking is not for them! Although it is very nice to believe in one's own perfection, the author has observed a large enough number of students and scientists over a sufficiently long period of time to know that this is just not reasonable behavior. We all make mistakes.

The best attitude for scientists to take is not that they *might* make a mistake, but that they *will* make a mistake;[1] the only problem is where to find it. Accept this suggestion or not, as you like. But remember this: At least the first few times that you run an analysis, the odds are about 99 to 1 that you will make an error. As you become more experienced, the odds might drop to about 10 to 1. The author once had occasion to talk with one of our most outstanding statisticians. To decide a matter it became necessary to run a simple statistical test. Our answer was obviously absurd, so we tried to discover the error. After several checks, however, the fault remained obscure. Finally, a third person, who could look at the problem from a fresh point of view, checked our computations and found the error. The statistician admitted that he was never very good in arithmetic and that he frequently made errors in addition and subtraction.

The first place that an error can be made occurs when you first start to obtain your data. More often than not the experimenter observes behavior and records it by writing it down, so let us take such a case as an example.

Suppose that you are running rats in a T-maze and that you are recording (1) their latency, (2) their running time, and (3) whether they turned left or right. You might take a large piece of paper on which you can identify your rat and have three columns for your three kinds of data, noting the data for each rat in the appropriate column. Once you indicate the time values and the direction the rat

---

[1] Unfortunately, since the publication of the first edition of this book there has been empirical documentation of this point for *articles already published in professional journals*. Wolins (1962) reanalyzed data obtained from authors who had published their work, and found several different kinds of errors, including the miscalculation of F-tests. For example, one author had reported F values to be reliable when in reality, according to Wolins, they were near one.

turned, you move on to your next animal; the event is over and there is no possibility for further checking. Hence, any error you make in writing down your data is uncorrectable. You should therefore be exceptionally careful in recording the correct value. You might fix the value firmly in mind, and then write it down, asking yourself all the time whether you are transcribing the right value. After it is written down, check yourself again to make sure that it is correct. If you find a value that seems particularly out of line, you might double-check it to see if it is right. After double-checking such an unusual datum, it is worthwhile to make a note that it is correct, for later on you might return to it with considerable doubt. For instance, if most of your rats take about 2 seconds to run the maze, and you write down that one rat had a running time of 57 seconds, take an extra look at the timer to make sure that this reading is correct. Then, if it is, make a little note beside "57 seconds," indicating that the value has been checked.

Frequently, experimenters transcribe the original records of behavior onto another sheet for their statistical analysis. Such a job is long and tedious, and therefore conducive to errors. In recopying data onto new sheets, considerable vigilance must be exercised. The finished job should be checked to make sure that no errors in transcription have been committed. (It is frequently possible to avoid this step. For instance, if you can plan your data sheet so that you can record the measures of behavior directly on the sheet that you will use for your statistical analysis, you will avoid errors of transcription.)

In writing data on a sheet, legibility is of utmost importance, for the reading of numbers is a frequent source of error. You may be surprised at the difficulty you might have in reading your own writing, particularly after a period of time. If you use a pencil, that pencil should be quite sharp and hard, to reduce smudging. If possible record your data in ink, and if you have to change a number, first make sure it is thoroughly erased or eradicated with ink eradicator.

Labeling of all aspects of your data sheet should be complete, since you may wish to refer to the data at some later time. You should label the experiment clearly, giving its title, the date, place of conduct, and so on. You should unambiguously label each source of data. Your three columns might be labeled, for example, "latency of response in leaving start box," "time in running from start box to close of goal box door," and "direction of turn." Each statistical operation should be clearly labeled. If you run a *t*-test, for instance, the top of your work sheet should state that it is a *t*-test between such and such conditions, using such and such a measure as the dependent variable. In short, label everything pertinent to the records and analysis so that you can return to your work years later and understand it readily.

The actual conduct of the statistical analysis is probably going to be the greatest source of error. It is thus advisable to check each step as you move along. For example, you will probably begin by computing the sums and sums of squares for your groups. Before you substitute these values into your equation you should check them. Otherwise, if they are in error, all of your later work will have to be redone. Similarly, each multiplication, division, subtraction, and addition should be checked just after it has been made, before you move on to the next operation that incorporates the result. After you have computed your statistical test, checking each step along the way, you should put it aside and do the entire example again, without looking at your previous work. If your second computation, performed independently of the first, checks with your first computation, the probability that you have erred is decreased (it is not eliminated, of course, for you may have made the error twice).

It is advantageous to have someone else conduct the same statistical analysis so that your results can be compared. Perhaps you might ask a friend to check you when the friend is criticizing the first draft of your write-up. It is also advisable to indicate when you have checked a number or operation. One way to accomplish this is to place a small dot above and to the right of the value (do not place it so low that the dot might be confused with a decimal point). The values of indicating a checked result are: (1) that you can better keep track of where you are in your work, and (2) that at some later time you will know whether or not the work has been checked. Concerning the statistical analysis, another source of errors deserves particular comment. When conducting the statistical analysis, some people can easily leave out steps, thus attempting to progress faster. For instance, if your equation calls for you to square a term and then divide that term by the number of participants, you might tend to do both of these operations at once, merely writing down the result. If you will try *not* to do this, not only will you find that your errors are reduced, but you will be able to check each step of your work more closely. In the above example, for instance, you should write down the square of the number and its divisor. Then write down the result of the division.

Ethics of Research

Sometimes psychologists are human in the worst sense of the word. As a consequence of a number of unethical acts, we have developed standards for the protection of research participants. Some examples may shock you, such as the student who was falsely told that her husband had been seriously injured in an automobile accident, the students who were falsely informed that their test scores indicated that they were not intelligent enough to be in college, or the students who were required to participate in research projects in order to be admitted to a low-cost rooming house.

To prevent such violation of ethics, the American Psychological Association studied a large number of critical incidents and developed a set of principles which, in and of themselves, are quite interesting to read (see *Ethical Principles in the Conduct of Research with Human Participants,* 1973). The general guideline is that the psychologist assumes obligations for the welfare of research participants. Having assumed that responsibility, the psychologist should seriously consider each step in the course of the research necessary to maintain the dignity and welfare of those participants. This includes the following steps.

1. The investigator should inform the individual of all features of the research that might be relevant to his or her willingness to participate, and should explain all aspects of the research about which the participant inquires.

2. Openness and honesty should characterize the relationship between the investigator and the research participant. Should it be necessary to conceal or deceive the participant for the purposes of the research, the investigator should make sure that the participant understands the reasons for this action and should explain all relevant aspects of the research at the conclusion of the data collection phase.

3. The participant should be free to decline to participate in the research or to discontinue participation at any time.

4. The investigator should honor all promises and commitments made to the participant.

5. The researcher should take special pains to protect the participants from physical and mental discomfort, harm, and danger. If any such risk exists, the investigator should inform the participant of that fact, secure the participant's consent before proceeding, and take all possible measures to minimize distress. A research procedure that might cause serious and lasting harm should never be employed.

6. At the conclusion of the data collection phase, the participant should be thoroughly "debriefed"—there should be a full clarification of the nature of the study and any misconceptions that may have arisen should be removed.

7. Information obtained about the participant during the course of the study is confidential.

Such principles seem obvious enough, and it is amazing that we needed to develop them at all. It's almost as if anyone who needs to be told such things shouldn't do research anyway. In addition, some have held that the principles restrict the freedom of already ethical scientists. Our country seems to be at a point in which some absurdly extreme procedures are being followed to protect human rights—the danger is that in going "overboard" with excessively restrictive procedures to protect human rights, we can prevent the normal conduct of our everyday business. An excessively rigid interpretation of ethical principles might indeed prevent some valuable research. For instance, Culliton (1974) discussed the issue of the National Research Act banning fetal research. Steiner (1972) makes this point in an interesting article entitled "The evils of research: Or what my mother didn't tell me about the sins of academia." Steiner modified our code of research ethics and applied it to protect students from such outrageous abuses as taking examinations. Pointing out that while a participant in an experiment knows that his or her reactions will be evaluated, the personal consequences are unimportant; Steiner holds that the same is not true for the student taking an examination—the "perils" that confront the student who takes an examination are very great indeed. Consequently, because giving examinations is so much more dangerous than doing research, Steiner believes that the ethical principles that guide the former must be far more stringent. As he put it: "Consequently, I have formulated a single principle that covers, in one fell swoop, all possible eventualities. I call it Principle 0.000.

> No examination may be given unless all parties agree that no evaluative purpose will be served and that no one but the student will ever see the examination paper. Whenever there is reason to doubt that these procedures will be followed, the professor must clear his examination with an ethics board consisting of two football coaches, one custodian, and Abbie Hoffman.

At this point I must apologize for having injected the matter of examinations into our discussion of research ethics" (p. 767). Steiner then goes on to suggest that the violations of our ethical principles are few indeed. It is thus erroneously assumed that all experimenters are guilty until proven innocent, thus erecting so many barriers in the path of the researcher that the research is immobilized. Actually he believes that the most flagrant abuse of human participants is one for which the principles of ethics offers no real cure:

> I have encountered a few studies in which I think human subjects have been injured, but I have read many studies in which I think the time and energy of human subjects have been wasted. Perhaps three years of reading manuscripts for the *Journal of Personality and Social Psychology* sensitized me to this problem. But when 80% of the papers submitted to leading journals are rejected—usually because of poor conceptualization, faulty methodology, or the triviality of the hypotheses—it is

apparent that a lot of human effort is being squandered. Of course, mistakes occur in the best of families, and even the most astute researcher is going to run off a few duds. But I think the evidence suggests that some of the people who are doing psychological research ought to be doing something else instead. By changing their occupational specialty, they might save themselves a lot of headaches and disappointments, simplify the task of journal editors, and, most importantly, avoid imposing on the precious time of the subjects who serve in their experiments. If psychological research has a bad name, it is probably not because we injure a lot of subjects but because we involve subjects in trivial, ill-conceived, or clumsily executed studies. What I am proposing is that every potential researcher be allotted a quota of studies during which he must demonstrate that he is not wasting his own and his subjects' time. Those who exhaust the allotted quota without producing anything worth while should have their hunting licenses withdrawn. Such a system might make all of us a little more careful about bestowing the PhD mantle on candidates of doubtful research competence and ingenuity. Perhaps I can call the compulsory withdrawal of hunting licenses Principle 0.001 (p. 768).

Guide to Use of Statistical Analyses[2]

In planning your experiment you, of course, must specify your statistical analysis. Perhaps the flow chart on page 416 might provide you with an overview of the methods of analysis that have been treated piecemeal in the design chapters.

Retain Your Original Data

Your research data are, as you should now be well aware, obtained through the expenditure of a considerable amount of energy and time. They are, therefore, valuable and should be preserved. Furthermore, once the experiment has been published, the data are public and should be made available on request for further advancement of science. Three good reasons for retaining data (with interesting illustrations of each reason) have been suggested by Johnson: "(a) it may be desirable to reanalyze the data from a different point of view; (b) it may be possible to compare data with additional data collected on the same . . . [participants] . . . at a later point in time; and (c) it may prove feasible to loan the data to other investigators for their research use" (1964, p. 350).

[2] This is a modification of a flow chart developed by Ronald L. Webster. I very much appreciate his letting me use this, as I do so many other things he has done.

Do you have two
groups or two
treatment conditions?

NO   YES

→ Are the groups
matched or does
each subject re-
ceive more than
one treatment?

NO   YES

→ Use the A-test or
t-test for correlated
data.

You should have two
independent groups.
If so, use the t-test
for independent
(randomized) groups.
If not, you have a problem.
See the instructor.

Do you have three or
more groups along *one*
independent variable
dimension?

NO   YES

→ Use Duncan's Multiple
Range Test.

Do you have two
independent variables
with two levels for
each variable?

NO   YES

→ This is a 2 X 2 factorial design. Independent
variable A has two levels (e.g., hi and low).
Independent variable B has two levels (large
and small). (If there were
three levels of A and two
of B, you would have a
3 X 2 factorial design.)
Use analysis of variance.

	A	
	Hi	Low
Large		
B Small		

You may have three or
more independent
variables with two or
more levels of each
variable, e.g., if there
are two levels of each
independent variable,
you have a 2 X 2 X 2
factorial design. Use
analysis of variance.

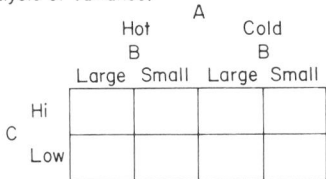

	A			
	Hot		Cold	
	B		B	
	Large	Small	Large	Small
Hi				
C Low				

Traditionally, data were often not retained by professional psychologists. For example, Wolins (1962) contacted 37 authors whose articles appeared in journals of the American Psychological Association. Of 32 who replied, 21 reported that their data had been misplaced, lost, or inadvertently destroyed. More recently, however, Craig and Reese (1973) reported considerable improvement. They received replies from 45 of 53 requests, with only 9 of the 45 refusing to share their data.

Finally, when you plan to retain your data recall a point made in the previous section. That is, if you look at them years hence, will you be able to understand them? Did you clearly label your data sheets?

Combining Tests of Reliability

Experimenters frequently have available two or more sets of experimental results that test the same hypothesis. Now, since the experiments are independent of each other, it is possible to combine the results of the statistical tests. Although this procedure may be used for several reasons, one particularly advantageous one is that in neither of the separate experiments was it possible to reject the null hypothesis. Yet the means of the two groups might have been in the same direction, and their differences sufficiently great so that they were strongly suggestive. In such cases it is possible to combine the tests of reliability in order to obtain a sound test of the empirical hypothesis. Incidentally, the experiments whose results of statistical tests are to be combined, need not be identical in design and procedure and can differ in various respects other than in terms of the participants.

A number of techniques for combining two or more statistical tests are available (Lindquist, 1953; Mosteller and Bush, 1954, chapter 8). Although we cannot possibly go into the advantages and disadvantages of each technique, one approach is extremely easy to use, although it is applicable only to the case where there are but two experiments. To illustrate, say that in one experiment the probability that the null hypothesis is false is 0.07, whereas in the second experiment it is 0.10. Clearly, in neither experiment was it possible to reject the null hypothesis, assuming the probability level was set at 0.05. On the further assumption that the means of the two groups in both experiments were in the same direction (e.g., the experimental group had the higher mean in both cases), we can combine these findings by referring to Figure 16-1. Thus, we locate 0.07 on the horizontal axis, and 0.10 on the vertical axis (we could, of course, reverse these if we wished). Reading up and across until the lines for the two values of P intersect, we obtain the combined probability. In this example it is less than 0.05, so that, considering the two experiments as combined, we are able to reject the null hypothesis, whereas considering them separately this was not possible.

Should you desire a more powerful method, you might consider the additive method reported by Edgington (1972a). As he kindly pointed out to me, whereas the overall P in this example of combining Ps of 0.07 and 0.10 was just less than 0.05, the additive method yields a P of about 0.014. A computationally simpler approximation to the additive method is the normal curve method (Edgington, 1972b).

Assumptions Underlying the Use of Statistical Tests

In applying the statistical tests to the experimental designs presented in this book, one must make certain assumptions. In general, these are that: (1) the dependent variable values are independent; (2) the variances of the groups are equal (homogeneous); (3) the population distribution is normal; (4) the treatment effects and the error effects are additive. You should be able to get a rough idea of the nature of assumptions 2 and 3. To help you visualize the character of assumption 4, assume that any given dependent variable is a function of two classes of variables — your independent variable and the various extraneous variables. Now we may assume that the dependent variable values due to these two sources of variation can be expressed as an algebraic sum of the effect of one and the effect of the other, i.e., if R is the response measure used as the dependent variable, if I is the effect of the independent variable, and if E is the combined effect of all of the extraneous variables, then the additivity assumption says that $R = I + E$.

Various tests are available to determine whether or not your particular data allow you to regard the assumptions of (2) homogeneity of variance (e.g., Games, 1972; Shukla, 1972), of (3) normality (e.g., D'Agostino, 1970), and of (4) additivity (e.g., Rojas, 1973) as tenable. It does not seem feasible at the present level of approach, however, to elaborate these assumptions or the nature of the tests for determining whether or not they are satisfied in order to justify your statistical tests. Certainly you will consider these matters further in your courses in statistics. In addition it is often difficult to determine whether or

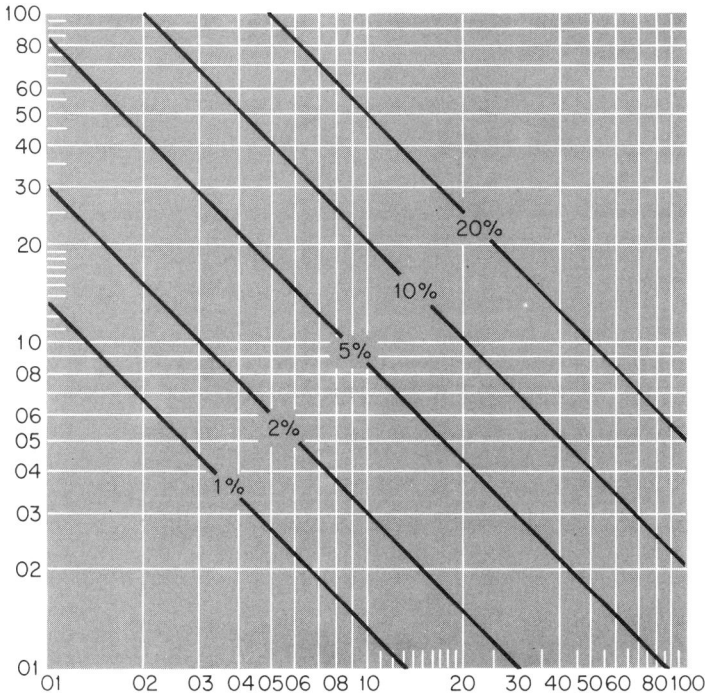

Figure 16-1. The probability of two combined tests of reliability. Find the value of one probability along a horizontal axis and the second probability along a vertical axis. Where the vertical line crosses the horizontal line one finds a diagonal that indicates the combined probability value. Five of the more commonly used significance levels are indicated here (from P. C. Baker, personal communication).

not the assumptions are exactly satisfied, i.e., the tests used for this purpose are rather insensitive. And the consensus is that rather sizable departures from assumptions 2, 3, and 4 can be tolerated still yielding valid statistical analysis (Lindquist, 1953; McNemar, 1962; Dixon & Massey, 1951; Cochran & Cox, 1957; Anderson & Bancroft, 1952; Boneau, 1960, etc.). Our statistical tests are quite robust in that they lead to proper conclusions often with deviations from these assumptions. We may add that the assumptions of normality and homogeneity may be violated with increasing security as number of participants per group increase (Scheffé, 1959). For further information, you should consult these or a number of other references that are easily available in the area of statistics.

The first assumption, however, is essential, since each dependent variable value must be independent of every other dependent variable value.[3] For example, if one value is 15, the determination that a second value is, say, 10 must in no way be influenced by, or related to, the fact that the first value is 15. If participants have been selected at random, and if one and only one value on each dependent variable is used for each participant, then the assumption of independence should be satisfied. In chapter 5 we considered how students frequently violate this assumption as follows: Suppose that a certain participant in a learning experiment yields values of 10, 8, 6, 5, and 4 on each of five trials. Some students might want to refer to each of these values as values of X in computing ΣX. Accordingly, they might say that $n = 5$. Now, this is a clear violation of the assumption of independence, for the second value is related to the first value, the third to the first and second, and so on. That is, all of these values were made by the same participant, and if that person happens to be capable, all the values will tend to be high. A second error that would be committed in this procedure is the artificial inflation of n. That is, these values are all for one participant; therefore, in this context n cannot possibly equal 5. The proper procedure would be to obtain one single value for this participant, say by adding them up and using the sum as the dependent variable value. Hence, the proper values are $\Sigma X = 33$ and $n = 1$.

We have not emphasized statistical assumptions, except for the one concerning independence, primarily because of the introductory nature of this book. As you progress in statistical and experimental work, you will want to learn more about the assumptions of homogeneity of variances and normality of the populations. With a more thorough knowledge of these assumptions and how you might determine when they are met, you will be in a better position to evaluate possible errors introduced by violating them. And, of course, when really serious violations of these assumptions occur, you will have to effect some remedy. In general, there are three possible remedies. One is to transform your data (e.g., to take the square root or reciprocal of each value) and then to continue the analysis in the usual way (cf. Edwards, 1968). The second would be to use a type of statistical test known as a nonparametric test, which does not require that you meet the assumptions of normality and homogeneity of variances (cf. Siegel, 1956). Nonparametric tests still require the assumption of independence, however. The third is to adjust your level of probability. The logic

---

[3] In the case of the matched groups design, the independence assumption takes a slightly different form; that is, that the values of D are independent. Hence, a more adequate statement of this assumption would be that the treatment effects and the error are independent. That is, in terms of the symbols used for the fourth assumption, I and E are independent.

here is that your P value for your statistical test is not a true one when the assumptions for it have been seriously violated; but it is possible to adjust that value so that it more closely approximates the true possibility level (cf. Horsnell, 1953).

Reducing Error Variance

The purpose of this section is to consider ways in which the experimenter can increase the chances of rejecting the null hypothesis, *if in fact it should be rejected.* The point may be illustrated by taking two extremes. First, if an experimenter conducts a "sloppy" experiment (e.g., the controls are poor) the chances of rejecting a null hypothesis that is false are quite low. On the other hand, if one conducts a highly refined (e.g., well-controlled) experiment, the chances of rejecting a false null hypothesis are increased. In short, if there is a lot of "noise" in the experimental situation, the dependent variable values are going to vary for a lot of reasons other than the fact that the independent variable varied, thus obscuring any systematic relationship between the independent and dependent variable. There are two general ways in which an experimenter can increase the chances of rejecting a null hypothesis that really should be rejected. To understand them, let us get the basic equation for the t-test for the two-randomized-groups design before us:[4]

$$(16\text{-}1) \qquad t = \frac{\overline{X}_1 - \overline{X}_2}{\sqrt{\dfrac{s_1^2}{n_1} + \dfrac{s_2^2}{n_2}}}$$

Now, we know that the larger the value of t, the greater the likelihood that we will be able to reject the null hypothesis. Hence, our question is: how can we design an experiment such that the value of t can legitimately be increased? In other words, how can we increase the numerator of Equation (16-1) and decrease the denominator? We have previously seen that the numerator can often be increased by exaggerating the difference in the independent variable values that you administer to your two groups. For instance, if we are seeking to determine whether amount of practice affects amount learned, we are more likely to obtain a reliable difference between two groups if we have them practice 100 trials versus 10 trials, than if we have them practice 15 trials versus 10 trials. This is so because we would probably increase the difference between the means of the dependent vari-

---

[4] See footnote 12, p. 217.

able of the two groups and, as we said, the greater the mean difference, the larger the value of t. Let us now consider the denominator of Equation (16-1).

In every experiment there is a certain error variance, and in our statistical analysis we obtain an estimate of it. In the two-groups designs, the error variance is the denominator of the t ratio. Where we use Duncan's range test, our estimate of the error variance is provided by Equation (9-5) (p. 250), which is an equation for computing the square root of the error variance. And in factorial designs, the error variance is the denominator of the F ratio. Basically, these three estimates of the error variance are measures of the same thing—the extent to which participants treated alike exhibit variability in their dependent variable values. There are many reasons why we obtain different values for participants treated alike. For one, organisms are all "made" differently, and they will all react somewhat differently to the same experimental treatment. For another, it simply is impossible to treat all participants in the same group precisely the same; we always have a number of extraneous variables operating on our participants in a random manner, resulting in differential effects. And finally, some of our error variance is due to imperfections in our measuring devices. No device can provide a completely "true" score, nor can we as humans make completely accurate and consistent readings of the measuring device.

In many ways, it is unfortunate that dependent variable values for participants treated alike show so much variability, but we must learn to live with the fact that this variability is with us. The best we can do is attempt to reduce it. Let us emphasize the reason that we want to reduce the error variance in an experiment. The illustration will be in terms of the t-test as used in a two-groups design, but what we have to say may be considered applicable to other designs and their statistical tests. Let us say that we know that the difference between the means of two groups is 5. Consider two situations, one where the error variance is rather large, and one where it is rather small. For example, say that the error variance is 5 in the first case, but 2 in the second. For the first case, then, our computed t would be: $t = 5/5 = 1.0$, and for the second it would be $t = 5/2 = 2.5$. Clearly, in the first case we would *fail* to reject our null hypothesis, while in the second case we are likely to reject it. For both situations we have the same mean difference. Such a difference in our two experiments is, of course, of the utmost importance. We naturally seek to reject the null hypothesis, if it "should" be rejected. What we mean is that the null hypothesis specifies that there is no mean difference between our two groups. However, in a given experiment we find that the means of our groups do differ. The question is whether or not the difference is sufficiently large to allow us to reject the null hypothesis. Now if our error variance is sufficiently large, we shall not. But it may be rejected if the error variance is sufficiently small. Hence we seek to obtain an error variance that is sufficiently small to allow us to reject our null hypothesis. If after reducing it as much as possible, we still cannot reject our null hypothesis, then it seems reasonable to conclude that the null hypothesis should actually not be rejected.

This point emphasizes that we are *not* trying to find out how to increase our chances of rejecting the null hypothesis in a biased sort of way; we only want to increase our chances of rejecting the null hypothesis if it really should be rejected. Let us now consider ways in which we can reduce the error variance in our experiments. To do this we shall consider the denominator of Equation (16-1) in greater detail. First, we can see clearly that as the variances (i.e., s_1^2 and s_2^2) of the groups decrease, the size of t increases. To illustrate, assume that the mean difference is 5, that the variances are each 64, and that n_1 and n_2 are both 8. In this event,

$$t = \frac{5}{\sqrt{\frac{64}{8} + \frac{64}{8}}} = 1.25$$

Now let us say that the experiment is conducted again, except that this time we are able to reduce the variances to 16. In this case,

$$t = \frac{5}{\sqrt{\frac{16}{8} + \frac{16}{8}}} = 2.50$$

Granting, then, that it is highly advisable to reduce the variances of our groups, how can we accomplish this? There are several possibilities. First, recall that our participants when they enter the experimental situation, are all different, and that the larger such differences, the greater the variances of our groups. Therefore, one obvious way to reduce the variances of our groups, and hence the error variance, is to reduce the extent to which our participants are different. Psychologists frequently increase the homogeneity of their groups by selection. For example, we work with a number of different strains of rats. In any given experiment, however, all the rats are usually taken from a single strain — the Wistar strain, the Sprague-Dawley strain, or whatever. If a psychologist randomly assigns rats from several different strains to groups, variances are probably going to increase. Working with humans is more difficult, but even here the selection of participants that are similar is a frequent practice, and should be considered. For example, using college students as participants undoubtedly results in smaller variances than if we selected participants at random from the general population of humans. But you could even be selective in your college population; you might use only females, only students with IQs above 120, only those with low anxiety scores, and so on.

At this point, one serious objection that comes to mind is that by selection of participants you thus restrict the extent to which you can generalize your results. Thus, if you sample only high-IQ students, you will certainly be in danger if you try to generalize your findings to low-IQ students, or to any other population that you have not sampled. For this reason, selection of homogeneous participants for only two groups in an experiment (e.g., experimental versus control groups) should be seriously pondered before it is adopted. For the greater the extent to which you select homogeneous participants, the less sound your basis for a broad generalization.

The solution to this problem could come from a systematic sampling of a population in which a number of different values for the population are incorporated in the experiment. In this case a factorial design is entailed, as in Table 14-2. For instance, you might classify your participants as high IQ, medium IQ, or low IQ (this is called *stratification*). In this case you have three homogeneous groups, and you can also generalize to your participant population as far as IQ is concerned. Furthermore, and this is a somewhat more advanced point, you can reduce your error variance by using this type of design, i.e., when you classify participants into levels (the vertical cells of the factorial), you compute the variation in the dependent variable due to levels (here, three levels of IQ); then this variation is "automatically" taken out of the error term, resulting in an increase in the precision of the experiment. As we said, this is a somewhat more advanced point than is really appropriate here, but perhaps by being alerted to it, your future work will be facilitated. In any event, what we have said is that: (1) error variance can be decreased by selecting homogenous participants; (2) but if you use only two groups you restrict the extent to which you can generalize; (3) therefore, you can stratify your participants by repeating your experiment at different levels of homogeneous participants, thus decreasing your error variance while maintaining the extent to which you can generalize.

A *second* way in which you can reduce your variances is in your experimental procedure. The ideal is to treat all participants in the same group as precisely alike as possible. We cannot emphasize this too strongly. We have counseled the use of a tape recorder for administering instructions, in order that all participants would receive precisely the same words, with precisely the same intonations. If you rather casually tell the participant what to do, varying the way in which you say it with different people, you are probably increasing your variances. Similarly, the greater the number of extraneous variables that are operating in a random fashion, the greater will be your variances. If, for example, noises are present in varying degrees for some individuals, but not present at all for others, your group variances are probably going to increase. Here again, however, you should recognize that if you eliminate extraneous variables to any great extent, you will have difficulty in generalizing to situations where they are present. For example, if all your participants are located in sound-deadened rooms, then you should not, strictly speaking, generalize to situations where noises are randomly present. But since we usually are not trying to generalize, at least not immediately, to such uncontrolled stimulus conditions, this general objection need not greatly disturb us.

A *third* way to reduce your variances concerns the errors that you might make — errors in reading your measuring instruments, in recording your data, and in your statistical analysis. The more errors that are present, the larger will be the variances, assuming that such errors are of a random nature. This point also relates to the matter of the reliability of your dependent variable, or perhaps more appropriately to how reliably you measured it, as discussed on pages 189–192. Hence, the more reliable your measures of the dependent variable, the less will be your error variance. One way in which the reliability of the dependent variable measure can be increased is to make more than one observation on each participant; if your experimental procedure allows this you would be wise to consider it.

The three techniques noted above are ways of reducing the error variance by reducing the variances of your groups. Another possible technique for reducing the error variance concerns the design that you select. The clearest example for the designs that we have considered would be to replace the two-randomized-groups design with the matched-groups design for two groups, providing that there is a substantial correlation between the independent variable and dependent variable. As you may recall, the error variance is reduced in accordance with the size of the correlation (p. 224). The factorial design can also be used to decrease your error variance. For example, you might incorporate an otherwise extraneous variable in your design and remove the variance attributable to that variable from your error variance. This was the point made above for IQs of participants (p. 424).

A technique that is frequently effective in reducing error variance is the "analysis of covariance." Briefly, this technique enables you to obtain a measure of what you think is a particularly relevant extraneous variable that you are not controlling. This usually has to do with some characteristic of your participants. For instance, if you are conducting a study of the effect of certain psychological variables on weight, you might use as your measure the weight of your participants before you administer your experimental treatments. Through analysis of covariance, you then can "statistically control" this variable; that is, you can remove the effect of initial weight from your dependent variable scores, thus decreasing your error variance. We might note that the degree of success in reducing error variance with the analysis of covariance depends on the size of the correlation between your extraneous variable and your dependent variable. The application of this statistical technique, however, is not always simple: It can be seriously misused and one cannot be assured that it can "save" a shoddy experiment (cf. Harris, Bisbee, and Evans, 1971; Maxwell and Cramer, 1975). Some researchers overuse this method as in the instance of a person I once overheard asking of a researcher "where is your analysis of covariance"—the understanding in his department was that it is *always* used in experimentation. In your future study of experimentation and statistics you should attempt to learn how this technique is applied.

We may note that Ray (1960, ch. 17) offered an excellent discussion of the use of various designs for reducing error variance. Let us summarize his presentation by starting with the notion of *precision* of a design: The more precise a design, the less the error variance resulting from its use. Ray considered five different designs: (1) a randomized-groups design (as in chapters 5, 9, and 10) which is analyzed by means of an analysis of variance; (2) a gains design in which there is an initial measure and a post training measure of proficiency for each participant; the dependent variable measure is then the amount of gain for each participant; the dependent variable measure is then the amount of gain for each participant, which value is obtained by subtracting the posttraining score from the pretraining score (as discussed on pp. 327–29); (3) an analysis of covariance design; (4) a matching design (chapter 8); and (5) a repeated treatments design—in this case the same participant is exposed to several experimental treatments, as in the use of counterbalancing (see pp. 161–164). Ray took as his standard the randomized-groups design and ascertained the effectiveness (in terms of amount of precision) of each of the other designs relative to that of the randomized-groups design. His conclusions were (1) all of the latter four designs are more precise than the randomized-groups design; (2) of the latter four designs, the repeated treatments design is generally superior to the gains design, to the covariance design, and to the matching design; and (3) the differences in precision between a matching and a covariance design are not very great. For more specific conclusions, you should refer to Ray; in particular, note that the superiority in precision of some designs depends on the value of the correlation that obtains in your experiment (e.g., between your matching and your dependent variable). The general point here is that error variance *can* be reduced by judicious selection of the design that you use.

Referring back to Equation (16-1) (p. 423) we have seen that as the variances of our groups decrease, the error variance decreases, and the size of t increases. The other factor in the denominator is n. As the size of n increases, the error variance decreases. This occurs for two reasons: because we use n in the computation of variances, and because n is also used otherwise. We might comment that increasing the number of participants per group is probably one of the easiest ways to decrease, usually very sharply, the error variance. More will be said shortly about the number of participants in an experiment.

We have now considered a number of ways in which the error variance of an experiment can be reduced. They should all provide useful hints for you as you develop your experimental plan. Some of the methods suggested should be quite easy to incorporate, while others will no doubt be unrealistic for your particular experiment. Furthermore, some of the procedures are generally more effective than others. One of the more interesting and promising developments in statistics and experimental design is the attempt to ascertain the relatively more effective procedures for decreasing error variance. Even though this topic is somewhat more advanced than is appropriate here, perhaps a word or two will sensitize you to the issue for your future learning. Overall and Dalal (1965), for example, presented a very valuable analysis of the factors that contribute to error variance. In their terms, the problem is to maximize the power of the experiment; more completely, they attempt to ascertain the relative effects of several variables on increasing the treatment effects relative to the experimental error. Among the variables that they consider are: (1) the number of independent observations made on each participant; (2) the number of participants per group; (3) the number of levels of a control variable (e.g., the number of values of some variable of secondary interest that is incorporated into a factorial design as in chapter 10); (4) the number of locations at which the experiment is repeated (e.g., the same experiment may be conducted in several different hospitals or schools); and (5) the number of treatment conditions in the experiment may be determined as a function of the particular value assumed by such variables. That is, as these kinds of variables vary (e.g., as the number of participants per group varies) the relative power of the experiment can be ascertained; in this way the experimenter can enter the tables with the particular values planned on, and determine the relative power of the experiment. One can also see whether certain changes in an experiment (e.g., increasing the number of participants by a certain amount) will have a noticeable effect on the power. Among the general conclusions offered by Overall and Dalal, some are particularly relevant (if unsurprising) here. For example, the number of observations made on each participant, the number of participants per group, and the number of locations at which the experiment was conducted all increase the likelihood of rejecting the null hypothesis (if it should be rejected). But, other things being equal, repeating the experiment in different locations has a relatively greater effect in increasing the estimated value of F (in the F-test). Another conclusion concerns the relative value of increasing the number of participants as against increasing the number of observations on each participant. Their answer is that "... it is unquestionably better to use more ... [participants] ... and test each one only once ... no matter how unreliable the measurement involved" (Overall & Dalal, 1965, p. 347).

These are but illustrations of the way in which our knowledge in the area of experimental design is increasing. As the results of continued research on this topic increase we can well expect that the power of experiments of the future will dwarf those of today.

We have tried in this section to indicate the importance of the reduction of error variance in experimentation and to suggest some of the ways that it might be accomplished. Unfortunately, it is not possible to provide an exhaustive coverage of the available techniques, because of both lack of space and complexities that would take us beyond our present level of discussion. Excellent treatments of this topic have been given elsewhere, although they require a somewhat advanced knowledge of experimentation and statistics (e.g., Cochran & Cox, 1957; Fisher, 1953).

Let us conclude this most important topic by summarizing the more important points that have been made. First, the likelihood of rejecting the null hypothesis can be increased by increasing the difference between the values of the independent variable administered to the groups in the experiment, and by decreasing the error variance. Specific ways that one is likely to decrease error variance are: (1) classify (stratify) participants into homogeneous levels according to their scores on some relevant measure; (2) standardize, in a strict fashion, the experimental procedures used; (3) reduce errors in observing and recording the dependent variable values (and make more than one measurement on each participant if practicable); (4) select a relatively precise design; (5) increase the number of participants per group; (6) replicate the experiment.

Number of Participants per Group

"How many participants should I have in my groups?" is a question that students usually ask in a beginning course in experimental psychology. We have touched on this question above, but there are several additional points that can profitably be made. First let us note a rather traditional procedure often used by experimenters. This is to observe a number of participants, more or less arbitrarily determined, and see how the results turn out. If the groups differ reliably, the experimenter may be satisfied with that number of participants, or additional participants may be studied to confirm the reliable findings. On the other hand, if the groups do not differ reliably but the differences are promising, more participants may be used in the hope that the additional data will produce reliability.[5]

Although we cannot adequately answer the student's question, we can offer some guiding considerations. First, we have seen that the larger the number of participants run, the more reliably can be estimated the mean difference (if such exists) between the groups. This is a true and sure statement, but it does not help very much. We can clearly say that 100 participants per group is better than 50. You may want to know if 20 participants per group is enough. That depends, first of all, on the "true" (population) difference between your groups, and second, on the size of the variances of your groups. What we can say is that the larger the true difference between groups, the smaller the number of participants required for the experiment; and the smaller the group variances, the fewer participants required. Now if you know what the differences are, and also what the variances are,

---

[5] This latter procedure cannot legitimately (strictly speaking) be defended in other than a preliminary investigation. Clearly, one who keeps adding participants until a reliable difference is obtained may capitalize on chance. For example, if one runs 10 participants per group and obtains a *t* value that approaches a reliability level of 0.05, perhaps 10 more participants might be added to each group. Assume that the *t* is now reliable. But the results of these additional participants might be due merely to chance. The experiment is stopped and success proclaimed. If still more participants were studied, however, reliability would be lost, and the experimenter would never know this fact. If such an experiment is to be cross-validated (replicated), this procedure is, of course, legitimate.

the number of participants required can be estimated. Unfortunately, experimenters do not usually have this information, or if they have it they do not consider the matter worth the effort required to answer the question. We shall not attempt to judge what should or should not be done in this respect but shall illustrate the procedure for determining the minimum number of participants required, given these two bits of information. (Possible sources of this information include: (1) an experiment reported in the literature similar to the one you want to run, from which you can abstract the necessary information, or better, (2) a pilot study conducted by yourself to yield estimates of the information that is needed.)

In any event, let us suppose that you are going to conduct a two-randomized-groups experiment. You estimate (on the basis of previously collected data) that the mean score of Condition A is 10, and that the mean of Condition B is 15. The difference between these means is 5. You also estimate that the variances of your two groups are both 75. Say that you set your probability level at 0.05, in which case the value of t that you will need to reject the null hypothesis is *approximately* 2 (you may be more precise if you like). Assume that you want an equal number of participants in both groups. Now we have this information:

$$\overline{X}_1 - \overline{X}_2 = 5$$

$$s_1^2 \text{ and } s_2^2 \text{ both } = 75$$

$$t = 2$$

Let us solve Equation (16-1) for n instead of for t. By simple algebraic manipulation we find that, on the above assumptions, Equation (16-1) becomes:

$$(16\text{-}2) \qquad n = \frac{2t^2s^2}{(\overline{X}_1 - \overline{X}_2)^2}$$

Substituting the above values in Equation (16-2) and solving for n, we find:

$$n = \frac{2(2)^2(75)}{(15 - 10)^2} = \frac{600}{25} = 24$$

We can say, therefore, that with this true mean difference, and with these variances for our two groups, and using the 5 percent level of reliability, we need a minimum of 24 participants per group to reject the null hypothesis. We have only approximated the value of t necessary at the 5 percent level, however, and we have not allowed for any possible increase in the variance of our two groups. Therefore, we should underline the word *minimum*. To be safe, then, we should probably run somewhat more than 24 participants per group; 30

would seem reasonable in this case.[6] This number has traditionally been more or less a rule of thumb in experimentation, with very many exceptions. After considering four aspects of a statistical test (alpha level, strength of the hypothesized relationship, power, and sample size) Cowles (1974) approached that value with the suggestion that we adopt the number of 35 participants per group for our common statistical tests such as *t*.

We have illustrated a technique for estimating the minimum number of participants necessary to reject the null hypothesis, if in fact you should reject it, for the case of the two-randomized-groups design. The technique can also be extended, with a little industry, to the equations for analyzing the other designs that we have taken up.

Replication

In the history of science many, many astounding findings have been erroneously reported. Quite recently a major psychologist, who had attained such fame that he had been knighted by the queen, was reported to be scientifically dishonest in his research implicating heredity in intelligence—nobody had ever repeated his research and it was reported to be a hoax, though it had formed the basis for important social movements. We have several times emphasized the self-correcting feature of science—if a scientist errs, for whatever reason, the error will be discovered, at least if it has any importance. The basic criterion in science for evaluating the validity of our conclusions is that research is repeated. The technical term for repeating an experiment is *replication*. By replication, we mean that the *methods* employed by a researcher are repeated in an effort to confirm or disconfirm the findings obtained. We say that an experiment was replicated when the experimental procedures were repeated on a new sample of participants. The replication may be by the original researcher, or preferably by another in a different laboratory—the latter would result in a higher probability of the conclusions (if they conformed to the original ones) than the former because any bias would be reduced, independent apparatus would be used, etc. Note an important distinction: replication refers to repeating the experiment, not to confirming the original findings. Hence, the replication of an experiment may either confirm or disconfirm (be consonant with or contradictory to) the findings of the earlier experiment. Unfortunately there is a

---

[6] This procedure is offered only as a rough guide, for we are neglecting power considerations of the statistical test. This procedure has a minimal power for rejecting the null hypothesis.

tendency for some to say that they "failed to replicate an experiment," which literally means that they failed to repeat the original methodology—what they mean to say is that they *did* replicate the experiment but failed to duplicate the findings. The distinction is an important one, analogous to the one between obtaining "no results" when what is meant was "negative results."

Though it is at the heart of our science, replication is relatively rare. This is understandable, but unfortunate. Priority for publishable space in our journals is given to original research. We thus probably retain many untruths that were the result of chance five times out of a 100 that the null hypothesis was erroneously rejected). Several solutions are possible. One is to include an earlier experiment in a new one and extend it by including another independent variable, such as discussed under the topic of factorial designs. Another is to encourage student research (master's theses, etc.) as replications, perhaps with short sections of our journals devoted to such studies. The problem should be *very* seriously considered.

A Look to the Future

This concludes our presentation. You have finished a book on experimental psychology, but the topic itself is endless. Among those who have studied this book, some will go on to become talented researchers; we hope that those who do will themselves discover some new and interesting characteristics of behavior. For all, we hope that an increased appreciation for sound psychological knowledge was gained.

Review of Chapter 16

1. You might consider reviewing your data sheets for an experiment together with your statistical analyses and relate those items to the discussion starting on page 409. Were you systematic in collecting and recording your data? Were your statistical analyses neatly and accurately carried out? Did you check yourself on each step or have a colleague doublecheck you? (If your work was not accurate, you probably could have saved yourself the time in even conducting your work.)

2. Did you adequately "debrief" your participants in your studies (if they were human)? Did you adequately carry out other items referred to under "research ethics" in this chapter?

3. Do you adequately understand the nature of statistical assumptions, especially that of independence? For this purpose you might review

the section on page 140 in chapter 5 entitled "A Statistical Error to Avoid."

4. How might you redesign an experiment that you conduct to achieve greater efficiency by reducing your error variance?

5. Define "replication." Consider the role of replication in science.

6. How might the emphasis on publishing only "positive results" in our journals negatively influence our science?

Appendix A
statistical tables

Table A-1 Table of t

df	P	0.9	0.8	0.7	0.6	0.5	0.4	0.3	0.2	0.1	0.05	0.02	0.01
1		0.158	0.325	0.510	0.727	1.000	1.376	1.963	3.078	6.314	12.706	31.821	63.657
2		0.142	0.289	0.445	0.617	0.816	1.061	1.386	1.886	2.920	4.303	6.965	9.925
3		0.137	0.277	0.424	0.584	0.765	0.978	1.250	1.638	2.353	3.182	4.541	5.841
4		0.134	0.271	0.414	0.589	0.741	0.941	1.190	1.533	2.132	2.776	3.747	4.604
5		0.132	0.267	0.408	0.559	0.727	0.920	1.156	1.476	2.015	2.571	3.365	4.032
6		0.131	0.265	0.404	0.553	0.718	0.906	1.134	1.440	1.943	2.447	3.143	3.707
7		0.130	0.263	0.402	0.549	0.711	0.896	1.119	1.415	1.895	2.365	2.998	3.499
8		0.130	0.262	0.399	0.546	0.706	0.889	1.108	1.397	1.860	2.306	2.896	3.355
9		0.129	0.261	0.398	0.543	0.703	0.883	1.100	1.383	1.833	2.262	2.821	3.250
10		0.129	0.260	0.397	0.542	0.700	0.879	1.093	1.372	1.812	2.228	2.764	3.169
11		0.129	0.260	0.396	0.540	0.697	0.876	1.088	1.363	1.796	2.201	2.718	3.106
12		0.128	0.259	0.395	0.539	0.695	0.873	1.083	1.356	1.782	2.179	2.681	3.055
13		0.128	0.259	0.394	0.538	0.694	0.870	1.079	1.350	1.771	2.160	2.650	3.012
14		0.128	0.258	0.393	0.537	0.692	0.868	1.076	1.345	1.761	2.145	2.624	2.977
15		0.128	0.258	0.393	0.536	0.691	0.866	1.074	1.341	1.753	2.131	2.602	2.947
16		0.128	0.258	0.392	0.535	0.690	0.865	1.071	1.337	1.746	2.120	2.583	2.921
17		0.128	0.257	0.392	0.534	0.689	0.863	1.069	1.333	1.740	2.110	2.567	2.898
18		0.127	0.257	0.392	0.534	0.688	0.862	1.067	1.330	1.734	2.101	2.552	2.878
19		0.127	0.257	0.391	0.533	0.688	0.861	1.066	1.328	1.729	2.093	2.539	2.861
20		0.127	0.257	0.391	0.533	0.687	0.860	1.064	1.325	1.725	2.086	2.528	2.845
21		0.127	0.257	0.391	0.532	0.686	0.859	1.063	1.323	1.721	2.080	2.518	2.831
22		0.127	0.256	0.390	0.532	0.686	0.858	1.061	1.321	1.717	2.074	2.508	2.819
23		0.127	0.256	0.390	0.532	0.685	0.858	1.060	1.319	1.714	2.069	2.500	2.807
24		0.127	0.256	0.390	0.531	0.685	0.857	1.059	1.318	1.711	2.064	2.492	2.797
25		0.127	0.256	0.390	0.531	0.684	0.856	1.058	1.316	1.708	2.060	2.485	2.787
26		0.127	0.256	0.390	0.531	0.684	0.856	1.058	1.315	1.706	2.056	2.479	2.779
27		0.127	0.256	0.389	0.531	0.684	0.855	1.057	1.314	1.703	2.052	2.473	2.771
28		0.127	0.256	0.389	0.530	0.683	0.855	1.056	1.313	1.701	2.048	2.467	2.763
29		0.127	0.256	0.389	0.530	0.683	0.854	1.055	1.311	1.699	2.045	2.462	2.756
30		0.127	0.256	0.389	0.530	0.683	0.854	1.055	1.310	1.697	2.042	2.457	2.750
∞		0.12566	0.25335	0.38532	0.52440	0.67449	0.84162	1.03643	1.28155	1.64485	1.95996	2.32634	2.57582

Table A-1 is reprinted from Table IV of Fisher: *Statistical Methods for Research Workers*, published by Oliver and Boyd Ltd., Edinburgh, by permission of the author and publishers.

Table A-2 Table of *A*

For any given value of n − 1, the table shows the values of A corresponding to various levels of probability. A is significant at a given level if it is equal to or less than the value shown in the table.

			PROBABILITY			
$n - 1$	0.10	0.05	0.02	0.01	0.001	$n - 1$
1	0.5125	0.5031	0.50049	0.50012	0.5000012	1
2	0.412	0.369	0.347	0.340	0.334	2
3	0.385	0.324	0.286	0.272	0.254	3
4	0.376	0.304	0.257	0.238	0.211	4
5	0.372	0.293	0.240	0.218	0.184	5
6	0.370	0.286	0.230	0.205	0.167	6
7	0.369	0.281	0.222	0.196	0.155	7
8	0.368	0.278	0.217	0.190	0.146	8
9	0.368	0.276	0.213	0.185	0.139	9
10	0.368	0.274	0.210	0.181	0.134	10
11	0.368	0.273	0.207	0.178	0.130	11
12	0.368	0.271	0.205	0.176	0.126	12
13	0.368	0.270	0.204	0.174	0.124	13
14	0.368	0.270	0.202	0.172	0.121	14
15	0.368	0.269	0.201	0.170	0.119	15
16	0.368	0.268	0.200	0.169	0.117	16
17	0.368	0.268	0.199	0.168	0.116	17
18	0.368	0.267	0.198	0.167	0.114	18
19	0.368	0.267	0.197	0.166	0.113	19
20	0.368	0.266	0.197	0.165	0.112	20
21	0.368	0.266	0.196	0.165	0.111	21
22	0.368	0.266	0.196	0.164	0.110	22
23	0.368	0.266	0.195	0.163	0.109	23
24	0.368	0.265	0.195	0.163	0.108	24
25	0.368	0.265	0.194	0.162	0.108	25
26	0.368	0.265	0.194	0.162	0.107	26
27	0.368	0.265	0.193	0.161	0.107	27
28	0.368	0.265	0.193	0.161	0.106	28
29	0.368	0.264	0.193	0.161	0.106	29
30	0.368	0.264	0.193	0.160	0.105	30
40	0.368	0.263	0.191	0.158	0.102	40
60	0.369	0.262	0.189	0.155	0.099	60
120	0.369	0.261	0.187	0.153	0.095	120
∞	0.370	0.260	0.185	0.151	0.092	∞

Table A-3 Table of F

df ASSOCIATED WITH NUMERATOR

df Associated with Denominator	P	1	2	3	4	5	6	8	12	24	∞
1	0.01	4052	4999	5403	5625	5764	5859	5981	6106	6234	6366
	0.05	161.45	199.50	215.71	224.58	230.16	233.99	238.88	243.91	249.05	254.32
	0.10	39.86	49.50	53.59	55.83	57.24	58.20	59.44	60.70	62.00	63.33
	0.20	9.47	12.00	13.06	13.73	14.01	14.26	14.59	14.90	15.24	15.58
2	0.01	98.49	99.00	99.17	99.25	99.30	99.33	99.36	99.42	99.46	99.50
	0.05	18.51	19.00	19.16	19.25	19.30	19.33	19.37	19.41	19.45	19.50
	0.10	8.53	9.00	9.16	9.24	9.29	9.33	9.37	9.41	9.45	9.49
	0.20	3.56	4.00	4.16	4.24	4.28	4.32	4.36	4.40	4.44	4.48
3	0.01	34.12	30.81	29.46	28.71	28.24	27.91	27.49	27.05	26.60	26.12
	0.05	10.13	9.55	9.28	9.12	9.01	8.94	8.84	8.74	8.64	8.53
	0.10	5.54	5.46	5.39	5.34	5.31	5.28	5.25	5.22	5.18	5.13
	0.20	2.68	2.89	2.94	2.96	2.97	2.97	2.98	2.98	2.98	2.98
4	0.01	21.20	18.00	16.69	15.98	15.52	15.21	14.80	14.37	13.93	13.46
	0.05	7.71	6.94	6.59	6.39	6.26	6.16	6.04	5.91	5.77	5.63
	0.10	4.54	4.32	4.19	4.11	4.05	4.01	3.95	3.90	3.83	3.76
	0.20	2.35	2.47	2.48	2.48	2.48	2.47	2.47	2.46	2.44	2.43
5	0.01	16.26	13.27	12.06	11.39	10.97	10.67	10.29	9.89	9.47	9.02
	0.05	6.61	5.79	5.41	5.19	5.05	4.95	4.82	4.68	4.53	4.36
	0.10	4.06	3.78	3.62	3.52	3.45	3.40	3.34	3.27	3.19	3.10
	0.20	2.18	2.26	2.25	2.24	2.23	2.22	2.20	2.18	2.16	2.13
6	0.01	13.74	10.92	9.78	9.15	8.75	8.47	8.10	7.72	7.31	6.88
	0.05	5.99	5.14	4.76	4.53	4.39	4.28	4.15	4.00	3.84	3.67
	0.10	3.78	3.46	3.29	3.18	3.11	3.05	2.98	2.90	2.82	2.72
	0.20	2.07	2.13	2.11	2.09	2.08	2.06	2.04	2.02	1.99	1.95

Table A-3 (Cont'd)

df Associated with Denominator	P	1	2	3	4	5	6	8	12	24	∞
7											
	0.01	12.25	9.55	8.45	7.85	7.46	7.19	6.84	6.47	6.07	5.65
	0.05	5.59	4.74	4.35	4.12	3.97	3.87	3.73	3.57	3.41	3.23
	0.10	3.59	3.26	3.07	2.96	2.88	2.83	2.75	2.67	2.58	2.47
	0.20	2.00	2.04	2.02	1.99	1.97	1.96	1.93	1.91	1.87	1.83
8											
	0.01	11.26	8.65	7.59	7.01	6.63	6.37	6.03	5.67	5.28	4.86
	0.05	5.32	4.46	4.07	3.84	3.69	3.58	3.44	3.28	3.12	2.93
	0.10	3.46	3.11	2.92	2.81	2.73	2.67	2.59	2.50	2.40	2.29
	0.20	1.95	1.98	1.95	1.92	1.90	1.88	1.86	1.83	1.79	1.74
9											
	0.01	10.56	8.02	6.99	6.42	6.06	5.80	5.47	5.11	4.73	4.31
	0.05	5.12	4.26	3.86	3.63	3.48	3.37	3.23	3.07	2.90	2.71
	0.10	3.36	3.01	2.81	2.69	2.61	2.55	2.47	2.38	2.28	2.16
	0.20	1.91	1.94	1.90	1.87	1.85	1.83	1.80	1.76	1.72	1.67
10											
	0.01	10.04	7.56	6.55	5.99	5.64	5.39	5.06	4.71	4.33	3.91
	0.05	4.96	4.10	3.71	3.48	3.33	3.22	3.07	2.91	2.74	2.54
	0.10	3.28	2.92	2.73	2.61	2.52	2.46	2.38	2.28	2.18	2.06
	0.20	1.88	1.90	1.86	1.83	1.80	1.78	1.75	1.72	1.67	1.62
11											
	0.01	9.65	7.20	6.22	5.67	5.32	5.07	4.74	4.40	4.02	3.60
	0.05	4.84	3.98	3.59	3.36	3.20	3.09	2.95	2.79	2.61	2.40
	0.10	3.23	2.86	2.66	2.54	2.45	2.39	2.30	2.21	2.10	1.97
	0.20	1.86	1.87	1.83	1.80	1.77	1.75	1.72	1.68	1.63	1.57
12											
	0.01	9.33	6.93	5.95	5.41	5.06	4.82	4.50	4.16	3.78	3.36
	0.05	4.75	3.88	3.49	3.26	3.11	3.00	2.85	2.69	2.50	2.30
	0.10	3.18	2.81	2.61	2.48	2.39	2.33	2.24	2.15	2.04	1.90
	0.20	1.84	1.85	1.80	1.77	1.74	1.72	1.69	1.65	1.60	1.54

13										
	0.01	9.07	6.70	5.74	5.20	4.86	4.62	4.30	3.96	3.59	3.16
	0.05	4.67	3.80	3.41	3.18	3.02	2.92	2.77	2.60	2.42	2.21
	0.10	3.14	2.76	2.56	2.43	2.35	2.28	2.20	2.10	1.98	1.85
	0.20	1.82	1.88	1.78	1.75	1.72	1.69	1.66	1.62	1.57	1.51
14										
	0.01	8.86	6.51	5.56	5.08	4.69	4.46	4.14	3.80	3.43	3.00
	0.05	4.60	3.74	3.34	3.11	2.96	2.85	2.70	2.53	2.35	2.13
	0.10	3.10	2.73	2.52	2.39	2.31	2.24	2.15	2.05	1.94	1.80
	0.20	1.81	1.81	1.76	1.78	1.70	1.67	1.64	1.60	1.55	1.48
15										
	0.01	8.68	6.36	5.42	4.89	4.56	4.32	4.00	3.67	3.29	2.87
	0.05	4.54	3.68	3.29	3.06	2.90	2.79	2.64	2.48	2.29	2.07
	0.10	3.07	2.70	2.49	2.36	2.27	2.21	2.12	2.02	1.90	1.76
	0.20	1.80	1.79	1.75	1.71	1.68	1.66	1.62	1.58	1.53	1.46
16										
	0.01	8.53	6.23	5.29	4.77	4.44	4.20	3.89	3.55	3.18	2.75
	0.05	4.49	3.63	3.24	3.01	2.85	2.74	2.59	2.42	2.24	2.01
	0.10	3.05	2.67	2.46	2.33	2.24	2.18	2.09	1.99	1.87	1.72
	0.20	1.79	1.78	1.74	1.70	1.67	1.64	1.61	1.56	1.51	1.43
17										
	0.01	8.40	6.11	5.18	4.67	4.34	4.10	3.79	3.45	3.08	2.65
	0.05	4.45	3.59	3.20	2.96	2.81	2.70	2.55	2.38	2.19	1.96
	0.10	3.03	2.64	2.44	2.31	2.22	2.15	2.06	1.96	1.84	1.69
	0.20	1.78	1.77	1.72	1.68	1.65	1.63	1.59	1.55	1.49	1.42
18										
	0.01	8.28	6.01	5.09	4.58	4.25	4.01	3.71	3.37	3.00	2.57
	0.05	4.41	3.55	3.16	2.93	2.77	2.66	2.51	2.34	2.15	1.92
	0.10	3.01	3.62	2.42	2.29	2.20	2.13	2.04	1.93	1.81	1.66
	0.20	1.77	1.76	1.71	1.67	1.64	1.62	1.58	1.53	1.48	1.40
19										
	0.01	8.18	5.93	5.01	4.50	4.17	3.94	3.63	3.30	2.92	2.49
	0.05	4.38	3.52	3.13	2.90	2.74	2.63	2.48	2.31	2.11	1.88
	0.10	2.99	2.61	2.40	2.27	2.18	2.11	2.02	1.91	1.79	1.63
	0.20	1.76	1.75	1.70	1.66	1.63	1.61	1.57	1.52	1.46	1.39

Table A-3 (Cont'd)

df Associated with Denominator	P	1	2	3	4	5	6	8	12	24	∞
20											
	0.01	8.10	5.85	4.94	4.43	4.10	3.87	3.56	3.23	2.86	2.42
	0.05	4.35	3.49	3.10	2.87	2.71	2.60	2.45	2.28	2.08	1.84
	0.10	2.97	2.59	2.38	2.25	2.16	2.09	2.00	1.89	1.77	1.61
	0.20	1.76	1.75	1.70	1.65	1.62	1.60	1.56	1.51	1.45	1.37
21											
	0.01	8.02	5.78	4.87	4.37	4.04	3.81	3.51	3.17	2.80	2.36
	0.05	4.32	3.47	3.07	2.84	2.68	2.57	2.42	2.25	2.05	1.81
	0.10	2.96	2.57	2.36	2.23	2.14	2.08	1.98	1.88	1.75	1.59
	0.20	1.75	1.74	1.69	1.65	1.61	1.59	1.55	1.50	1.44	1.36
22											
	0.01	7.94	5.72	4.82	4.31	3.99	3.76	3.45	3.12	2.75	2.31
	0.05	4.30	3.44	3.05	2.82	2.66	2.55	2.40	2.23	2.03	1.78
	0.10	2.95	2.56	2.35	2.22	2.13	2.06	1.97	1.86	1.73	1.57
	0.20	1.75	1.73	1.68	1.64	1.61	1.58	1.54	1.49	1.43	1.35
23											
	0.01	7.88	5.66	4.76	4.26	3.94	3.71	3.41	3.07	2.70	2.26
	0.05	4.28	3.42	3.03	2.80	2.64	2.53	2.38	2.20	2.00	1.76
	0.10	2.94	2.55	2.34	2.21	2.11	2.05	1.95	1.84	1.72	1.55
	0.20	1.74	1.73	1.68	1.63	1.60	1.57	1.53	1.49	1.42	1.34
24											
	0.01	7.82	5.61	4.72	4.22	3.90	3.67	3.36	3.03	2.66	2.21
	0.05	4.26	3.40	3.01	2.78	2.62	2.51	2.36	2.18	1.98	1.73
	0.10	2.93	2.54	2.33	2.19	2.10	2.04	1.94	1.83	1.70	1.53
	0.20	1.74	1.72	1.67	1.63	1.59	1.57	1.53	1.48	1.42	1.33
25											
	0.01	7.77	5.57	4.68	4.18	3.86	3.63	3.32	2.99	2.62	2.17
	0.05	4.24	3.38	2.99	2.76	2.60	2.49	2.34	2.16	1.96	1.71
	0.10	2.92	2.53	2.32	2.18	2.09	2.02	1.93	1.82	1.69	1.52
	0.20	1.73	1.72	1.66	1.62	1.59	1.56	1.52	1.47	1.41	1.32

Table A-3 (Cont'd)

df	α										
26	0.01	7.72	5.53	4.64	4.14	3.82	3.59	3.29	2.96	2.58	2.13
	0.05	4.22	3.37	2.98	2.74	2.59	2.47	2.32	2.15	1.95	1.69
	0.10	2.91	2.52	2.31	2.17	2.08	2.01	1.92	1.81	1.68	1.50
	0.20	1.73	1.71	1.66	1.62	1.58	1.56	1.52	1.47	1.40	1.31
27	0.01	7.68	5.49	4.60	4.11	3.78	3.56	3.26	2.93	2.55	2.10
	0.05	4.21	3.35	2.96	2.73	2.57	2.46	2.30	2.13	1.93	1.67
	0.10	2.90	2.51	2.30	2.17	2.07	2.00	1.91	1.80	1.67	1.49
	0.20	1.73	1.71	1.66	1.61	1.58	1.55	1.51	1.46	1.40	1.30
28	0.01	7.64	5.45	4.57	4.07	3.75	3.53	3.23	2.90	2.52	2.06
	0.05	4.20	3.34	2.95	2.71	2.56	2.44	2.29	2.12	1.91	1.65
	0.10	2.89	2.50	2.29	2.16	2.06	2.00	1.90	1.79	1.66	1.48
	0.20	1.72	1.71	1.65	1.61	1.57	1.55	1.51	1.46	1.39	1.30
29	0.01	7.60	5.42	4.54	4.04	3.73	3.50	3.20	2.87	2.49	2.03
	0.05	4.18	3.33	2.93	2.70	2.54	2.43	2.28	2.10	1.90	1.64
	0.10	2.89	2.50	2.28	2.15	2.06	1.99	1.89	1.78	1.65	1.47
	0.20	1.72	1.70	1.65	1.60	1.57	1.54	1.50	1.45	1.39	1.29
30	0.01	7.56	5.39	4.51	4.02	3.70	3.47	3.17	2.84	2.47	2.01
	0.05	4.17	3.32	2.92	2.69	2.53	2.42	2.27	2.09	1.89	1.62
	0.10	2.88	2.49	2.28	2.14	2.05	1.98	1.88	1.77	1.64	1.46
	0.20	1.72	1.70	1.64	1.60	1.57	1.54	1.50	1.45	1.38	1.28
40	0.01	7.31	5.18	4.31	3.83	3.51	3.29	2.99	2.66	2.29	1.80
	0.05	4.08	3.23	2.84	2.61	2.45	2.34	2.18	2.00	1.79	1.51
	0.10	2.84	2.44	2.23	2.09	2.00	1.93	1.83	1.71	1.57	1.38
	0.20	1.70	1.68	1.62	1.57	1.54	1.51	1.47	1.41	1.34	1.24
60	0.01	7.08	4.98	4.13	3.65	3.34	3.12	2.82	2.50	2.12	1.60
	0.05	4.00	3.15	2.76	2.52	2.37	2.25	2.10	1.92	1.70	1.39
	0.10	2.79	2.39	2.18	2.04	1.95	1.87	1.77	1.66	1.51	1.29
	0.20	1.68	1.65	1.59	1.55	1.51	1.48	1.44	1.38	1.31	1.18

Table A-3 (Cont'd)

df Associated with Denominator						*df* ASSOCIATED WITH NUMERATOR					
	P	*1*	*2*	*3*	*4*	*5*	*6*	*8*	*12*	*24*	*∞*
120											
	0.01	6.85	4.79	3.95	3.48	3.17	2.96	2.66	2.34	1.95	1.38
	0.05	3.92	3.07	2.68	2.45	2.29	2.17	2.02	1.83	1.61	1.25
	0.10	2.75	2.35	2.13	1.99	1.90	1.82	1.72	1.60	1.45	1.19
	0.20	1.66	1.63	1.57	1.52	1.48	1.45	1.41	1.35	1.27	1.12
∞											
	0.01	6.64	4.60	3.78	3.32	3.02	2.80	2.51	2.18	1.79	1.00
	0.05	3.84	2.99	2.60	2.37	2.21	2.09	1.94	1.75	1.52	1.00
	0.10	2.71	2.30	2.08	1.94	1.85	1.77	1.67	1.55	1.38	1.00
	0.20	1.64	1.61	1.55	1.50	1.46	1.43	1.38	1.32	1.23	1.00

*Table A-3 is abridged from Table V of Fisher and Yates: *Statistical Tables of Biological, Agricultural, and Medical Research*, published by Oliver and Boyd Ltd., Edinburgh, by permission of the author and publishers.

Table A-4 Values of r_p for Duncan's Range Test (reliability level = 5 percent)

df	\multicolumn{16}{c}{NUMBER OF GROUPS}
	2	3	4	5	6	7	8	9	10	12	14	16	18	20	50	100
1	18.0	18.0	18.0	18.0	18.0	18.0	18.0	18.0	18.0	18.0	18.0	18.0	18.0	18.0	18.0	18.0
2	6.09	6.09	6.09	6.09	6.09	6.09	6.09	6.09	6.09	6.09	6.09	6.09	6.09	6.09	6.09	6.09
3	4.50	4.50	4.50	4.50	4.50	4.50	4.50	4.50	4.50	4.50	4.50	4.50	4.50	4.50	4.50	4.50
4	3.93	4.01	4.02	4.02	4.02	4.02	4.02	4.02	4.02	4.02	4.02	4.02	4.02	4.02	4.02	4.02
5	3.64	3.74	3.79	3.83	3.83	3.83	3.83	3.83	3.83	3.83	3.83	3.83	3.83	3.83	3.83	3.83
6	3.46	3.58	3.64	3.68	3.68	3.68	3.68	3.68	3.68	3.68	3.68	3.68	3.68	3.68	3.68	3.68
7	3.35	3.47	3.54	3.58	3.60	3.61	3.61	3.61	3.61	3.61	3.61	3.61	3.61	3.61	3.61	3.61
8	3.26	3.39	3.47	3.52	3.55	3.56	3.56	3.56	3.56	3.56	3.56	3.56	3.56	3.56	3.56	3.56
9	3.20	3.34	3.41	3.47	3.50	3.52	3.52	3.52	3.52	3.52	3.52	3.52	3.52	3.52	3.52	3.52
10	3.15	3.30	3.37	3.43	3.46	3.47	3.47	3.47	3.47	3.47	3.47	3.47	3.47	3.48	3.48	3.48
11	3.11	3.27	3.35	3.39	3.43	3.44	3.45	3.46	3.46	3.46	3.46	3.46	3.47	3.48	3.48	3.48
12	3.08	3.23	3.33	3.36	3.40	3.42	3.44	3.44	3.46	3.46	3.46	3.46	3.47	3.48	3.48	3.48
13	3.06	3.21	3.30	3.35	3.38	3.41	3.42	3.44	3.45	3.45	3.46	3.46	3.47	3.47	3.47	3.47
14	3.03	3.18	3.27	3.33	3.37	3.39	3.41	3.42	3.44	3.45	3.46	3.46	3.47	3.47	3.47	3.47
15	3.01	3.16	3.25	3.31	3.36	3.38	3.40	3.42	3.43	3.44	3.45	3.46	3.47	3.47	3.47	3.47
16	3.00	3.15	3.23	3.30	3.34	3.37	3.39	3.41	3.43	3.44	3.45	3.46	3.47	3.47	3.47	3.47
17	2.98	3.13	3.22	3.28	3.33	3.36	3.38	3.40	3.42	3.44	3.45	3.46	3.47	3.47	3.47	3.47
18	2.97	3.12	3.21	3.27	3.32	3.35	3.37	3.39	3.41	3.43	3.45	3.46	3.47	3.47	3.47	3.47
19	2.96	3.11	3.19	3.26	3.31	3.35	3.37	3.39	3.41	3.43	3.44	3.46	3.47	3.47	3.47	3.47
20	2.95	3.10	3.18	3.25	3.30	3.34	3.36	3.38	3.40	3.43	3.44	3.46	3.46	3.47	3.47	3.47
22	2.93	3.08	3.17	3.24	3.29	3.32	3.35	3.37	3.39	3.42	3.44	3.45	3.46	3.47	3.47	3.47
24	2.92	3.07	3.15	3.22	3.28	3.31	3.34	3.37	3.38	3.41	3.44	3.45	3.46	3.47	3.47	3.47
26	2.91	3.06	3.14	3.21	3.27	3.30	3.34	3.36	3.38	3.41	3.43	3.45	3.46	3.47	3.47	3.47
28	2.90	3.04	3.13	3.20	3.26	3.30	3.33	3.35	3.37	3.40	3.43	3.45	3.46	3.47	3.47	3.47
30	2.89	3.04	3.12	3.20	3.25	3.29	3.32	3.35	3.37	3.40	3.43	3.44	3.46	3.47	3.47	3.47
40	2.86	3.01	3.10	3.17	3.22	3.27	3.30	3.33	3.35	3.39	3.42	3.44	3.46	3.47	3.47	3.47
60	2.83	2.98	3.08	3.14	3.20	3.24	3.28	3.31	3.33	3.37	3.40	3.43	3.45	3.47	3.48	3.48
100	2.80	2.95	3.05	3.12	3.18	3.22	3.26	3.29	3.32	3.36	3.40	3.42	3.45	3.47	3.53	3.53
∞	2.77	2.92	3.02	3.09	3.15	3.19	3.23	3.26	3.29	3.34	3.38	3.41	3.44	3.47	3.61	3.67

Table A-5 Values of r_p for Duncan's Range Test (reliability level = 1 percent)

df	2	3	4	5	6	7	8	9	10	12	14	16	18	20	50	100
1	90.0	90.0	90.0	90.0	90.0	90.0	90.0	90.0	90.0	90.0	90.0	90.0	90.0	90.0	90.0	90.0
2	14.0	14.0	14.0	14.0	14.0	14.0	14.0	14.0	14.0	14.0	14.0	14.0	14.0	14.0	14.0	14.0
3	8.26	8.5	8.6	8.7	8.8	8.9	8.9	9.0	9.0	9.0	9.1	9.2	9.3	9.3	9.3	9.3
4	6.51	6.8	6.9	7.0	7.1	7.1	7.2	7.2	7.3	7.3	7.4	7.4	7.5	7.5	7.5	7.5
5	5.70	5.96	6.11	6.18	6.26	6.33	6.40	6.41	6.5	6.6	6.6	6.7	6.7	6.8	6.8	6.8
6	5.24	5.51	5.65	5.73	5.81	5.88	5.95	6.00	6.0	6.1	6.2	6.2	6.3	6.3	6.3	6.3
7	4.95	5.22	5.37	5.45	5.53	5.61	5.69	5.73	5.8	5.8	5.9	5.9	6.0	6.0	6.0	6.0
8	4.74	5.00	5.14	5.23	5.32	5.40	5.47	5.51	5.5	5.6	5.7	5.7	5.8	5.8	5.8	5.8
9	4.60	4.86	4.99	5.08	5.17	5.25	5.32	5.36	5.4	5.5	5.5	5.6	5.7	5.7	5.7	5.7
10	4.48	4.73	4.88	4.96	5.06	5.13	5.20	5.24	5.28	5.36	5.42	5.48	5.54	5.55	5.55	5.55
11	4.39	4.63	4.77	4.86	4.94	5.01	5.06	5.12	5.15	5.24	5.28	5.34	5.38	5.39	5.39	5.39
12	4.32	4.55	4.68	4.76	4.84	4.92	4.96	5.02	5.07	5.13	5.17	5.22	5.24	5.26	5.26	5.26
13	4.26	4.48	4.62	4.69	4.74	4.84	4.88	4.94	4.98	5.04	5.08	5.13	5.14	5.15	5.15	5.15
14	4.21	4.42	4.55	4.63	4.70	4.78	4.83	4.87	4.91	4.96	5.00	5.04	5.06	5.07	5.07	5.07
15	4.17	4.37	4.50	4.58	4.64	4.72	4.77	4.81	4.84	4.90	4.94	4.97	4.99	5.00	5.00	5.00
16	4.13	4.34	4.45	4.54	4.60	4.67	4.72	4.76	4.79	4.84	4.88	4.91	4.93	4.94	4.94	4.94
17	4.10	4.30	4.41	4.50	4.56	4.63	4.68	4.72	4.75	4.80	4.83	4.86	4.88	4.89	4.89	4.89
18	4.07	4.27	4.38	4.46	4.53	4.59	4.64	4.68	4.71	4.76	4.79	4.82	4.84	4.85	4.85	4.85
19	4.05	4.24	4.35	4.43	4.50	4.56	4.61	4.64	4.67	4.72	4.76	4.79	4.81	4.82	4.82	4.82
20	4.02	4.22	4.33	4.40	4.47	4.53	4.58	4.61	4.65	4.69	4.73	4.76	4.78	4.79	4.79	4.79
22	3.99	4.17	4.28	4.36	4.42	4.48	4.53	4.57	4.60	4.65	4.68	4.71	4.74	4.75	4.75	4.75
24	3.96	4.14	4.24	4.33	4.39	4.44	4.49	4.53	4.57	4.62	4.64	4.67	4.70	4.72	4.74	4.74
26	3.93	4.11	4.21	4.30	4.36	4.41	4.46	4.50	4.53	4.58	4.62	4 65	4.67	4.69	4.73	4.73
28	3.91	4.08	4.18	4.28	4.34	4.39	4.43	4.47	4.51	4.56	4.60	4.62	4.65	4.67	4.72	4.72
30	3.89	4.06	4.16	4.22	4.32	4.36	4.41	4.45	4.48	4.54	4.58	4.61	4.63	4.65	4.71	4.71
40	3.82	3.99	4.10	4.17	4.24	4.30	4.34	4.37	4.41	4.46	4.51	4.54	4.57	4.59	4.69	4.69
60	3.76	3.92	4.03	4.12	4.17	4.23	4.27	4.31	4.34	4.39	4.44	4.47	4.50	4.53	4.66	4.66
100	3.71	3.86	3.98	4.06	4.11	4.17	4.21	4.25	4.29	4.35	4.38	4.42	4.45	4.48	4.64	4.65
∞	3.64	3.80	3.90	3.98	4.04	4.09	4.14	4.17	4.20	4.26	4.31	4.34	4.38	4.41	4.60	4.63

NUMBER OF GROUPS

Table A-6 Squares and Square Roots

N	N^2	$\sqrt{N}$	$\sqrt{10N}$
1.00	1.0000	1.00000	3.16228
1.01	1.0201	1.00499	3.17805
1.02	1.0404	1.00995	3.19374
1.03	1.0609	1.01489	3.20936
1.04	1.0816	1.01980	3.22490
1.05	1.1025	1.02470	3.24037
1.06	1.1236	1.02956	3.25576
1.07	1.1449	1.03441	3.27109
1.08	1.1664	1.03923	3.28634
1.09	1.1881	1.04403	3.30151
1.10	1.2100	1.04881	3.31662
1.11	1.2321	1.05357	3.33167
1.12	1.2544	1.05830	3.34664
1.13	1.2769	1.06301	3.36155
1.14	1.2996	1.06771	3.37639
1.15	1.3225	1.07238	3.39116
1.16	1.3456	1.07703	3.40588
1.17	1.3689	1.08167	3.42053
1.18	1.3924	1.08628	3.43511
1.19	1.4161	1.09087	3.44964
1.20	1.4400	1.09545	3.46410
1.21	1.4641	1.10000	3.47851
1.22	1.4884	1.10454	3.49285
1.23	1.5129	1.10905	3.50714
1.24	1.5376	1.11355	3.52136
1.25	1.5625	1.11803	3.53553
1.26	1.5876	1.12250	3.54965
1.27	1.6129	1.12694	3.56371
1.28	1.6384	1.13137	3.57771
1.29	1.6641	1.13578	3.59166
1.30	1.6900	1.14018	3.60555
1.31	1.7161	1.14455	3.61939
1.32	1.7424	1.14891	3.63318
1.33	1.7689	1.15326	3.64692
1.34	1.7956	1.15758	3.66060
1.35	1.8225	1.16190	3.67423
1.36	1.8496	1.16619	3.68782
1.37	1.8769	1.17047	3.70135
1.38	1.9044	1.17473	3.71484
1.39	1.9321	1.17898	3.72827
1.40	1.9600	1.18322	3.74166
1.41	1.9881	1.18743	3.75500
1.42	2.0164	1.19164	3.76829
1.43	2.0449	1.19583	3.78153
1.44	2.0736	1.20000	3.79473
1.45	2.1025	1.20416	3.80789
1.46	2.1316	1.20830	3.82099
1.47	2.1609	1.21244	3.83406
1.48	2.1904	1.21655	3.84708
1.49	2.2201	1.22066	3.86005
1.50	2.2500	1.22474	3.87298
N	N^2	$\sqrt{N}$	$\sqrt{10N}$

N	N^2	$\sqrt{N}$	$\sqrt{10N}$
1.50	2.2500	1.22474	3.87298
1.51	2.2801	1.22882	3.88587
1.52	2.3104	1.23288	3.89872
1.53	2.3409	1.23693	3.91152
1.54	2.3716	1.24097	3.92428
1.55	2.4025	1.24499	3.93700
1.56	2.4336	1.24900	3.94968
1.57	2.4649	1.25300	3.96232
1.58	2.4964	1.25698	3.97492
1.59	2.5281	1.26095	3.98748
1.60	2.5600	1.26491	4.00000
1.61	2.5921	1.26886	4.01248
1.62	2.6244	1.27279	4.02492
1.63	2.6569	1.27671	4.03733
1.64	2.6896	1.28062	4.04969
1.65	2.7225	1.28452	4.06202
1.66	2.7556	1.28841	4.07431
1.67	2.7889	1.29228	4.08656
1.68	2.8224	1.29615	4.09878
1.69	2.8561	1.30000	4.11096
1.70	2.8900	1.30384	4.12311
1.71	2.9241	1.30767	4.13521
1.72	2.9584	1.31149	4.14729
1.73	2.9929	1.31529	4.15933
1.74	3.0276	1.31909	4.17133
1.75	3.0625	1.32288	4.18330
1.76	3.0976	1.32665	4.19524
1.77	3.1329	1.33041	4.20714
1.78	3.1684	1.33417	4.21900
1.79	3.2041	1.33791	4.23084
1.80	3.2400	1.34164	4.24264
1.81	3.2761	1.34536	4.25441
1.82	3.3124	1.34907	4.26615
1.83	3.3489	1.35277	4.27785
1.84	3.3856	1.35647	4.28952
1.85	3.4225	1.36015	4.30116
1.86	3.4596	1.36382	4.31277
1.87	3.4969	1.36748	4.32435
1.88	3.5344	1.37113	4.33590
1.89	3.5721	1.37477	4.34741
1.90	3.6100	1.37840	4.35890
1.91	3.6481	1.38203	4.37035
1.92	3.6864	1.38564	4.38178
1.93	3.7249	1.38924	4.39318
1.94	3.7636	1.39284	4.40454
1.95	3.8025	1.39642	4.41588
1.96	3.8416	1.40000	4.42719
1.97	3.8809	1.40357	4.43847
1.98	3.9204	1.40712	4.44972
1.99	3.9601	1.41067	4.46094
2.00	4.0000	1.41421	4.47214
N	N^2	$\sqrt{N}$	$\sqrt{10N}$

Table A-6 (Cont'd)

N	N^2	$\sqrt{N}$	$\sqrt{10N}$	N	N^2	$\sqrt{N}$	$\sqrt{10N}$
2.00	4.0000	1.41421	4.47214	**2.50**	6.2500	1.58114	5.00000
2.01	4.0401	1.41774	4.48330	2.51	6.3001	1.58430	5.00999
2.02	4.0804	1.42127	4.49444	2.52	6.3504	1.58745	5.01996
2.03	4.1209	1.42478	4.50555	2.53	6.4009	1.59060	5.02991
2.04	4.1616	1.42829	4.51664	2.54	6.4516	1.59374	5.03984
2.05	4.2025	1.43178	4.52769	2.55	6.5025	1.59687	5.04975
2.06	4.2436	1.43527	4.53872	2.56	6.5536	1.60000	5.05964
2.07	4.2849	1.43875	4.54973	2.57	6.6049	1.60312	5.06952
2.08	4.3264	1.44222	4.56070	2.58	6.6564	1.60624	5.07937
2.09	4.3681	1.44568	4.57165	2.59	6.7081	1.60935	5.08920
2.10	4.4100	1.44914	4.58258	**2.60**	6.7600	1.61245	5.09902
2.11	4.4521	1.45258	4.59347	2.61	6.8121	1.61555	5.10882
2.12	4.4944	1.45602	4.60435	2.62	6.8644	1.61864	5.11859
2.13	4.5369	1.45945	4.61519	2.63	6.9169	1.62173	5.12835
2.14	4.5796	1.46287	4.62601	2.64	6.9696	1.62481	5.13809
2.15	4.6225	1.46629	4.63681	2.65	7.0225	1.62788	5.14782
2.16	4.6656	1.46969	4.64758	2.66	7.0756	1.63095	5.15752
2.17	4.7089	1.47309	4.65833	2.67	7.1289	1.63401	5.16720
2.18	4.7524	1.47648	4.66905	2.68	7.1824	1.63707	5.17687
2.19	4.7961	1.47986	4.67974	2.69	7.2361	1.64012	5.18652
2.20	4.8400	1.48324	4.69042	**2.70**	7.2900	1.64317	5.19615
2.21	4.8841	1.48661	4.70106	2.71	7.3441	1.64621	5.20577
2.22	4.9284	1.48997	4.71169	2.72	7.3984	1.64924	5.21536
2.23	4.9729	1.49332	4.72229	2.73	7.4529	1.65227	5.22494
2.24	5.0176	1.49666	4.73286	2.74	7.5076	1.65529	5.23450
2.25	5.0625	1.50000	4.74342	2.75	7.5625	1.65831	5.24404
2.26	5.1076	1.50333	4.75395	2.76	7.6176	1.66132	5.25357
2.27	5.1529	1.50665	4.76445	2.77	7.6729	1.66433	5.26308
2.28	5.1984	1.50997	4.77493	2.78	7.7284	1.66733	5.27257
2.29	5.2441	1.51327	4.78539	2.79	7.7841	1.67033	5.28205
2.30	5.2900	1.51658	4.79583	**2.80**	7.8400	1.67332	5.29150
2.31	5.3361	1.51987	4.80625	2.81	7.8961	1.67631	5.30094
2.32	5.3824	1.52315	4.81664	2.82	7.9524	1.67929	5.31037
2.33	5.4289	1.52643	4.82701	2.83	8.0089	1.68226	5.31977
2.34	5.4756	1.52971	4.83735	2.84	8.0656	1.68523	5.32917
2.35	5.5225	1.53297	4.84768	2.85	8.1225	1.68819	5.33854
2.36	5.5696	1.53623	4.85798	2.86	8.1796	1.69115	5.34790
2.37	5.6169	1.53948	4.86826	2.87	8.2369	1.69411	5.35724
2.38	5.6644	1.54272	4.87852	2.88	8.2944	1.69706	5.36656
2.39	5.7121	1.54596	4.88876	2.89	8.3521	1.70000	5.37587
2.40	5.7600	1.54919	4.89898	**2.90**	8.4100	1.70294	5.38516
2.41	5.8081	1.55242	4.90918	2.91	8.4681	1.70587	5.39444
2.42	5.8564	1.55563	4.91935	2.92	8.5264	1.70880	5.40370
2.43	5.9049	1.55885	4.92950	2.93	8.5849	1.71172	5.41295
2.44	5.9536	1.56205	4.93964	2.94	8.6436	1.71464	5.42218
2.45	6.0025	1.56525	4.94975	2.95	8.7025	1.71756	5.43139
2.46	6.0516	1.56844	4.95984	2.96	8.7616	1.72047	5.44059
2.47	6.1009	1.57162	4.96991	2.97	8.8209	1.72337	5.44977
2.48	6.1504	1.57480	4.97996	2.98	8.8804	1.72627	5.45894
2.49	6.2001	1.57797	4.98999	2.99	8.9401	1.72916	5.46809
2.50	6.2500	1.58114	5.00000	**3.00**	9.0000	1.73205	5.47723
N	N^2	$\sqrt{N}$	$\sqrt{10N}$	N	N^2	$\sqrt{N}$	$\sqrt{10N}$

Table A-6 (Cont'd)

N	N²	√N	√10N	N	N²	√N	√10N
3.00	9.0000	1.73205	5.47723	**3.50**	12.2500	1.87083	5.91608
3.01	9.0601	1.73494	5.48635	3.51	12.3201	1.87350	5.92453
3.02	9.1204	1.73781	5.49545	3.52	12.3904	1.87617	5.93296
3.03	9.1809	1.74069	5.50454	3.53	12.4609	1.87883	5.94138
3.04	9.2416	1.74356	5.51362	3.54	12.5316	1.88149	5.94979
3.05	9.3025	1.74642	5.52268	3.55	12.6025	1.88414	5.95819
3.06	9.3636	1.74929	5.53173	3.56	12.6736	1.88680	5.96657
3.07	9.4249	1.75214	5.54076	3.57	12.7449	1.88944	5.97495
3.08	9.4864	1.75499	5.54977	3.58	12.8164	1.89209	5.98331
3.09	9.5481	1.75784	5.55878	3.59	12.8881	1.89473	5.99166
3.10	9.6100	1.76068	5.56776	**3.60**	12.9600	1.89737	6.00000
3.11	9.6721	1.76352	5.57674	3.61	13.0321	1.90000	6.00833
3.12	9.7344	1.76635	5.58570	3.62	13.1044	1.90263	6.01664
3.13	9.7969	1.76918	5.59464	3.63	13.1769	1.90526	6.02495
3.14	9.8596	1.77200	5.60357	3.64	13.2496	1.90788	6.03324
3.15	9.9225	1.77482	5.61249	3.65	13.3225	1.91050	6.04152
3.16	9.9856	1.77764	5.62139	3.66	13.3956	1.91311	6.04979
3.17	10.0489	1.78045	5.63028	3.67	13.4689	1.91572	6.05805
3.18	10.1124	1.78326	5.63915	3.68	13.5424	1.91833	6.06630
3.19	10.1761	1.78606	5.64801	3.69	13.6161	1.92094	6.07454
3.20	10.2400	1.78885	5.65685	**3.70**	13.6900	1.92354	6.08276
3.21	10.3041	1.79165	5.66569	3.71	13.7641	1.92614	6.09098
3.22	10.3684	1.79444	5.67450	3.72	13.8384	1.92873	6.09918
3.23	10.4329	1.79722	5.68331	3.73	13.9129	1.93132	6.10737
3.24	10.4976	1.80000	5.69210	3.74	13.9876	1.93391	6.11555
3.25	10.5625	1.80278	5.70088	3.75	14.0625	1.93649	6.12372
3.26	10.6276	1.80555	5.70964	3.76	14.1376	1.93907	6.13188
3.27	10.6929	1.80831	5.71839	3.77	14.2129	1.94165	6.14003
3.28	10.7584	1.81108	5.72713	3.78	14.2884	1.94422	6.14817
3.29	10.8241	1.81384	5.73585	3.79	14.3641	1.94679	6.15630
3.30	10.8900	1.81659	5.74456	**3.80**	14.4400	1.94936	6.16441
3.31	10.9561	1.81934	5.75326	3.81	14.5161	1.95192	6.17252
3.32	11.0224	1.82209	5.76194	3.82	14.5924	1.95448	6.18061
3.33	11.0889	1.82483	5.77062	3.83	14.6689	1.95704	6.18870
3.34	11.1556	1.82757	5.77927	3.84	14.7456	1.95959	6.19677
3.35	11.2225	1.83030	5.78792	3.85	14.8225	1.96214	6.20484
3.36	11.2896	1.83303	5.79655	3.86	14.8996	1.96469	6.21289
3.37	11.3569	1.83576	5.80517	3.87	14.9769	1.96723	6.22093
3.38	11.4244	1.83848	5.81378	3.88	15.0544	1.96977	6.22896
3.39	11.4921	1.84120	5.82237	3.89	15.1321	1.97231	6.23699
3.40	11.5600	1.84391	5.83095	**3.90**	15.2100	1.97484	6.24500
3.41	11.6281	1.84662	5.83952	3.91	15.2881	1.97737	6.25300
3.42	11.6964	1.84932	5.84808	3.92	15.3664	1.97990	6.26099
3.43	11.7649	1.85203	5.85662	3.93	15.4449	1.98242	6.26897
3.44	11.8336	1.85472	5.86515	3.94	15.5236	1.98494	6.27694
3.45	11.9025	1.85742	5.87367	3.95	15.6025	1.98746	6.28490
3.46	11.9716	1.86011	5.88218	3.96	15.6816	1.98997	6.29285
3.47	12.0409	1.86279	5.89067	3.97	15.7609	1.99249	6.30079
3.48	12.1104	1.86548	5.89915	3.98	15.8404	1.99499	6.30872
3.49	12.1801	1.86815	5.90762	3.99	15.9201	1.99750	6.31664
3.50	12.2500	1.87083	5.91608	**4.00**	16.0000	2.00000	6.32456
N	**N²**	**√N**	**√10N**	**N**	**N²**	**√N**	**√10N**

Table A-6 (Cont'd)

N	N²	√N	√10N	N	N²	√N	√10N
4.00	16.0000	2.00000	6.32456	**4.50**	20.2500	2.12132	6.70820
4.01	16.0801	2.00250	6.33246	4.51	20.3401	2.12368	6.71565
4.02	16.1604	2.00499	6.34035	4.52	20.4304	2.12603	6.72309
4.03	16.2409	2.00749	6.34823	4.53	20.5209	2.12838	6.73053
4.04	16.3216	2.00998	6.35610	4.54	20.6116	2.13073	6.73795
4.05	16.4025	2.01246	6.36396	4.55	20.7025	2.13307	6.74537
4.06	16.4836	2.01494	6.37181	4.56	20.7936	2.13542	6.75278
4.07	16.5649	2.01742	6.37966	4.57	20.8849	2.13776	6.76018
4.08	16.6464	2.01990	6.38749	4.58	20.9764	2.14009	6.76757
4.09	16.7281	2.02237	6.39531	4.59	21.0681	2.14243	6.77495
4.10	16.8100	2.02485	6.40312	**4.60**	21.1600	2.14476	6.78233
4.11	16.8921	2.02731	6.41093	4.61	21.2521	2.14709	6.78970
4.12	16.9744	2.02978	6.41872	4.62	21.3444	2.14942	6.79706
4.13	17.0569	2.03224	6.42651	4.63	21.4369	2.15174	6.80441
4.14	17.1396	2.03470	6.43428	4.64	21.5296	2.15407	6.81175
4.15	17.2225	2.03715	6.44205	4.65	21.6225	2.15639	6.81909
4.16	17.3056	2.03961	6.44981	4.66	21.7156	2.15870	6.82642
4.17	17.3889	2.04206	6.45755	4.67	21.8089	2.16102	6.83374
4.18	17.4724	2.04450	6.46529	4.68	21.9024	2.16333	5.84105
4.19	17.5561	2.04695	6.47302	4.69	21.9961	2.16564	6.84836
4.20	17.6400	2.04939	6.48074	**4.70**	22.0900	2.16795	6.85565
4.21	17.7241	2.05183	6.48845	4.71	22.1841	2.17025	6.86294
4.22	17.8084	2.05426	6.49615	4.72	22.2784	2.17256	6.87023
4.23	17.8929	2.05670	6.50384	4.73	22.3729	2.17486	6.87750
4.24	17.9776	2.05913	6.51153	4.74	22.4676	2.17715	6.88477
4.25	18.0625	2.06155	6.51920	4.75	22.5625	2.17945	6.89202
4.26	18.1476	2.06398	6.52687	4.76	22.6576	2.18174	6.89928
4.27	18.2329	2.06640	6.53452	4.77	22.7529	2.18403	6.90652
4.28	18.3184	2.06882	6.54217	4.78	22.8484	2.18632	6.91375
4.29	18.4041	2.07123	6.54981	4.79	22.9441	2.18861	6.92098
4.30	18.4900	2.07364	6.55744	**4.80**	23.0400	2.19089	6.92820
4.31	18.5761	2.07605	6.56506	4.81	23.1361	2.19317	6.93542
4.32	18.6624	2.07846	6.57267	4.82	23.2324	2.19545	6.94262
4.33	18.7489	2.08087	6.58027	4.83	23.3289	2.19773	6.94982
4.34	18.8356	2.08327	6.58787	4.84	23.4256	2.20000	6.95701
4.35	18.9225	2.08567	6.59545	4.85	23.5225	2.20227	6.96419
4.36	19.0096	2.08806	6.60303	4.86	23.6196	2.20454	6.97137
4.37	19.0969	2.09045	6.61060	4.87	23.7169	2.20681	6.97854
4.38	19.1844	2.09284	6.61816	4.88	23.8144	2.20907	6.98570
4.39	19.2721	2.09523	6.62571	4.89	23.9121	2.21133	6.99285
4.40	19.3600	2.09762	6.63325	**4.90**	24.0100	2.21359	7.00000
4.41	19.4481	2.10000	6.64078	4.91	24.1081	2.21585	7.00714
4.42	19.5364	2.10238	6.64831	4.92	24.2064	2.21811	7.01427
4.43	19.6249	2.10476	6.65582	4.93	24.3049	2.22036	7.02140
4.44	19.7136	2.10713	6.66333	4.94	24.4036	2.22261	7.02851
4.45	19.8025	2.10950	6.67083	4.95	24.5025	2.22486	7.03562
4.46	19.8916	2.11187	6.67832	4.96	24.6016	2.22711	7.04273
4.47	19.9809	2.11424	6.68581	4.97	24.7009	2.22935	7.04982
4.48	20.0704	2.11660	6.69328	4.98	24.8004	2.23159	7.05691
4.49	20.1601	2.11896	6.70075	4.99	24.9001	2.23383	7.06399
4.50	20.2500	2.12132	6.70820	**5.00**	25.0000	2.23607	7.07107
N	N²	√N	√10N	N	N²	√N	√10N

Table A-6 (Cont'd)

N	N²	√N	√10N	N	N²	√N	√10N
5.00	25.0000	2.23607	7.07107	**5.50**	30.2500	2.34521	7.41620
5.01	25.1001	2.23830	7.07814	5.51	30.3601	2.34734	7.42294
5.02	25.2004	2.24054	7.08520	5.52	30.4704	2.34947	7.42967
5.03	25.3009	2.24277	7.09225	5.53	30.5809	2.35160	7.43640
5.04	25.4016	2.24499	7.09930	5.54	30.6916	2.35372	7.44312
5.05	25.5025	2.24722	7.10634	5.55	30.8025	2.35584	7.44983
5.06	25.6036	2.24944	7.11337	5.56	30.9136	2.35797	7.45654
5.07	25.7049	2.25167	7.12039	5.57	31.0249	2.36008	7.46324
5.08	25.8064	2.25389	7.12741	5.58	31.1364	2.36220	7.46994
5.09	25.9081	2.25610	7.13442	5.59	31.2481	2.36432	7.47663
5.10	26.0100	2.25832	7.14143	**5.60**	31.3600	2.36643	7.48331
5.11	26.1121	2.26053	7.14843	5.61	31.4721	2.36854	7.48999
5.12	26.2144	2.26274	7.15542	5.62	31.5844	2.37065	7.49667
5.13	26.3169	2.26495	7.16240	5.63	31.6969	2.37276	7.50333
5.14	26.4196	2.26716	7.16938	5.64	31.8096	2.37487	7.50999
5.15	26.5225	2.26936	7.17635	5.65	31.9225	2.37697	7.51665
5.16	26.6256	2.27156	7.18331	5.66	32.0356	2.37908	7.52330
5.17	26.7289	2.27376	7.19027	5.67	32.1489	2.38118	7.52994
5.18	26.8324	2.27596	7.19722	5.68	32.2624	2.38328	7.53658
5.19	26.9361	2.27816	7.20417	5.69	32.3761	2.38537	7.54321
5.20	27.0400	2.28035	7.21110	**5.70**	32.4900	2.38747	7.54983
5.21	27.1441	2.28254	7.21803	5.71	32.6041	2.38956	7.55645
5.22	27.2484	2.28473	7.22496	5.72	32.7184	2.39165	7.56307
5.23	27.3529	2.28692	7.23187	5.73	32.8329	2.39374	7.56968
5.24	27.4576	2.28910	7.23878	5.74	32.9476	2.39583	7.57628
5.25	27.5625	2.29129	7.24569	5.75	33.0625	2.39792	7.58288
5.26	27.6676	2.29347	7.25259	5.76	33.1776	2.40000	7.58947
5.27	27.7729	2.29565	7.25948	5.77	33.2929	2.40208	7.59605
5.28	27.8784	2.29783	7.26636	5.78	33.4084	2.40416	7.60263
5.29	27.9841	2.30000	7.27324	5.79	33.5241	2.40624	7.60920
5.30	28.0900	2.30217	7.28011	**5.80**	33.6400	2.40832	7.61577
5.31	28.1961	2.30434	7.28697	5.81	33.7561	2.41039	7.62234
5.32	28.3024	2.30651	7.29383	5.82	33.8724	2.41247	7.62889
5.33	28.4089	2.30868	7.30068	5.83	33.9889	2.41454	7.63544
5.34	28.5156	2.31084	7.30753	5.84	34.1056	2.41661	7.64199
5.35	28.6225	2.31301	7.31437	5.85	34.2225	2.41868	7.64853
5.36	28.7296	2.31517	7.32120	5.86	34.3396	2.42074	7.65506
5.37	28.8369	2.31733	7.32803	5.87	34.4569	2.42281	7.66159
5.38	28.9444	2.31948	7.33485	5.88	34.5744	2.42487	7.66812
5.39	29.0521	2.32164	7.34166	5.89	34.6921	2.42693	7.67463
5.40	29.1600	2.32379	7.34847	**5.90**	34.8100	2.42899	7.68115
5.41	29.2681	2.32594	7.35527	5.91	34.9281	2.43105	7.68765
5.42	29.3764	2.32809	7.36206	5.92	35.0464	2.43311	7.69415
5.43	29.4849	2.33024	7.36885	5.93	35.1649	2.43516	7.70065
5.44	29.5936	2.33238	7.37564	5.94	35.2836	2.43721	7.70714
5.45	29.7025	2.33452	7.38241	5.95	35.4025	2.43926	7.71362
5.46	29.8116	2.33666	7.38918	5.96	35.5216	2.44131	7.72010
5.47	29.9209	2.33880	7.39594	5.97	35.6409	2.44336	7.72658
5.48	30.0304	2.34094	7.40270	5.98	35.7604	2.44540	7.73305
5.49	30.1401	2.34307	7.40945	5.99	35.8801	2.44745	7.73951
5.50	30.2500	2.34521	7.41620	**6.00**	36.0000	2.44949	7.74597
N	**N²**	**√N**	**√10N**	**N**	**N²**	**√N**	**√10N**

Table A-6 (Cont'd)

N	N²	√N	√10N		N	N²	√N	√10N
6.00	36.0000	2.44949	7.74597		**6.50**	42.2500	2.54951	8.06226
6.01	36.1201	2.45153	7.75242		6.51	42.3801	2.55147	8.06846
6.02	36.2404	2.45357	7.75887		6.52	42.5104	2.55343	8.07465
6.03	36.3609	2.45561	7.76531		6.53	42.6409	2.55539	8.08084
6.04	36.4816	2.45764	7.77174		6.54	42.7716	2.55734	8.08703
6.05	36.6025	2.45967	7.77817		6.55	42.9025	2.55930	8.09321
6.06	36.7236	2.46171	7.78460		6.56	43.0336	2.56125	8.09938
6.07	36.8449	2.46374	7.79102		6.57	43.1649	2.56320	8.10555
6.08	36.9664	2.46577	7.79744		6.58	43.2964	2.56515	8.11172
6.09	37.0881	2.46779	7.80385		6.59	43.4281	2.56710	8.11788
6.10	37.2100	2.46982	7.81025		**6.60**	43.5600	2.56905	8.12404
6.11	37.3321	2.47184	7.81665		6.61	43.6921	2.57099	8.13019
6.12	37.4544	2.47386	7.82304		6.62	43.8244	2.57294	8.13634
6.13	37.5769	2.47588	7.82943		6.63	43.9569	2.57488	8.14248
6.14	37.6996	2.47790	7.83582		6.64	44.0896	2.57682	8.14862
6.15	37.8225	2.47992	7.84219		6.65	44.2225	2.57876	8.15475
6.16	37.9456	2.48193	7.84857		6.66	44.3556	2.58070	8.16088
6.17	38.0689	2.48395	7.85493		6.67	44.4889	2.58263	8.16701
6.18	38.1924	2.48596	7.86130		6.68	44.6224	2.58457	8.17313
6.19	38.3161	2.48797	7.86766		6.69	44.7561	2.58650	8.17924
6.20	38.4400	2.48998	7.87401		**6.70**	44.8900	2.58844	8.18535
6.21	38.5641	2.49199	7.88036		6.71	45.0241	2.59037	8.19146
6.22	38.6884	2.49399	7.88670		6.72	45.1584	2.59230	8.19756
6.23	38.8129	2.49600	7.89303		6.73	45.2929	2.59422	8.20366
6.24	38.9376	2.49800	7.89937		6.74	45.4276	2.59615	8.20975
6.25	39.0625	2.50000	7.90569		6.75	45.5625	2.59808	8.21584
6.26	39.1876	2.50200	7.91202		6.76	45.6976	2.60000	8.22192
6.27	39.3129	2.50400	7.91833		6.77	45.8329	2.60192	8.22800
6.28	39.4384	2.50599	7.92465		6.78	45.9684	2.60384	8.23408
6.29	39.5641	2.50799	7.93095		6.79	46.1041	2.60576	8.24015
6.30	39.6900	2.50998	7.93725		**6.80**	46.2400	2.60768	8.24621
6.31	39.8161	2.51197	7.94355		6.81	46.3761	2.60960	8.25227
6.32	39.9424	2.51396	7.94984		6.82	46.5124	2.61151	8.25833
6.33	40.0689	2.51595	7.95613		6.83	46.6489	2.61343	8.26438
6.34	40.1956	2.51794	7.96241		6.84	46.7856	2.61534	8.27043
6.35	40.3225	2.51992	7.96869		6.85	46.9225	2.61725	8.27647
6.36	40.4496	2.52190	7.97496		6.86	47.0596	2.61916	8.28251
6.37	40.5769	2.52389	7.98123		6.87	47.1969	2.62107	8.28855
6.38	40.7044	2.52587	7.98749		6.88	47.3344	2.62298	8.29458
6.39	40.8321	2.52784	7.99375		6.89	47.4721	2.62488	8.30060
6.40	40.9600	2.52982	8.00000		**6.90**	47.6100	2.62679	8.30662
6.41	41.0881	2.53180	8.00625		6.91	47.7481	2.62869	8.31264
6.42	41.2164	2.53377	8.01249		6.92	47.8864	2.63059	8.31865
6.43	41.3449	2.53574	8.01873		6.93	48.0249	2.63249	8.32466
6.44	41.4736	2.53772	8.02496		6.94	48.1636	2.63439	8.33067
6.45	41.6025	2.53969	8.03119		6.95	48.3025	2.63629	8.33667
6.46	41.7316	2.54165	8.03741		6.96	48.4416	2.63818	8.34266
6.47	41.8609	2.54362	8.04363		6.97	48.5809	2.64008	8.34865
6.48	41.9904	2.54558	8.04984		6.98	48.7204	2.64197	8.35464
6.49	42.1201	2.54755	8.05605		6.99	48.8601	2.64386	8.36062
6.50	42.2500	2.54951	8.06226		**7.00**	49.0000	2.64575	8.36660
N	**N²**	**√N**	**√10N**		**N**	**N²**	**√N**	**√10N**

Table A-6 (Cont'd)

N	N^2	$\sqrt{N}$	$\sqrt{10N}$	N	N^2	$\sqrt{N}$	$\sqrt{10N}$
7.00	49.0000	2.64575	8.36660	**7.50**	56.2500	2.73861	8.66025
7.01	49.1401	2.64764	8.37257	7.51	56.4001	2.74044	8.66603
7.02	49.2804	2.64953	8.37854	7.52	56.5504	2.74226	8.67179
7.03	49.4209	2.65141	8.38451	7.53	56.7009	2.74408	8.67756
7.04	49.5616	2.65330	8.39047	7.54	56.8516	2.74591	8.68332
7.05	49.7025	2.65518	8.39643	7.55	57.0025	2.74773	8.68907
7.06	49.8436	2.65707	8.40238	7.56	57.1536	2.74955	8.69483
7.07	49.9849	2.65895	8.40833	7.57	57.3049	2.75136	8.70057
7.08	50.1264	2.66083	8.41427	7.58	57.4564	2.75318	8.70632
7.09	50.2681	2.66271	8.42021	7.59	57.6081	2.75500	8.71206
7.10	50.4100	2.66458	8.42615	**7.60**	57.7600	2.75681	8.71780
7.11	50.5521	2.66646	8.43208	7.61	57.9121	2.75862	8.72353
7.12	50.6944	2.66833	8.43801	7.62	58.0644	2.76043	8.72926
7.13	50.8369	2.67021	8.44393	7.63	58.2169	2.76225	8.73499
7.14	50.9796	2.67208	8.44985	7.64	58.3696	2.76405	8.74071
7.15	51.1225	2.67395	8.45577	7.65	58.5225	2.76586	8.74643
7.16	51.2656	2.67582	8.46168	7.66	58.6756	2.76767	8.75214
7.17	51.4089	2.67769	8.46759	7.67	58.8289	2.76948	8.75785
7.18	51.5524	2.67955	8.47349	7.68	58.9824	2.77128	8.76356
7.19	51.6961	2.68142	8.47939	7.69	59.1361	2.77308	8.76926
7.20	51.8400	2.68328	8.48528	**7.70**	59.2900	2.77489	8.77496
7.21	51.9841	2.68514	8.49117	7.71	59.4441	2.77669	8.78066
7.22	52.1284	2.68701	8.49706	7.72	59.5984	2.77849	8.78635
7.23	52.2729	2.68887	8.50294	7.73	59.7529	2.78029	8.79204
7.24	52.4176	2.69072	8.50882	7.74	59.9076	2.78209	8.79773
7.25	52.5625	2.69258	8.51469	7.75	60.0625	2.78388	8.80341
7.26	52.7076	2.69444	8.52056	7.76	60.2176	2.78568	8.80909
7.27	52.8529	2.69629	8.52643	7.77	60.3729	2.78747	8.81476
7.28	52.9984	2.69815	8.53229	7.78	60.5284	2.78927	8.82043
7.29	53.1441	2.70000	8.53815	7.79	60.6841	2.79106	8.82610
7.30	53.2900	2.70185	8.54400	**7.80**	60.8400	2.79285	8.83176
7.31	53.4361	2.70370	8.54985	7.81	60.9961	2.79464	8.83742
7.32	53.5824	2.70555	8.55570	7.82	61.1524	2.79643	8.84308
7.33	53.7289	2.70740	8.56154	7.83	61.3089	2.79821	8.84873
7.34	53.8756	2.70924	8.56738	7.84	61.4656	2.80000	8.85438
7.35	54.0225	2.71109	8.57321	7.85	61.6225	2.80179	8.86002
7.36	54.1696	2.71293	8.57904	7.86	61.7796	2.80357	8.86566
7.37	54.3169	2.71477	8.58487	7.87	61.9369	2.80535	8.87130
7.38	54.4644	2.71662	8.59069	7.88	62.0944	2.80713	8.87694
7.39	54.6121	2.71846	8.59651	7.89	62.2521	2.80891	8.88257
7.40	54.7600	2.72029	8.60233	**7.90**	62.4100	2.81069	8.88819
7.41	54.9081	2.72213	8.60814	7.91	62.5681	2.81247	8.89382
7.42	55.0564	2.72397	8.61394	7.92	62.7264	2.81425	8.89944
7.43	55.2049	2.72580	8.61974	7.93	62.8849	2.81603	8.90505
7.44	55.3536	2.72764	8.62554	7.94	63.0436	2.81780	8.91067
7.45	55.5025	2.72947	8.63134	7.95	63.2025	2.81957	8.91628
7.46	55.6516	2.73130	8.63713	7.96	63.3616	2.82135	8.92188
7.47	55.8009	2.73313	8.64292	7.97	63.5209	2.82312	8.92749
7.48	55.9504	2.73496	8.64870	7.98	63.6804	2.82489	8.93308
7.49	56.1001	2.73679	8.65448	7.99	63.8401	2.82666	8.93868
7.50	56.2500	2.73861	8.66025	**8.00**	64.0000	2.82843	8.94427
N	**N^2**	**$\sqrt{N}$**	**$\sqrt{10N}$**	**N**	**N^2**	**$\sqrt{N}$**	**$\sqrt{10N}$**

Table A-6 (Cont'd)

N	N²	√N	√10N	N	N²	√N	√10N
8.00	64.0000	2.82843	8.94427	**8.50**	72.2500	2.91548	9.21954
8.01	64.1601	2.83019	8.94986	8.51	72.4201	2.91719	9.22497
8.02	64.3204	2.83196	8.95545	8.52	72.5904	2.91890	9.23038
8.03	64.4809	2.83373	8.96103	8.53	72.7609	2.92062	9.23580
8.04	64.6416	2.83549	8.96660	8.54	72.9316	2.92233	9.24121
8.05	64.8025	2.83725	8.97218	8.55	73.1025	2.92404	9.24662
8.06	64.9636	2.83901	8.97775	8.56	73.2736	2.92575	9.25203
8.07	65.1249	2.84077	8.98332	8.57	73.4449	2.92746	9.25743
8.08	65.2864	2.84253	8.98888	8.58	73.6164	2.92916	9.26283
8.09	65.4481	2.84429	8.99444	8.59	73.7881	2.93087	9.26823
8.10	65.6100	2.84605	9.00000	**8.60**	73.9600	2.93258	9.27362
8.11	65.7721	2.84781	9.00555	8.61	74.1321	2.93428	9.27901
8.12	65.9344	2.84956	9.01110	8.62	74.3044	2.93598	9.28440
8.13	66.0969	2.85132	9.01665	8.63	74.4769	2.93769	9.28978
8.14	66.2596	2.85307	9.02219	8.64	74.6496	2.93939	9.29516
8.15	66.4225	2.85482	9.02774	8.65	74.8225	2.94109	9.30054
8.16	66.5856	2.85657	9.03327	8.66	74.9956	2.94279	9.30591
8.17	66.7489	2.85832	9.03881	8.67	75.1689	2.94449	9.31128
8.18	66.9124	2.86007	9.04434	8.68	75.3424	2.94618	9.31665
8.19	67.0761	2.86182	9.04986	8.69	75.5161	2.94788	9.32202
8.20	67.2400	2.86356	9.05539	**8.70**	75.6900	2.94958	9.32738
8.21	67.4041	2.86531	9.06091	8.71	75.8641	2.95127	9.33274
8.22	67.5684	2.86705	9.06642	8.72	76.0384	2.95296	9.33809
8.23	67.7329	2.86880	9.07193	8.73	76.2129	2.95466	9.34345
8.24	67.8976	2.87054	9.07744	8.74	76.3876	2.95635	9.34880
8.25	68.0625	2.87228	9.08295	8.75	76.5625	2.95804	9.35414
8.26	68.2276	2.87402	9.08845	8.76	76.7376	2.95973	9.35949
8.27	68.3929	2.87576	9.09395	8.77	76.9129	2.96142	9.36483
8.28	68.5584	2.87750	9.09945	8.78	77.0884	2.96311	9.37017
8.29	68.7241	2.87924	9.10494	8.79	77.2641	2.96479	9.37550
8.30	68.8900	2.88097	9.11043	**8.80**	77.4400	2.96648	9.38083
8.31	69.0561	2.88271	9.11592	8.81	77.6161	2.96816	9.38616
8.32	69.2224	2.88444	9.12140	8.82	77.7924	2.96985	9.39149
8.33	69.3889	2.88617	9.12688	8.83	77.9689	2.97153	9.39681
8.34	69.5556	2.88791	9.13236	8.84	78.1456	2.97321	9.40213
8.35	69.7225	2.88964	9.13783	8.85	78.3225	2.97489	9.40744
8.36	69.8896	2.89137	9.14330	8.86	78.4996	2.97658	9.41276
8.37	70.0569	2.89310	9.14877	8.87	78.6769	2.97825	9.41807
8.38	70.2244	2.89482	9.15423	8.88	78.8544	2.97993	9.42338
8.39	70.3921	2.89655	9.15969	8.89	79.0321	2.98161	9.42868
8.40	70.5600	2.89828	9.16515	**8.90**	79.2100	2.98329	9.43398
8.41	70.7281	2.90000	9.17061	8.91	79.3881	2.98496	9.43928
8.42	70.8964	2.90172	9.17606	8.92	79.5664	2.98664	9.44458
8.43	71.0649	2.90345	9.18150	8.93	79.7449	2.98831	9.44987
8.44	71.2336	2.90517	9.18695	8.94	79.9236	2.98998	9.45516
8.45	71.4025	2.90689	9.19239	8.95	80.1025	2.99166	9.46044
8.46	71.5716	2.90861	9.19783	8.96	80.2816	2.99333	9.46573
8.47	71.7409	2.91033	9.20326	8.97	80.4609	2.99500	9.47101
8.48	71.9104	2.91204	9.20869	8.98	80.6404	2.99666	9.47629
8.49	72.0801	2.91376	9.21412	8.99	80.8201	2.99833	9.48156
8.50	72.2500	2.91548	9.21954	**9.00**	81.0000	3.00000	9.48683
N	N²	√N	√10N	N	N²	√N	√10N

Table A-6 (Cont'd)

N	N²	√N	√10N
9.00	81.0000	3.00000	9.48683
9.01	81.1801	3.00167	9.49210
9.02	81.3604	3.00333	9.49737
9.03	81.5409	3.00500	9.50263
9.04	81.7216	3.00666	9.50789
9.05	81.9025	3.00832	9.51315
9.06	82.0836	3.00998	9.51840
9.07	82.2649	3.01164	9.52365
9.08	82.4464	3.01330	9.52890
9.09	82.6281	3.01496	9.53415
9.10	82.8100	3.01662	9.53939
9.11	82.9921	3.01828	9.54463
9.12	83.1744	3.01993	9.54987
9.13	83.3569	3.02159	9.55510
9.14	83.5396	3.02324	9.56033
9.15	83.7225	3.02490	9.56556
9.16	83.9056	3.02655	9.57079
9.17	84.0889	3.02820	9.57601
9.18	84.2724	3.02985	9.58123
9.19	84.4561	3.03150	9.58645
9.20	84.6400	3.03315	9.59166
9.21	84.8241	3.03480	9.59687
9.22	85.0084	3.03645	9.60208
9.23	85.1929	3.03809	9.60729
9.24	85.3776	3.03974	9.61249
9.25	85.5625	3.04138	9.61769
9.26	85.7476	3.04302	9.62289
9.27	85.9329	3.04467	9.62808
9.28	86.1184	3.04631	9.63328
9.29	86.3041	3.04795	9.63846
9.30	86.4900	3.04959	9.64365
9.31	86.6761	3.05123	9.64883
9.32	86.8624	3.05287	9.65401
9.33	87.0489	3.05450	9.65919
9.34	87.2356	3.05614	9.66437
9.35	87.4225	3.05778	9.66954
9.36	87.6096	3.05941	9.67471
9.37	87.7969	3.06105	9.67988
9.38	87.9844	3.06268	9.68504
9.39	88.1721	3.06431	9.69020
9.40	88.3600	3.06594	9.69536
9.41	88.5481	3.06757	9.70052
9.42	88.7364	3.06920	9.70567
9.43	88.9249	3.07083	9.71082
9.44	89.1136	3.07246	9.71597
9.45	89.3025	3.07409	9.72111
9.46	89.4916	3.07571	9.72625
9.47	89.6809	3.07734	9.73139
9.48	89.8704	3.07896	9.73653
9.49	90.0601	3.08058	9.74166
9.50	90.2500	3.08221	9.74679
N	N²	√N	√10N

N	N²	√N	√10N
9.50	90.2500	3.08221	9.74679
9.51	90.4401	3.08383	9.75192
9.52	90.6304	3.08545	9.75705
9.53	90.8209	3.08707	9.76217
9.54	91.0116	3.08869	9.76729
9.55	91.2025	3.09031	9.77241
9.56	91.3936	3.09192	9.77753
9.57	91.5849	3.09354	9.78264
9.58	91.7764	3.09516	9.78775
9.59	91.9681	3.09677	9.79285
9.60	92.1600	3.09839	9.79796
9.61	92.3521	3.10000	9.80306
9.62	92.5444	3.10161	9.80816
9.63	92.7369	3.10322	9.81326
9.64	92.9296	3.10483	9.81835
9.65	93.1225	3.10644	9.82344
9.66	93.3156	3.10805	9.82853
9.67	93.5089	3.10966	9.83362
9.68	93.7024	3.11127	9.83870
9.69	93.8961	3.11288	9.84378
9.70	94.0900	3.11448	9.84886
9.71	94.2841	3.11609	9.85393
9.72	94.4784	3.11769	9.85901
9.73	94.6729	3.11929	9.86408
9.74	94.8676	3.12090	9.86914
9.75	95.0625	3.12250	9.87421
9.76	95.2576	3.12410	9.87927
9.77	95.4529	3.12570	9.88433
9.78	95.6484	3.12730	9.88939
9.79	95.8441	3.12890	9.89444
9.80	96.0400	3.13050	9.89949
9.81	96.2361	3.13209	9.90454
9.82	96.4324	3.13369	9.90959
9.83	96.6289	3.13528	9.91464
9.84	96.8256	3.13688	9.91968
9.85	97.0225	3.13847	9.92472
9.86	97.2196	3.14006	9.92975
9.87	97.4169	3.14166	9.93479
9.88	97.6144	3.14325	9.93982
9.89	97.8121	3.14484	9.94485
9.90	98.0100	3.14643	9.94987
9.91	98.2081	3.14802	9.95490
9.92	98.4064	3.14960	9.95992
9.93	98.6049	3.15119	9.96494
9.94	98.8036	3.15278	9.96995
9.95	99.0025	3.15436	9.97497
9.96	99.2016	3.15595	9.97998
9.97	99.4009	3.15753	9.98499
9.98	99.6004	3.15911	9.98999
9.99	99.8001	3.16070	9.99500
10.00	100.000	3.16228	10.0000
N	N²	√N	√10N

Appendix B
answers to problems

In the event that your answers to the computed values of the statistical tests approximate, but do not precisely equal, those given below, you should first consider the "number of places" used in the different computations in order to understand discrepancies.

Chapter 5

1. With 26 df, a t of 2.14 is reliable beyond the 0.05 level. Hence the null hypothesis may be rejected.

2. With 30 df, a t of 2.20 is reliable beyond the 0.05 level. Since this was the criterion set for rejecting the null hypothesis, and since the direction of the means is that specified by the empirical hypothesis, it may be concluded that the empirical hypothesis was confirmed—that the independent variable influenced the dependent variable.

3. With 13 df, the computed t of 4.30 is reliable beyond the 1 percent level. Since the group that received the tranquilizer had the lesser mean psychotic tendency it may be concluded that the drug produces the advertised effect.

4. The computed t of 0.51 is not reliable. Since the experimenter could not reject the null hypothesis, the empirical hypothesis was not confirmed.

5. The suspicion is not confirmed — the computed t is 0.10.

Chapter 6

1. The confounding in this study is especially atrocious. The participants in the two groups undoubtedly differ in a large number of respects other than type of method. For instance, there may be differences in intelligence, opportunity to study, socioeconomic level, as well as differences in reading proficiency prior to learning by either method, and certainly there were different teachers. The proper approach would be to randomly assign participants from the same class in a given school to two groups, and then to randomly determine which group is taught by each method, both groups being taught by the same instructor.

2. The characteristics of the individual tanks and targets are confounded with the independent variable. It may be that one tank gun is more accurate than the other, and that one set of targets is easier to hit than the other. To control these variables one might have all participants fire from the same tank (continually checking the calibration of the gun) on the same set of targets. Or half of the participants from each group could fire from each tank onto each set of targets.

3. The conclusion reached in this study is limited to the effects of class from which the children came. Undoubtedly these classes differ in a number of respects, among which is age at which they are toilet trained. The dependent variable results may thus be due to some other differential experience of the groups such as amount of social stimulation, or amount of money spent on family needs. The obvious, but difficult, way to conduct this experiment in order to establish a causal relation would be to randomly select a group of children, randomly assign them to two groups, and then randomly determine the age at which each group is toilet trained.

4. The control group should also be operated on, except that the hypothalamus should not be impaired. It could be that some structure other than the hypothalamus is disturbed during the operation, and this other structure may be responsible for the "missing" behavior.

5. There may be other reasons for not reporting an emotionally loaded word than that it is not perceived. For instance "sex" may actually be perceived by a participant, but the individual waits until he or she is absolutely sure that that is the word, possibly saving the person from a "social blunder." In addition, the frequency with which the loaded and neutral words are used in everyday life undoubtedly differs, thus affecting the threshold for recognition of the words. A better approach would be to start with a number of words that are emotionally neutral (or with nonsense syllables), and make some of them emotionally loaded (such as associating an electric shock with them). The loaded and neutral words should be equated for frequency of use.

6. One should not accept this conclusion because there is no control for the effects of suggestion. The control group should have experienced some treatment similar to that of the experimental group, such as having the same pattern of needles placed on their body but not actually penetrating the skin. Thus controlling for the effect of suggestion one would expect no difference between the experimental (acupuncture) and control (placebo) groups with regard to improvement in shoulder discomfort, a finding actually reported by Moore and Burke (1976). Other scientific studies independently have confirmed that mere suggestion that pain will be reduced through experimental techniques is sufficient to lead patients to report decreased pain (cf. especially the works of Felix Mann and of Ronald Melzack, as reported in the *Medical Tribune*, May 16, 1973).

7. This research does not allow us to draw any sound conclusions because of its faulty methodology in a number of respects. The two groups of students taught under the two different methods were in all likelihood different before the research started. Not having randomly assigned students to the two classes but allowing them to be selected on the basis of class hours produces a serious confound. With the same instructor having taught both classes it is possible that there is an interaction between instructor's belief and the relative efficacy of the pass-fail method and performance of students in the two classes. Hence, the students being taught by the pass-fail method may have out-performed what would have been normal for them. The most serious methodological criticism is that failure to reject the null hypothesis is not equivalent to accepting the null hypothesis. Thus, failure to find that performance under the two methods did not differ significantly does not allow one to conclude that the two methods are equally effective. There are an infinite number of reasons why two conditions in a study may *not* differ significantly, only one of which is that the (population) means on the dependent variable scores of the two groups are equal. A much more likely reason for failing to reject the null hypothesis is that there is excessive experimental error in the conduct of research; typically this is due to poor control methodology. Finally, casual observation of a difference in "classroom atmosphere" hardly provides the kind of information upon which educational curricula should be based. Regardless, as an extension of this kind of conclusion, one could probably predict that a course in which there were no grades at all would result in *total* freedom from "grade-oriented tensions."

Chapter 8

1. With 7 *df* the computed *t* of 2.58 is reliable at the 5 percent level. Hence the null hypothesis may be rejected. However the participants who used the Eastern Grip had a higher mean score, from which we can conclude that the empirical hypothesis is not confirmed.

2. The computed *t* of 2.97 with 6 *df* is reliable beyond the 0.05 level. Since the experimental group had the higher mean score, the empirical hypothesis is confirmed.

3. With 19 *df*, the computed *t* of 6.93 is reliable beyond the 0.02 level. Since the group that used the training aid had the higher mean score, we may conclude that the training aid facilitated map reading proficiency.

Chapter 9

1. $R_2 = 1.04$ and $R_3 = 1.09$.

Mean Scores for Groups

English Majors	Art Majors	Chemistry Majors
2.17	5.50	9.33

All groups are reliably different from each other.

2. $R_2 = 2.42$, $R_3 = 2.54$, and $R_4 = 2.62$.

Mean Scores for Groups

III	II	I	IV
2.00	2.38	6.25	7.12

Groups II and III both have reliably lower means than do

Groups I and IV. It might be added that, for greatest proficiency, these fictitious data indicate that considerable practice or extremely little practice are most beneficial (a U-shaped curve).

3. $R_2 = 2.28$, $R_3 = 2.40$, $R_4 = 2.47$ and $R_5 = 2.53$

Mean Scores for Groups

1	2	3	4	5
3.45	3.82	4.18	6.64	8.45

The order of means increases systematically with the independent variable. The higher two means are reliably superior to the lower three means. In general, the hypothesis was confirmed. It would have been more desirable, however, to have obtained a reliable difference between each of the groups, a goal that might have been achieved had a larger number of participants per group been studied.

4. R_2 for the comparison between Groups B and C is 7.36, between Groups A and B it is 8.33, and R_3 for the comparison between Groups A and C is 8.35.

Mean Scores for Groups Taught by:

Method C	Method B	Method A
27.23	29.90	46.62

Method A is to be preferred since it led to reliably greater proficiency than did the other two methods.

Chapter 10

1.

2.   *Analysis of Variance*

Source of Variation	Sum of Squares	df	Mean Square	F
Over-all Among	(91.13)	(3)		
Between Drugs	89.28	1	89.28	39.33
Between Psychoses	1.28	1	1.28	.56
D × P	.57	1	.57	.25
Within Groups	54.58	24	2.27	
Total	145.71	27		

Since the F for "between drugs" is reliable, variation of this independent variable is effective. The mean score for the participants who received drugs is higher than that for those who did not receive drugs. Hence we may conclude that administration of the drug led to an increase in normality. The lack of reliable F's for the "between psychoses" and interaction sources of variation indicates that there is no difference in normality as a function of type of psychosis, nor that there is an interaction between the variables.

3.   Amount of drug administered

4.

Amount of drug administered

	None	2cc.	4cc.	6cc.
Paranoid				
Manic depressive				
Schizophrenic				
Normal				

Type of participant

5.

Type of brand

	Old Zincs	Counts
Without filter		
With filter		

Filter

6. *Analysis of Variance*

Source of Variation	Sum of Squares	df	Mean Square	F
Over-all Among	(275.88)	(3)		
Between Brands	0.03	1	0.03	0.02
Between Filters	0.23	1	0.23	0.12
B × F	275.62	1	275.62	144.30
Within Groups	68.90	36	1.91	
Total	344.78	39		

Since neither variation of brands nor filters resulted in reliable differences, we may conclude that variation of these variables, considered by themselves, did not affect steadiness. However the interaction was reliable. From this we may conclude that whether or not brand affects steadiness depends on whether or not a filter was used — that smoking Old Zincs with a filter leads to greater steadiness than does smoking Counts with a filter, but that smoking Counts without a filter leads to greater steadiness than smoking Old Zincs without a filter. It would appear that putting a filter on Counts decreases steadiness, but putting a filter on Old Zincs increases steadiness. In fact, Counts without a filter lead to about the same amount of steadiness as Old Zincs with a filter, as a diagram of the interaction would show. But we don't recommend that you smoke either brand.

7. *Analysis of Variance*

Source of Variation	Sum of Squares	df	Mean Square	F
Over-all Among	(80.68)	(3)	26.89	
Between Opium	30.04	1	30.04	27.81
Between Marijuana	48.89	1	48.89	45.27
O × M	1.75	1	1.75	1.62
Within Groups	26.00	24	1.08	
Total	106.68	27		

Since the F's for "between Opium" and "between Marijuana" are both reliable, we can conclude that smoking both of them lead to hallucinatory activity, and that there is no interaction between these two variables since this latter source of variation is not reliable.

References

ALBRECHT, HEATHER A. Replication of pronoun satiation by prior verbal stimulation. Unpublished manuscript, 1965.

ALBRECHT, HEATHER A., and WEBSTER, R. L. The effect of prior verbal stimulation on associates to stimulus words. Unpublished manuscript, 1966. *Publication Manual of the American Psychological Association* (2nd ed.), 1974, p. 19.

ANDERSON, R. L., and BANCROFT, T. A. *Statistical theory in research.* New York: McGraw-Hill, 1952.

ASHER, R. Why are medical journals so dull? *British Medical Journal,* 1958, *2,* 502.

BABICH, F. R., JACOBSON, A. L., BUBASH, SUZANNE, and JACOBSON, ANN. Transfer of a response to naive rats by injection of ribonucleic acid extracted from trained rats. *Science,* 1965, *149,* 656–57.

BACHRACH, A. J. *Psychological research: An introduction* (2nd ed.). New York: Random House, 1965.

BARBER, B. Resistance by scientists to scientific discovery. *Science,* 1961, *134,* 596–602.

BARBER, T. X., and SILVER, M. J. Fact, fiction, and the experimenter bias effect. *Psychological Bulletin,* Monograph Supplement, 1968, *70,* 1–29.

BATKIN, S., WOODWARD, W. T., COLE, R. E., and HALL, J. B. RNA and actinomycin-D enhancement of learning in the carp. *Psychonomic Science*, 1966, *5*, 345–46.

BEATTY, WILLIAM W. How blind is blind? A simple procedure for observer naiveté. *Psychological Bulletin*, 1972, *78*, 70–71.

BERSH, P. J. The influence of two variables upon the establishment of a secondary reinforcer for operant responses. *Journal of Experimental Psychology*, 1951, *41*, 62–73.

BINDER, A., McCONNELL, D., and SJOHOLM, N. A. Verbal conditioning as a function of experimenter characteristics. *Journal of Abnormal and Social Psychology*, 1957, *55*, 309–14.

BOE, E. E. Effect of punishment during and intensity on the extinction of an instrumental response. *Journal of Experimental Psychology*, 1966, *72*, 125–31.

BONEAU, C. A. The effects of violation of assumptions underlying the *t*-test. *Psychological Bulletin*, 1960, *57*, 49–64.

BREAUX, ROBERT. Analysis of variance of one-, two-, and three-treatment designs for a PDP-8. *Behavior Research Methods and Instrumentation*, 1972, *4*, 271–72.

BRIDGEMAN, P. W. *The logic of modern physics.* New York: Macmillan, 1927.

BROWN, C. C., and SAUCER, R. T. *Electronic instrumentation for the behavioral sciences.* Springfield, Ill.: Charles C Thomas, 1958.

BRUNSWIK, E. *Perception and the representative design of psychological experiments.* Berkeley and Los Angeles: University of California Press, 1956.

BUGELSKI, B. R. *Experimental psychology.* New York: Henry Holt, 1951.

BYRNE, W. L. et al. Memory transfer. *Science*, 1966, *153*, 658–59.

BURCH, M. E., and MAGSDICK, W. K. The retention in rats of an incompletely learned maze solution for short intervals of time. *Journals of Comparative Psychology*, 1933, *16, 385–409.*

CALVIN, A. D., McGUIGAN, F. J., TYRRELL, S., and SOYARS, M. Manifest anxiety and the Palmar Perspiration Index. *Journal of Consulting Psychology*, 1956, *20*, 356.

CALVIN, A. D., PERKINS, M. J., and HOFFMAN, F. K. The effect of nondifferential reward and non-reward on discriminative learning in children. *Child Development*, 1956, *27*, 439–46.

CALVIN, A. D., SCRIVEN, M., GALLAGHER, J. J., HANLEY, C., McCONNELL, J. V., and McGUIGAN, F. J. *Psychology.* Boston: Allyn & Bacon, 1961.

CAMPBELL, DONALD T. Reforms as experiments. *American Psychologist*, 1969, *24*, 409–29.

CAMPBELL, DONALD T. Assessing the impact of planned social change.

Social Research and Public Policies. The Dartmouth/OECD Conference, 1975.

CAMPBELL, DONALD T., and BORUCH, ROBERT F. Making the case for randomized assignment to treatments by considering the alternatives: Six ways in which quasi-experimental evaluations in compensatory education tend to underestimate effects. *Evaluation and experiment: Some critical issues in assessing social programs.* Seattle: Academic Press, Inc., 1975.

CAMPBELL, DONALD T. and STANLEY, JULIAN C. Experimental and quasi-experimental designs for research. *Handbook of research on teaching.* Chicago: Rand McNally, 1963.

CANNON, W. B. *The way of an investigator.* New York: Norton, 1945.

CANTRIL, H. *The invasion from Mars.* Princeton, N.J.: Princeton University Press, 1940.

COCHRAN, W. G., and COX, G. M. *Experimental designs.* New York: John Wiley, 1957.

COHEN, M. R., and NAGEL, E. *Logic and scientific method.* New York: Harcourt Brace Jovanovich, 1934.

CONANT, JAMES B. *On understanding science: An historical approach* New York: New American Library, 1951.

COOK, THOMAS D. and CAMPBELL, DONALD T. The design and conduct of quasi-experiments and true experiments in field settings. *Handbook of industrial and organizational psychology.* Chicago: Rand McNally College Publishing Co., 1976.

CORNSWEET, T. N. *The design of electric circuits in the behavioral sciences.* New York: John Wiley, 1963.

COWLES, M. P. $N = 35$: A rule of thumb for psychological researchers. *Perceptual and Motor Skills,* 1974, *38,* 1135–38.

CRAIG, JAMES R., and REESE, SANDRA C. Psychology in action. Retention of raw data: A problem revisited. *American Psychologist,* 1973, *28,* 723.

CRONBACH, L. J., and FURBY, L. How we should measure "change"—or should we? *Psychological Bulletin,* 1970, *74,* 68–80.

CULLITON, BARBARA J. National research act: restores training, bans fetal research. *Science,* 1974, *185,* 426–27.

D'AGOSTINO, RALPH B. Simple compact portable test of normality: Geary's test revisited. *Psychological Bulletin,* 1970, *74,* 138–40.

DAILEY, JOHN T., and PIKREL, EVAN W. Some psychological contributions to defenses against hijackers. *American Psychologist,* 1975, *30,* 161–65.

DEESE, J. *The psychology of learning.* New York: McGraw-Hill, 1952.

DICKINSON, T. L., and WOLINS, L. Analysis of repeated measures and other designs. *Multivariate Behavioral Research,* 1974 (July), 353–71.

Dixon, W. J., and Massey, F. J., Jr. *Statistical analysis.* New York: McGraw-Hill, 1951.

Doyle, A. C. *Sherlock Holmes.* Garden City, N.Y.: Garden City Press, 1938.

Dubs, H. H. *Rational induction.* Chicago: University of Chicago Press, 1930.

Duncan, D. B. Multiple range and multiple F-tests. *Biometrics,* 1955, *11*, 1–42.

Duncan, D. B. Multiple range tests for correlated and heteroscedastic means. *Biometrics,* 1957, *13*, 164–76.

Ebbinghaus, E. *Memory: A contribution to experimental psychology* (translated by H. A. Ruger and C. E. Busserius). New York: Columbia University Press, 1913.

Edgington, E. S. Statistical inference from $N = 1$ experiments. *The Journal of Psychology,* 1967, *65*, 195–99.

Edgington, E. S. An additive method for combining probability values from independent experiments. *Journal of Psychology,* 1972, *80*, 351–62. (a)

Edgington, E. S. A normal curve method for combining probability values from independent experiments. *Journal of Psychology,* 1972, *82*, 85–89. (b)

Edgington, E. S. $N = $ Experiments: hypothesis testing. *The Canadian Psychologist,* 1972, *13*, 121–34. (c)

Edwards, A. L. *Experimental design in psychological research* (3rd ed.). New York: Holt, Rinehart & Winston, 1968.

Erlebacher, A. Design and analysis of experiments contrasting the within- and between-subjects' manipulation of the independent variable. *Psychological Bulletin,* 1977, *84*, 212–19.

Farber, I. E., and Spence, K. W. Complex learning and conditioning as a function of anxiety. *Journal of Experimental Psychology,* 1953, *45*, 120–25.

Feigl, H., and Scriven, M. *Minnesota studies in the philosophy of science.* (Volume I: The foundations of science and the concepts of psychology and psychoanalysis.) Minneapolis: University of Minnesota Press, 1956.

Fisher, A. E. Chemical stimulation of the brain. *Psychobiology.* San Francisco: W. H. Freeman & Company, 1964. Copyright © 1964 by Scientific American, Inc. All rights reserved.

Fisher, R. A. *The design of experiments* (6th ed.). New York: Hafner, 1953.

Frank, P. G. (ed.). *The validation of scientific theories.* Boston: Beacon, 1956.

Gaito, J. Repeated measurements designs and counterbalancing. *Psychological Bulletin,* 1961, *58*, 46–54.

GAITO, J. Repeated measurements designs and tests of null hypotheses. *Educational and Psychological Measurement*, 1973, *33*, 69–75.

GAMES, P. A., WINKLER, H. B., and PROBERT, D. A. Robust tests for homogeneity of variance. *Educational and Psychological Measurement*, 1972, *32*, 887–909.

GAMPEL, DOROTHY H. Temporal factors in verbal satiation. *Journal of Experimental Psychology*, 1966, *72*, 201–6.

GILLIS, JOHN S. Participants instead of subjects. *American Psychologist*, 1976, *31*, 95–97.

GOTTMAN, J. M., McFALL, R. M., and BARNETT, J. T. Design and analysis of research using time series. *Psychological Bulletin*, 1969, *72*, 299–306.

GREENWALD, A. G. Within-subjects designs: To use or not to use? *Psychological Bulletin*, 1976, *83*, 314–20.

GRICE, G. R., and HUNTER, J. J. Stimulus intensity effects depend upon the type of experimental design. *Psychological Review*, 1964, *71*, 247–56.

GRINGS, W. W. *Laboratory instrumentation in psychology*. Palo Alto, Calif.: National Press, 1954.

GUILFORD, J. P. *Fundamental statistics in psychology and education* (4th ed.). New York: McGraw-Hill, 1965.

GUTHRIE, E. R. *The psychology of learning* (rev. ed.). New York: Harper & Row, 1952.

HAMMOND, K. R. Subject and object sampling: A note. *Psychological Bulletin*, 1948, *45*, 530–33.

HAMMOND, K. R. Representative vs. systematic design in clinical psychology. *Psychological Bulletin*, 1954, *51*, 150–59.

HARLOW, H. F., and ZIMMERMAN, R. R. The development of affectional responses in infant monkeys. *American Philosophical Society*, 1958, *102*, 501–9.

HARRIS, DAVID R., BISBEE, CHARLES T., and EVANS, SELBY H. Further comments—misuse of analysis of covariance. *Psychological Bulletin*, 1971, *75*, 220–22.

HARRIS, F. R., WOLF, M. M., and BAER, D. M. Effects of adult social reinforcement on child behavior. *Young Children*, 1964, *20*, 8–17.

HAYS, W. L. *Statistics for psychologists*. New York: Holt, Rinehart & Winston, 1963.

HEIDBREDER, E. *Seven psychologies*. New York: Appleton-Century-Crofts, 1933.

HEMPEL, C. G. Studies in the logic of confirmation. *Mind*, 1945, *54*, 1–26, 97–121.

HEMPEL, C. G. *Aspects of scientific explanations: And other essays in the philosophy of science*. New York: Free Press, 1965.

HEMPEL, C. G., and OPPENHEIM, P. The logic of explanation. *Philosophy of Science*, 1948, *15*, 135–75.

HERNÁNDEZ-PEON, R., SCHERRER, H., and JOUVET, M. Modification of electric activity in cochlear nucleus during "attention" in unanesthetized cats. *Science*, 1956, *123*, 331–32.

HESS, E. H. Attitude and pupil size. *Scientific American*, 1965, *212*, 46–54.

HIMMELFARB, SAMUEL, What do you do when the control group doesn't fit into the factorial design? *Psychological Bulletin*, 1975, *82*, 363–68.

HOENIG, S. A., and PAYNE, F. L. *How to build and use electronic devices without frustration, panic, mountains of money, or an engineering degree*. Boston: Little, Brown, 1973.

HORNBECK, FREDERICK W. Computers in behavioral science. Factorial analyses of variance with appended control groups. *Behavioral Science*, 1973, *18*, 213–20.

HORSNELL, G. The effect of unequal group variances on the F-test for the homogeneity of group means. *Biometrika*, 1953, *46*, 128–36.

HUCK, SCHUYLER W., and MCLEAN, ROBERT A. Using a repeated measures ANOVA to analyze the data from a pretest-posttest design: A potentially confusing task. *Psychological Bulletin*, 1975, *82*, 511–18.

HULL, C. L. *Principles of behavior*. New York: Appleton-Century-Crofts, 1943.

HULL, C. L. *A behavior system*. New Haven, Conn.: Yale University Press, 1952.

IGEL, G. J., and CALVIN, A. D. The development of affectional responses in infant dogs. *Journal of Comparative & Physiological Psychology*, 1960, *53*, 302–5.

JACKSON, HERBERT W. *Introduction to electrical circuits*, 4th ed. Englewood Cliffs, N.J.: Prentice-Hall, 1975.

JACOBSON, A. L., FRIED, C., and HOROWITZ, S. D. Planarians and memory. *Nature*, 1966, *209*, 599–601.

JAMES, W. *Principles of psychology*. Chicago: Encyclopædia Britannica, Inc., 1952, p. 809.

JENKINS, J. G., and DALLENBACH, K. M. Oblivescence during sleep and waking. *American Journal of Psychology*, 1924, *35*, 605–12.

JOHNSON, P., and BAILEY, D. E. Some determinants of the use of relationships in discrimination learning. *Journal of Experimental Psychology*, 1966, *71*, 365–72.

JOHNSON, R. W. Retain the original data! *American Psychologist*, 1964, *19*, 350–51.

JONES, F. P. Experimental method in antiquity. *American Psychologist*, 1964, *19*, 419–20.

KENNY, DAVID A. A quasi-experimental approach to assessing treatment effects in the nonequivalent control group design. *Psychological Bulletin*, 1975, *82*, 345–62.

KESELMAN, H. J., TOOTHAKER, L. E., and SHOOTER, M. An evaluation of two unequal n_k forms of the Tukey multiple comparison statistic. *Journal of the American Statistical Association*, 1975, *70*, 584–87.

KIESLER, D. J. Some myths of psychotherapy research and the search for a paradigm. *Psychological Bulletin*, 1966, *65*, 110–36.

KRAMER, C. Y. Extension of multiple range tests to group means with unequal numbers of replications. *Biometrics*, 1956, *12*, 307–10.

KRAMER, C. Y. Extension of multiple range tests to group correlated adjusted means. *Biometrics*, 1957, *13*, 13–18.

LABOUVIE, E. W., BARTSCH, T. W., NESSELROADE, J. R., and BALTES, P. B. On the internal and external validity of simple longitudinal designs. *Child Development*, 1974, *45*, 282–90.

LACHMAN, R., MEEHAN, J. T., and BRADLEY, ROSALEE. Observing response and word association in concept shifts: Two-choice and four-choice selective learning. *Journal of Psychology*, 1965, *59*, 349–57.

LEITENBERG, H. The use of single-case methodology in psychotherapy research. *Journal of Abnormal Psychology*, 1973, *82*, 87–101.

LEPLEY, W. M. The participation of implicit speech in acts of writing. *American Journal of Psychology*, 1952, *65*, 597–99.

LEVY, P. Psychological statistics: A teaching paradigm. *Bulletin of the British Psychological Society*, 1973, *26*, 12.

LEWIS, D. J. Rats and men. *American Journal of Sociology*, 1953, *59*, 131–35.

LI, J. C. R. *Introduction to statistical inference.* Ann Arbor, Mich.: Edwards Brothers, 1957.

LINDQUIST, E. F. *Design and analysis of experiments in psychology and education.* Boston: Houghton, Mifflin, 1953.

MAXWELL, SCOTT, and CRAMER, ELLIOT M. A note on analysis of covariance *Psychological Bulletin*, 1975, *82*, 187–90.

McGUIGAN, F. J. Confirmation of theories in psychology. *Psychological Review*, 1956, *63*, 98–104.

McGUIGAN, F. J. The effect of precision, delay and schedule of knowledge of results on performance. *Journal of Experimental Psychology*, 1959, *58*, 79–84.

McGUIGAN, F. J. The experimenter: A neglected stimulus object. *Psychological Bulletin*, 1963, *60*, 421–28.

McGuigan, F. J. Covert oral behavior and auditory hallucinations. *Psychophysiology*, 1966, *3*, 73–80.(a)

McGuigan, F. J. *Thinking: Studies of covert language processes.* New York: Appleton-Century-Crofts, 1966.(b)

McGuigan, F. J. The *G* statistic, an index of amount learned. *Journal of National Society for Programmed Instruction*, 1967, *6*, 14–16.

McGuigan, F. J. Covert oral behavior as a function of quality of handwriting. *The American Journal of Psychology*, 1970, *83*, 337–88.

McGuigan, F. J. Amount learned: An empirical basis for grading teachers and students. *Teaching of Psychology*, 1974, *1*, 10–15.

McGuigan, F. J. *Cognitive psychophysiology—principles of covert behavior.* New York: Plenum Publishing Corp., in press.

McGuigan, F. J., and Bailey, S. Covert response patterns during the processing of language stimuli. *Interamerican Journal of Psychology*, 1969, *3*, 289–99.

McGuigan, F. J., Calvin, A. D., and Richardson, Elizabeth C. Manifest anxiety, palmar perspiration index, and stylus maze-learning. *American Journal of Psychology*, 1959, *67*, 434–38.

McGuigan, F. J., Hutchens, Carolyn, Eason, Nancy and Reynolds, Teddy. The retroactive interference of motor activity with knowledge of results. *Journal of General Psychology*, 1964, *70*, 279–81.

McGuigan, F. J., and MacCaslin, E. F. The relationship between rifle steadiness and rifle markmanship and the effect of rifle training on rifle steadiness. *Journal of Applied Psychology*, 1955, *39*, 156–59.(a)

McGuigan, F. J., and MacCaslin, E. F. Whole and part methods in learning a perceptual motor skill. *American Journal of Psychology*, 1955, *47*, 658–61. (b)

McGuigan, F. J., and Peters, R. J., Jr. Assessing the effectiveness of programmed texts: Methodology and some findings. *Journal of Programmed Instruction*, 1965, *3*, 23–34.

McNemar, Q. *Psychological statistics* (3rd ed.). New York: John Wiley, 1962.

Meehl, Paul E. Some methodological reflections on the difficulties of psychoanalytic research. *Psychological Issues*, 1973, *8*, 104–17.

Miklich, D. R., Purcell, K., and Weiss, J. H. Methods and designs: Practical aspects of the use of radio telemetry in the behavioral sciences. *Behavior Research Methods and Instrumentation*, 1974, *6*, 461–66.

Moore, M. E., and Berk, S. N. Acupuncture for chronic shoulder pain. *Annals of Internal Medicine*, 1976, *84*, 381–84.

Morgan, C. L. *An introduction to comparative psychology* (2nd ed.). London: Walter Scott, 1906.

Mosteller, F., and Bush, R. R. Selected quantitative techniques. In

472

G. Lindzey (ed.), *Handbook of social psychology.* Cambridge, Mass.: Addison-Wesley, 1954.

NAGEL, ERNEST. *The structure of science: Problems in the logic of scientific explanation.* New York: Harcourt Brace Jovanovich, 1961.

NAMBOODIRI, N. KRISHMAN, Experimental designs in which each subject is used repeatedly. *Psychological Bulletin,* 1972, *77,* 54–64.

NEWBURY, E. Current interpretation and significance of Lloyd Morgan's canon. *Psychological Bulletin,* 1954, *51,* 70–74.

O'CONNOR, EDWARD F., JR. Extending classical test theory to the measurement of change. *Review of Educational Research,* 1972, *42,* 73–97.(a)

O'CONNOR, EDWARD F., JR. Response to Cronbach and Furby's "How we should measure 'change'—or should we?" *Psychological Bulletin,* 1972, *78,* 159–60.(b)

OVERALL, J. E., and DALAL, S. N. Design of experiments to maximize power relative to cost. *Psychological Bulletin,* 1965, *5,* 339–50.

PAGE, I. H. Serotonin. *Scientific American,* 1957, *197,* 52–56.

PAGE, STEWART, and YATES, ELIZABETH. Attitudes of psychologists toward experimenter controls in research. *The Canadian Psychologist,* 1973, *14,* 202–7.

PAVLIK, W. B., and CARLTON, P. L. A reversed partial-reinforcement effect. *Journal of Experimental Psychology,* 1965, *70,* 417–23.

PERLMUTTER, J., and MYERS, J. L. A comparison of two procedures for testing multiple contrasts. *Psychological Bulletin,* 1973, *79,* 181–84.

PETERS, C. C., and VANVOORHIS, W. R. *Statistical procedures and their mathematical bases.* New York: McGraw-Hill, 1940.

POOR, DAVID D. S. Analysis of variance for repeated measures designs: Two approaches. *Psychological Bulletin,* 1973, *80,* 204–9.

PORTER, PAUL B. A highly significant comment. *American Psychologist,* 1973, *28,* 189–90.

POULTON, E. C. Unwanted range effects from using within-subject experimental designs. *Psychological Bulletin,* 173, *80,* 113–21.

POULTON, E. C. Range effects are characteristic of a person serving in a within-subjects experimental design—a reply to Rothstein. *Psychological Bulletin,* 1974, *81,* 201–3.

POULTON, E. C., and FREEMAN, P. R. Unwanted asymmetrical transfer effects with balanced experimental designs. *Psychological Bulletin,* 1966, *66,* 1–8.

PREMACK, D. Reinforcement theory. In D. Levine (ed.), *Nebraska symposium on motivation,* 1965. Lincoln: University of Nebraska Press, 1965.

RAY, W. S. *An introduction to experimental design.* New York: Macmillan, 1960.

REICHENBACH, H. *Experience and prediction.* Chicago: University of Chicago Press, 1938.

REICHENBACK, H. *Elements of symbolic logic.* New York: Macmillan, 1947.

REICHENBACH, H. *The theory of probability* (2nd. ed.). Berkeley: University of California Press, 1949.

REVUSKY, SAMUEL H. Some statistical treatments compatible with individual organism methodology. *The Experimental Analysis of Behavior,* 1967, *10,* 319–30.

REYNOLDS, R. W., and MEEKER, M. R. Thiosemicarbazide injection followed by electric shock increases resistance to stress in rats. *Science,* 1966, *151,* 1101–2.

ROJAS, BASILIO A. On Tukey's test of additivity. *Biometrics,* 1973, *29,* 45–52.

ROSENTHAL, R. *Experimenter effects in behavioral research.* New York: Appleton-Century-Crofts, 1966.

ROTHSTEIN, L. D. Reply to Poulton. *Psychological Bulletin,* 1974, *81,* 199–201.

RYAN, T. A. Multiple comparisons in psychological research. *Psychological Bulletin,* 1959, *56,* 26–47.

SANDLER, J. A test of the significance of the difference between the means of correlated measures, based on a simplification of student's *t. British Journal of Psychology,* 1955, *46,* 225–26.

SCHOENFELD, W. N., ANTONITIS, J. J., and BERSH, P. J. A preliminary study of training conditions necessary for secondary reinforcement. *Journal of Experimental Psychology,* 1950, *40,* 40–45.

SCHEFFÉ, H. A method for judging all contrasts in the analysis of variance. *Biometrika,* 1953, *40,* 87–104.

SCHEFFÉ, H. *The analysis of variance.* New York: John Wiley, 1959.

SCRIVEN, M. Explanation and prediction in evolutionary theory. *Science,* 1959, *130,* 477–82.

SEIDEL, R. J., and ROTBERG, IRIS C. Effects of written verbalization and timing of information on problem solving in programed learning. *Journal of Educational Psychology,* 1966, *57,* 151–58.

SHINE, LESTER C. II, and BOWER, SAMUEL M. A one-way analysis of variance for single-subject designs. *Educational and Psychological Measurement,* 1971, *31,* 105–13.

SHUKLA, G. K. An invariant test for the homogeneity of variance in a two-way classification. *Biometrics,* 1972, *28,* 1063–72.

SIDMAN, M. *Tactics of scientific research.* New York: Basic Books, 1960.

SIDOWSKI, J. B. *Experimental methods and instrumentation in psychology.* New York: McGraw-Hill, 1966.

SIEGEL, S. *Nonparametric statistics for the behavioral sciences.* New York: McGraw-Hill, 1956.

SINGH, S. D. Effect of human environment on cognitive behavior in the Rhesus monkey. *Journal of Comparative & Physiological Psychology,* 1966, *61,* 280–83.

SKINNER, B. F. *Waldon two.* New York: Macmillan, 1948.

SKINNER, B. F. *Science and human behavior.* New York: Macmillan, 1953.

SKINNER, B. F. *Cumulative record.* New York: Appleton-Century-Crofts, 1959.

SKINNER, B. F. The design of cultures. *Daedalus,* 1961, pp. 534–46.

SOLOMON, R. L. An extension of control group design. *Psychological Bulletin,* 1949, *46,* 137–50.

SPENCE, K. W. The postulates and methods of "Behaviorism." *Psychological Review,* 1948, *55,* 67–78.

SPENCE, K. W., FARBER, I. E., and MCFANN, H. H. The relation of anxiety (drive) level to performance in competitional and non-competitional paired-associates learning. *Journal of Experimental Psychology,* 1956, *52,* 296–305.

SPIRES, A. M. Subject-experimenter interaction in verbal conditioning. Unpublished doctoral dissertation, New York University, 1960.

STEINER, IVAN D. The evils of research: Or what my mother didn't tell me about the sins of academia. *American Psychologist,* 1972, *27,* 766–68.

SULZBACHER, S. I. Effects of response mode and subject characteristics on learning in programed instruction. *National Society of Programed Instruction Journal,* 1967, *4,* 10–11.

TAYLOR, J. A. A personality scale of manifest anxiety. *Journal of Abnormal & Social Psychology,* 1953, *48,* 285–90.

TOCCI, RONALD J. *Fundamentals of electronic devices* (2nd ed.). Columbus, O.: Bobbs-Merrill, 1975.

TVERSKY, A., and EDWARDS, W. Information versus reward in binary choices. *Journal of Experimental Psychology,* 1966, *71,* 680–83.

UNDERWOOD, B. J. The effect of successive interpolations on retroactive and proactive inhibition. *Psychological Monographs,* 1945, *38,* 29–38.

UNDERWOOD, B. J. *Experimental psychology.* New York: Appleton-Century-Crofts, 1949.

UNDERWOOD, B. J. Interference and forgetting. *Psychological Review,* 1957, *64,* 49–60.(a)

UNDERWOOD, B. J. *Psychological research.* New York: Appleton-Century-Crofts, 1957.(b)

UNDERWOOD, B. J. *Experimental psychology.* New York: Appleton-Century-Crofts, 1966.(a)

UNDERWOOD, B. J. *Problems in experimental design and inference.* New York: Appleton, 1966.(b)

VALLE, F. P. Free and forced exploration in rats as a function of between-vs. within-Ss design. *Psychonomic Science*, 1972, *29*, 11–13.

VENABLES, P. H., and MARTIN, I. (eds.). *A manual of psychophysiological methods.* New York: John Wiley, 1967.

WEBSTER, R. L., and WEINGOLD, H. P. Suppression of pronoun choices as a function of prior verbal stimulation. Paper read at Southeastern Psychological Association meeting, 1965.

WHITE, LLOYD A. Why engineers can't write. *Research Development*, 1971, *22*, 24–26.

WHITE, M. A., and DUKER, J. Some unprinciples of psychological research. *American Psychologist*, 1971, *26*, 397–399.

WHITE, M. A., and DUKER, J. Suggested standards for children's samples. *American Psychologist*, 1973, *28*, 700–703.

WIKE, EDWARD L. Water beds and sexual satisfaction: Wike's law of low odd primes (WLLOP). *Psychological Reports*, 1973, *33*, 192–94.

WINE, R. L. *Statistics for scientists and engineers.* Englewood Cliffs, N.J.: Prentice-Hall, 1964.

WINE, R. L. *Beginning statistics.* Cambridge, Mass.: Winthrop, 1976.

WINER, B. J. *Statistical principles in experimental design.* New York: McGraw-Hill, 1971.

WOLINS, L. Responsibility for raw data. *American Psychologist*, 1962, *17*, 657–58.

WOODWORTH, R. S., and SCHLOSBERG, H. *Experimental psychology.* New York: Henry Holt, 1955.

YANOF, H. M. *Biomedical electronics.* Philadelphia: F. A. Davis Co., 1965.

ZUCKER, M. H. *Electronic circuits for the behavioral and biomedical sciences.* San Francisco: W. H. Freeman, 1969.

Glossary

ACCIDENTAL UNIVERSALITY: A relationship in which there is *no* element of necessity between the antecedent and consequent conditions (see Nomic Universality; p. 54).

ADDITIVITY (STATISTICAL ASSUMPTION): In using parametric statistics one assumes that the treatment and error effects in an experiment are additive (p. 418).

ANALYSIS OF VARIANCE: A method of analyzing the total variance present in an experiment into components. By computing ratios between those components (such as with the F test), various empirical questions can be answered such as determining whether two or more groups in an experiment reliably differ (pp. 260–272 and Chapter 9).

ANALYTIC STATEMENT: A statement that is always true (it cannot be false), but is empirically empty (p. 48).

ASYMMETRICAL TRANSFER: When, in a repeated treatments' design, there is differential transfer such that the transfer from one condition to a second is different from the second to the first, as in counterbalancing (p. 163).

BALANCING: A technique of controlling extraneous variables by assuring that they affect members of the experimental group equally with those of the control group (p. 155, p. 164).

BEHAVIOR: The organization of responses usually to accomplish some goal (p. 7 and Chapter 7).

BETWEEN-GROUPS DESIGNS: Experimental designs in which there are two or more independent groups of participants such that each group receives a different amount of the independent variable. Contrasted with within-groups designs (p. 326).

CENTRAL TENDENCY: A statistical measure of a representative value of all measures in a distribution. Includes the mean, median, and mode (p. 218).

CHANCE: A concept that indicates that an event is determined by unpredicted antecedents. In a completely deterministic world, chance would be an expression of our ignorance. To the extent to which there is lack of order (chaos) in the world, events are determined by unpredictable antecedents (see LaPlace's Superman).

CIRCULARITY, VICIOUS: Fallacious reasoning that occurs when an answer is based on a question and the question based on the answer, with no appeal to outside, independent information (p. 39).

CLINICAL METHOD: A method, also known as the "case history method" or "life history method," in which an attempt is made to help an individual solve personal problems or to collect information about an individual's history. The probability of an evidence report resulting from this method is extremely low (p. 70).

COMPARISON GROUP: See Control Group.

CONCATENATION: The characteristic of an interlocked, compatible system that adds an increment in probability to each component proposition within the system (p. 362).

CONFIRMATION: The process of subjecting a statement (hypothesis, theory, law, etc.) to empirical test. The consequences may be that the probability of the statement is decreased (disconfirmed, not supported) or increased (confirmed, supported). Distinguished from "replication" in that "replication" refers to the repetition of the methods of a scientific study (p. 15, p. 366 and Chapter 13).

CONFIRMATORY EXPERIMENTS: Those in which an explicit hypothesis is formed and subjected to test; contrasted with exploratory experiments (p. 75).

CONFOUNDING: The presence of an extraneous variable, systematically related to the independent variable, such that it might differentially affect the dependent variable values of the groups in the investigation (p. 148).

CONSTANCY OF CONDITIONS: A technique of control of extraneous variables by keeping them at the same value for all conditions throughout the experiment (p. 154).

CONTEXT OF DISCOVERY: Manner in which the scientist actually arrives at a hypothesis (p. 58).

CONTEXT OF JUSTIFICATION: The presentation of the proof that a hypothesis is probably true (p. 58)

CONTINUOUS VARIABLE: A variable that may change by any amount and thus may assume any fraction of a value—it may be represented by any point along a line (p. 8).

CONTRADICTORY STATEMENT: One that always assumes a truth value of false (p. 48).

CONTROL, EXPERIMENTAL: Techniques employed to insure that variation of a dependent variable is related to systematic manipulation of the independent variable and not to extraneous variables.

CONTROL GROUP(S): The group(s) that receive some normal or standard or zero treatment condition. A primary function is to establish a standard against which the experimental treatment administered to the experimental group(s) may be evaluated. It is important to contrast "control group" with "comparison group," in that the former identifies a component of an experiment versus a group of participants used in one of the nonexperimental methods (p. 7, and especially Chapters 4, 6, and 12).

CORRELATION: A measure of the extent to which two variables are related, most commonly measured by the Pearson Product Moment Coefficient of Correlation (r). The greater the absolute value of r the stronger the (linear) relationship between the two variables (p. 201).

COUNTERBALANCING: A technique for controlling progressive effects in an experiment such that each condition is presented to each participant an equal number of times, and each condition must occur an equal number of times at each practice session. In complete counterbalancing all possible orders of the variables must be presented, as distinguished from incomplete counterbalancing (p. 161).

DEDUCTIVE INFERENCE: An inference that conforms to the rules of deductive (*vs.* inductive) logic; consequently the conclusion is necessarily true, based on the premises (p. 362, p. 380).

DEGREES OF FREEDOM: A value required in ascertaining the computed probability of the results of a statistical test (like the *t*-test). There are different equations for each statistical test. Literally, the term specifies the number of independent observations that are possible for a source of variation minus the number of independent parameters that were used in estimating that variation computation. (p. 130 and throughout).

DEGREE OF PROBABILITY: A probability somewhere between 0 (absolutely false) and 1 (absolutely true).

DEPENDENT VARIABLE: Some well-defined aspect of behavior (a response) that is measured in a study. The value that it assumes is hypothesized to be dependent on the value assumed by the independent variable and thus is expected to be a change in behavior systematically related to the independent variable. The dependent variable is also dependent on extraneous variables, thus requiring control procedures (p. 7, p. 10, and Chapter 7 and p. 183).

DESIGN, EXPERIMENTAL: A specific plan used in assigning participants to conditions, as in a two-groups design (Chapter 5) or a factorial design (Chapter 10). The plan is for systematically varying independent variables and noting consequent changes in dependent variables. This definition distinguishes the design from the method of statistical analysis. "Experimental design" has also been used to include all of the steps of the experimental plan (p. 102).

DETERMINISM: The assumption that events are lawful so that we can establish relationships between and among variables that are lasting or permanent. While we cannot assert that the universe is completely deterministic, the assumption of determinism is necessary (though not sufficient) to obtain knowledge (p. 5).

DIRECT STATEMENT: A sentence (proposition) that refers to a limited set of phenomena that are immediately observable (p. 364).

DISCONFIRMED: See confirmation.

DISCONTINUOUS OR DISCRETE VARIABLE: A variable that can only assume numerical values that differ by clearly defined steps with no intermediate values possible, as in 1, 2, etc., but not 1.67 (p. 8).

ELIMINATION: A technique for controlling extraneous variables by removing them from the experimental situation (p. 153).

EMPIRICISM: The study of observable events in nature in an effort to solve problems. Empiricism is necessary for the formulation of a testable statement such as an empirical hypothesis or empirical law. Contrasted with nativism (p. 47).

ENVIRONMENT: The sum total of stimuli that may affect an organism's receptors. The stimuli may be in the *external* environment as lights that excite the eye, or in the *internal* environment as in muscular contractions that excite muscle spindle receptors for the kinesthetic modality. We assume that every receptor excitation results in some behavioral change, even though that change may be extremely slight, as in covert responses (p. 7).

ERROR VARIANCE: The error in an experiment, defined as the denominator of the t or F ratios. See experimental error (p. 421).

EVIDENCE REPORT: The summarized results of an empirical study that are used in the confirmation process for a hypothesis (p. 69, p. 361, and Chapter 13).

EXISTENTIAL HYPOTHESIS: A statement that asserts that a relationship holds for at least one case (p. 55, p. 369, and Chapter 13).

EXPERIMENTAL ERROR: The sum of all uncontrolled sources of variation that affect a dependent variable measure. It is the denominator of statistical tests such as the t-test and the F-test.

EXPERIMENTAL GROUP(s): The set or sets of individuals who receive the experimental treatment(s) (p. 6 and especially Chapter 5).

EXPERIMENTAL PLAN: The development of steps of an experiment necessary for the conduct of a study (p. 77).

EXPERIMENTAL TREATMENT(s): That set of special conditions whose effect on behavior is to be evaluated. The term "treatment" is broader than is defined here for psychology and probably derives from the early development of experimental designs and statistics in the field of agriculture, where plots of ground were literally "treated" with different combinations of plant-nurturing substances (p. 6).

EXPERIMENTATION: The application of the experimental method in which there is purposive manipulation of the independent variable (p. 71).

EXPLANATION: The process of deducing a specific statement from a more general statement, thus explaining the specific statement (p. 15, p. 24, p. 383, p. 404, and Chapter 15).

EXPLORATORY EXPERIMENTS: Contrasted with confirmatory experiments in that there is little knowledge about a given problem so that vague, poorly formed hypotheses are the only statements that can be used to guide the study (p. 75).

EXTRANEOUS VARIABLES: Variables in an experimental or nonexperimental study that may operate freely to influence the dependent variables. Such variables need to be controlled or otherwise seriously considered in order to not invalidate the study (p. 11 and Chapter 6).

EXTRANEOUS VARIABLE CONTROL: The regulation of the extraneous variables in a study (p. 147).

F STATISTIC (F TEST): A ratio between two variances. In the analysis of variance it is used to determine whether an experimental effect is statistically significant (most frequently, whether there is a reliable difference between the means of two groups) (p. 267).

FIXED EFFECTS MODEL: A factorial design in which all values of the independent variables about which inferences are to be made are specified in the design (p. 307).

FUNDAMENTAL DEFINITION: A general definition that encompasses specific definitions of a concept (p. 36).

G-RATIO: The ratio of amount actually learned to amount that could possibly have been learned. This ratio corrects for the amount of knowledge that students have prior to their participation in an experiment (p. 242).

GENERAL IMPLICATION: "If..., then...." If certain conditions hold, then certain other conditions should also hold (Chapter 3).

GENERALIZATION: An inductive inference from a specific set of data to a broader class of potential observations. We inductively infer that what is found to be true for a random sample is probably true for the universe for which the random sample is representative. Generalizations conform to the calculus of probability (p. 14, p. 380, p. 386, and Chapter 15).

HOMOGENEITY OF VARIANCE: Statistical assumption of equal variances (p. 418).

HYPOTHESIS: A testable statement that is advanced as a potential solution to a problem (p. 6 and Chapter 3).

INDEPENDENCE (STATISTICAL ASSUMPTION): The primary statistical assumption for a parametric analysis, viz., that each dependent variable value statistically analyzed must be independent of each other dependent variable value (p. 419).

INDEPENDENT VARIABLE: An aspect of the environment that is empirically investigated for the purpose of determining whether or not it influences behavior (p. 7 and Chapter 7; p. 178).

INDEPENDENT VARIABLE CONTROL: The variation of the independent variable in a known and specified manner, which is the defining characteristic of an experiment. The lack of such purposive manipulation is characteristic of the method of systematic observation and other quasi-experimental designs (p. 147).

INDIRECT STATEMENT: A sentence (proposition) that cannot be directly tested but must be reduced to a set of direct statements for confirmation (p. 365).

INDUCTIVE INFERENCE: An inference that is less than certain (P < 1.0), and conforms to the calculus of probability. "Probability inference" is a synonym (p. 362; p. 380).

INDUCTIVE SCHEMA: A concatenated representation of scientific reasoning processes, including inductive and deductive inferences, scientific generalizations, concatenation, explanation, and prediction (Chapter 14).

INDUCTIVE SIMPLICITY: The inference that the simplest relationship provides the preferred prediction (p. 236).

INTERACTION: There is an interaction between two or more independent variables if dependent variable values obtained under levels of one independent variable behave differently under different levels of the other independent variable. Dependent variable values related to one independent variable in a factorial design thus depend on, or vary systematically with, the values of another independent variable in that design (p. 285, p. 393).

INTERSUBJECTIVE RELIABILITY: See Objectivity.

KNOWLEDGE: A statement or collection of statements of relationships among variables that have been empirically confirmed. The statements may be strictly empirical ones or they may be theoretical (based on empirical findings) with varying amounts of systematic interrelations (see "concatenation"). The statements have been empirically confirmed with a reasonably high degree of probability (p. 24 and Chapter 13).

LAW: A statement of an empirical relationship between or among variables (in psychology, between independent and dependent variables) that has a high degree of probability.

LEVEL OF RELIABILITY (SIGNIFICANCE): The probability of rejecting the null hypothesis when in fact it is true. This is symbolized in statistics by $\propto$.

LOGICAL CONSTRUCT: A variable that is hypothesized to help account for unobserved events intervening between stimuli and responses (as used in psychology). Logical constructs may be of two varieties: (1) hypothetical constructs and (2) intervening variables. The distinction is that hypothetical constructs have reality status while intervening variables are convenient fictions.

MATERIALISM: Traditionally, a monistic position on the mind-body problem in which it is asserted that there are only physicalistic events in the universe. Materialism is a necessary assumption for solving any problem and constitutes a foundation of science (p. 4).

MEANING: A statement has meaning (is meaningful) if it is testable. Otherwise, it is meaningless or untestable (see Metaphysics) (p. 27).

MECHANISM: A classic position opposed to vitalism which asserts a mechanical model for understanding behavior. Organisms thus behave according to the principles of physics (p. 5).

METAPHYSICS: Disciplines, especially those of classical philosophy and religion, that pose some (not all) questions that are empirically unsolvable. The word "questions" is thus misused here and could better be replaced by "pseudo-questions" or "pseudo-problems." (p. 2).

METHOD OF SYSTEMATIC OBSERVATION: A nonexperimental method of the quasi-experimental variety in which events as they naturally occur are systematically studied (p. 71 and Chapter 12).

MIXED MODEL: A factorial design in which some independent variable values are fixed while others are random samples from the larger population of possible independent variable values (p. 309).

MORGAN'S CANON: The principle advanced by C. Lloyd Morgan stating that animal activity should be interpreted in terms of processes which are lower in the scale of psychological evolution and development in preference to higher psychological processes, if that is possible (p. 62).

NATURALISTIC OBSERVATION: Empirical inquiry where there is no intervention with the phenomena being studied. Rather, there is only systematic gathering of data on behavior as it naturally occurs; consequently there is no effort to control extraneous variables (p. 357).

NOMIC UNIVERSALITY: A statement that expresses some element of necessity between the antecedent and consequent conditions, as distinguished from accidental universality (p. 54).

NONPARAMETRIC STATISTICS: Statistics for testing null hypotheses that do not require satisfying the standard assumptions for parametric statistics, as specified in Chapter 16.

NORMALITY (STATISTICAL ASSUMPTION): The assumption that population distributions dealt with in an empirical study are normal (p. 418).

NULL HYPOTHESIS: A statement specifying characteristics of parameters (versus sample values) that are statistically evaluated. In this book the only form of a null hypotheses has been specifying that differences between parametric means are zero (p. 128).

OBJECTIVITY: A necesssary characteristic of scientific observations wherein a number of observers can agree that a given event with specified characteristics has occurred. It is the demand that knowledge be based on reliable, repeatable observations. The formal statement of objectivity is in a principle of intersubjective reliability (p. 4).

OCCAM'S RAZOR: The principle advanced by William of Occam that entities should not be multiplied without necessity. An instance of the principle of parsimony (p. 62).

OPERATIONAL DEFINITION: One that indicates that a certain phenomenon exists by specifying the precise methods by which the phenomenon is measured (p. 35).

ORDER EFFECTS: The effects on behavior of presenting two or more experimental treatments to the same participant. Such order effects would not occur with the use of a between-groups design in which each participant receives only one treatment. See also counterbalancing (p. 322).

PARAMETER: A measure ascertained from all possible observations in a population or universe, e.g., a population mean (μ) is distinguished from a sample mean $\overline{X}$.

PARSIMONY: The principle that the simplest explanation should be preferred to more complex explanations (p. 62).

PILOT EXPERIMENT: The initial stages of experimentation in which a small number of participants are studied in order to test and modify the experimental procedures (p. 76).

POPULATION (OR UNIVERSE): All possible observations made according to specified rules (p. 387).

PREDICTION: To inductively infer the nature of an unobserved event on the basis of prior knowledge. Post-diction employs precisely the same inductive processes as prediction. The distinction is that in prediction the event has not yet occurred; when one post-dicts, the event has already occurred but has not been systematically observed. One might thus predict some characteristics of the world in 1999, but post-dict that Caesar visited Ireland (p. 14, p. 384, and p. 407).

PROBABILITY: As used in this context, the empirically determined likelihood that an event did or will occur. We may empirically determine the degree of probability for a statement as varying between 0.0 and 1.0 such that no empirical statement may have a probability of 0.0 (false) or 1.0 (true) (p. 29).

PROBABILITY THEORY OF TESTABILITY (MEANING): A proposition is testable if, and only if, it is possible to determine a degree of probability for it (p. 27).

PROBLEMS: Questions that initiate scientific inquiries. Hypotheses that are empirically testable are posed as solutions to problems. "Problems" are to be distinguished from pseudo-problems ("unsolvable problems") that may appear to be real questions but on further examination cannot be attacked through empirical means (p. 3 and Chapter 2).

PSEUDO-PROBLEM: See Problems.

PURPOSIVE MANIPULATION: Systematic control of an independent variable in an experiment. Contrasted with selection of independent variable values. See Independent Variable Control (p. 181).

QUASI-EXPERIMENTAL DESIGNS: Non-experimental designs in which there is always confounding (p. 351).

RANDOM EFFECTS MODEL: A factorial design model in which the independent variable values represent a random sample from the population of possible independent variable values (p. 308).

RANDOMIZATION: A process by which each member of a universe has an equal probability of being selected, as in a table of random numbers (p. 86).

RANDOM SAMPLE: A sample (in psychology usually of participants) that is selected from a population such that all possible individuals have equal probabilities of being selected.

RELIABILITY: The extent to which the same observations are obtained in repeated studies.

REPLICATION: The conduct of an additional study in which the method of the first (usually an experiment) is precisely repeated. The term is sometimes used to indicate that the results of the second experiment confirmed the first, though this is a confusion of confirmation and replication (repeating). In confirming, one repeats *and* obtains the same findings. Similar, but different, uses occur in statistics (p. 14, p. 431).

RESPONSE: The contraction of muscle (skeletal, smooth, or cardiac) or the secretion of glands that constitute the components of behavior (p. 7 and Chapter 7).

SAMPLE: A subset of observations from a population (p. 387).

SCIENCE: The application of the scientific method to solvable problems (p. 4). For other definitions see p. 1.

SCIENCE (APPLIED, TECHNOLOGICAL, PRACTICAL): The activity of solving problems of immediate need in everyday life. Many of these problems can be solved by the application of the fruits of pure science while others require the conduct of empirical research "in the field." (p. 347, p. 386).

SCIENCE (PURE, BASIC): The conduct of systematic inquiries only for the sake of knowledge itself, without regard to the solution of practical problems. Our universities are the primary agents for the search for knowledge for its own sake, and for the retention of that knowledge for society (p. 347, p. 386).

SCIENTIFIC METHOD: A serial process by which the sciences obtain answers to their questions (p. 3).

SERENDIPITY: The finding of something unexpected that is more valuable than the original purpose of the research (p. 64).

SIGNIFICANT: Used in statistics to indicate reliability. We say a statistical test, such as the t-test, may be statistically significant (reliable) when the null hypothesis is rejected, thus indicating that the phenomenon being studied is a reliable one (p. 13 and Chapter 5).

STANDARD DEVIATION: A measure of variability that indicates how the values in a distribution of scores are spread out or concentrated (p. 218).

STATISTIC: A value that is computed from observations of a sample from a population, e.g., a sample mean is $\overline{X}$.

SYNTHETIC STATEMENT: One that may be *either* true or false, or more precisely, it has a probability varying between 0.0 and 1.0 (p. 48).

TECHNOLOGY: See Science (applied).

TESTABLE: A statement that may be directly (or indirectly as with theories) subjected to empirical confirmation. A testable statement is one for which a determinable degree of probability may be assigned to it, in accordance with the probability theory of testability (p. 28.)

UNILATERAL VERIFIABILITY: The characteristic of a hypothesis as being verifiable as false but not as true or vice versa (p. 369).

UNIVERSAL HYPOTHESIS: A statement that asserts that a given relationship holds for all variables contained therein, for all time and at all places (p. 55, p. 367, and Chapter 13).

VARIABLE: Anything that can change in value and thus assume different numerical values (p. 7).

VERIFICATION: A process of determining that a hypothesis is strictly true or false, as contrasted with confirmation (p. 367).

WITHIN-GROUPS DESIGNS: Experimental designs in which participants receive two or more values of the independent variable. Contrasted with between-groups designs (Chapter 11).

Index